Lecture Notes in Mathematics

Edited by A. Dold, B. Eckmann and F. Takens

1439

Emilio Bujalance
José J. Etayo
José M. Gamboa
Grzegorz Gromadzki

Automorphism Groups of Compact Bordered Klein Surfaces

A Combinatorial Approach

Springer-Verlag
Berlin Heidelberg New York London
Paris Tokyo Hong Kong Barcelona

Authors

Emilio Bujalance
Dpto. de Matemáticas Fundamentales
Facultad de Ciencias
Universidad a Distancia (UNED)
28040 Madrid, Spain

José Javier Etayo
José Manuel Gamboa
Dpto. de Algebra
Facultad de Matemáticas
Universidad Complutense
28040 Madrid, Spain

Grzegorz Gromadzki
Institute of Mathematics, WSP
Chodkiewicza 30
85-064 Bydgoszcz, Poland

Mathematics Subject Classification (1980): 14H; 20H; 30F

ISBN 3-540-52941-1 Springer-Verlag Berlin Heidelberg New York
ISBN 0-387-52941-1 Springer-Verlag New York Berlin Heidelberg

Printing and binding: Druckhaus Beltz, Hemsbach/Bergstr.
2146/3140-543210 – Printed on acid-free paper

To Raquel (and our daughters
Raquel, M. Teresa, Carla and Lidia)
To Almudena
To Alicia
To Terenia

INTRODUCTION

Classical results on automorphism groups of complex algebraic curves.

Given a complex algebraic curve by means of its polynomial equations, it is very difficult to get information about its birational automorphisms, unless the curve is either rational or elliptic. For curves C of genus $p \geq 2$, Schwarz proved in 1879 the finiteness of the group Aut(C) of automorphisms of C, [111]. Afterwards Hurwitz, applying his famous ramification formula, showed that $|\mathrm{Aut}(C)| \leq 84(p-1)$, [69]. By means of classical methods of complex algebraic geometry, Klein showed that $|\mathrm{Aut}(C)| \leq 48$ if $p=2$. Then, Gordan proved that $|\mathrm{Aut}(C)| \leq 120$ for $p=4$, [51], and Wiman, who carefully studied the cases $2 \leq p \leq 6$, in particular established that $|\mathrm{Aut}(C)| \leq 192$ for $p=5$ and $|\mathrm{Aut}(C)| < 420$ for $p=6$, [127], [128]. Hence in case $p=2,4,5,6$, Hurwitz's bound is not attained. In case $p=3$, there is only one curve of genus 3 with $168=84(3-1)$ automorphisms: Klein's quartic $x^3y+y^3+x=0$.

Apart from these and other facts on curves of low genus, further results of a more general nature were known by the end of the last century. For instance, Wiman proved that the order of each automorphism of C is always $\leq 2(2p+1)$; he also studied hyperelliptic curves in detail. Every curve of genus $p=2$ is hyperelliptic, and so it admits an involution. On contrary, for $p \geq 3$, curves with non-trivial automorphisms are exceptional; they constitute the singular locus of the moduli space of curves of genus p, and this explains the interest of this topic.

Complex algebraic curves and Riemann surfaces.

As observed by Riemann, the birational geometry of (irreducible) complex algebraic curves can be studied in a transcendental way. Indeed, algebraic function fields in one variable over $\mathbb{C}$ are nothing but fields of meromorphic functions on compact Riemann surfaces. Hence, groups of birational automorphisms of complex algebraic curves are the same as automorphism groups of compact Riemann surfaces.

This new point of view propelled the theory ahead, in the early sixties, with the remarkable work of Macbeath. After Poincare **[106]**, it was well known that each compact Riemann surface S of (algebraic) genus $p \geq 2$ can be represented as an orbit space H/Γ of the upper half complex plane H. Here H is endowed with the conformal structure induced by the group Ω of Möbius transformations, and the acting group Γ is a fuchsian group, *i.e.*, a discrete subgroup of Ω. The group Γ can be chosen with no elements of finite order. With this representation at hand, Macbeath proved that a finite group G is a group of automorphisms of S if and only if $G=\Gamma'/\Gamma$ for another fuchsian group Γ', **[82]**. This was a new, combinatorial, topological, group-theoretical method, to adress questions on groups of automorphisms.

The general strategy for a better understanding of automorphism groups of a compact Riemann surface S of genus $p \geq 2$ is explained by Macbeath **[82]**, and paraphrased by Accola, **[1]**. First one looks for some group G of automorphisms of S. Let T_0 be the set of branching points in $S_0=S/G$ of the n-sheeted covering $\pi:S \longrightarrow S_0$, and let F_0 be the fundamental group of $S_0 \setminus T_0$. Let K be the kernel of the natural homomorphism from F_0 into the n-th symmetric group. Then a given automorphism $f_0 \in Aut(S_0)$ lifts to some $f \in Aut(S)$ if and only if f_0 restricts to a permutation on T_0 and $f_0^*(K)=K$. As a consequence, if $Aut(S_0)$ contains a subgroup with m elements verifying these conditions, then Aut(S) contains a subgroup of order mn. Notice that this method heavily relies on a good choice of G in order to work well with $Aut(S_0)$. The techniques used here are those of Weierstrass points and of fuchsian groups. A large number of results can be proved with the aid of the first method, but many others require the second one.

The set W of Weierstrass points of S is finite and each $f \in Aut(S)$ restricts to a permutation of W. This permutation determines completely f when S is not hyperelliptic and completely, up to the canonical involution, when S is hyperelliptic. This gives another proof of the finiteness of Aut(S). The method of Weierstrass points is specially fruitful to provide geometrical information about the behaviour of an automorphism h of S. For example, there is an equality relating the order and number of fixed points of h with the genera of S and S/h, **[7]**, **[42]**, **[47]**, **[80]**.

The theory of fuchsian groups is the most powerful in order to investigate the structure of automorphism groups of complex algebraic curves. However, the situation is not fully satisfactory, in the following sense: given such a curve C we can represent its associated Riemann surface S_C in the form H/Γ_C. Unfortunately, except for its algebraic structure, we do not have a good knowledge of the fuchsian group Γ_C. In particular, although Γ_C reflects some geometrical properties of C as hyperellipticity or q-gonality, there is no known relation between Γ_C and any algebraic equations defining C. This explains why results on groups of automorphisms of complex algebraic curves obtained by means of fuchsian groups theory are not effective. Sometimes it is possible to prove the existence of a curve C with certain geometric properties whose group of automorphisms is of a given type, but we cannot find explicit algebraic equations for C.

Some results on automorphisms of Riemann surfaces.

Let us retrieve some significant facts in this area, obtained by means of fuchsian group theory. Hurwitz's ramification formula can be read as $|\Gamma'/\Gamma|=\mu(\Gamma)/\mu(\Gamma')$, where Γ is a surface fuchsian normal subgroup of the fuchsian group Γ' and μ represents the area of a fundamental region of the corresponding group. If $S=H/\Gamma$ has genus p, then $\mu(\Gamma)=4\pi(p-1)$ while, by Siegel's theorem [113], $\mu(\Gamma')\geq\pi/21$. Hence $|Aut(S)|\leq 84(p-1)$ as announced and the equality is only attained if Γ' is the fuchsian triangular group $M=(2,3,7)$. In such a way, the study of Hurwitz groups (=groups of order $84(p-1)$ acting on surfaces of genus $p\geq 2$), becomes the study of finite factors of M.

Using this, Macbeath proved the existence of Hurwitz groups for infinitely many values of p - for $p=7$ in [85] and for $p=2m^6+1$ in [83]. The non-existence of Hurwitz groups for infinitely many values of p was also proved in the latter paper. The same kind of arguments led Accola [1] and Maclachlan [89] to show that for every $p\geq 2$ there is a surface of genus p with $8(p+1)$ automorphisms; actually, there are infinitely many of such surfaces. Besides, Greenberg showed that every finite group is the group of automorphisms of some surface [53], and Harvey computed the minimum genus of the surfaces admitting an automorphism of a given order, [61]. As a consequence, he found a new proof of Wiman's bound. Among the vast literature on the subject we must quote here the papers by Cohen, [37], [38], Greenberg [54], Sah [110] and Zomorrodian [131].

Real algebraic curves and Klein surfaces.

Up to now we only have been concerned with *complex* algebraic geometry. However, what happens if the ground field is the field $\mathbb{R}$ of real numbers?. This was overlooked for a long time but in the last two decades many beautiful mathematics have been developed in the real setting, demanding specific tools for this new field of research (*cf.* the books of Brumfiel [10], Delfs-Knebusch [41], Bochnak-Coste-Roy [8] and Knebusch [74]).

A particular question in this area is the study of groups of birational automorphisms of real algebraic curves. As in the complex case, very few is known about Aut(C) from the algebraic equations defining C except when C has genus 0 - Lüroth's theorem - or genus 1, see Alling [4]. Even the recent methods in effective algebra and computational geometry do not give yet a satisfactory answer to this question.

In order to work in an analogous way to the complex case, one is forced to enlarge the category of Riemann surfaces. This raises two different problems. Firstly, since nonorientable surfaces do not admit any analytic structure, a more general notion is needed. The suitable one is dianalyticity, which includes analytic and antianalytic maps; the former preserve the orientation, the latter reverse it and both are conformal. Then, orientable and nonorientable surfaces, with or without boundary, admit a dianalytic structure. These are Klein surfaces, introduced by Alling and Greenleaf [5], following up ideas by Klein. Of course, Riemann surfaces are precisely the orientable, unbordered Klein surfaces.

Secondly, it was necessary to represent algebraic function fields in one variable over R as fields of "meromorphic functions" on some objects. These objects are exactly Klein surfaces. Moreover, *real* fields (=fields in which -1 is not a sum of squares) are the meromorphic function fields of *bordered* surfaces. Thus the categories of compact bordered Klein surfaces and of real (irreducible) algebraic curves are equivalent. This functorial equivalence is necessary to apply the methods described before to problems in real geometry. We explain it in pure real algebra terms in the appendix at the end of the book.

Hence, for the study of birational automorphisms of real algebraic curves we can focus on compact Klein surfaces S of algebraic genus $p \geq 2$. The notion of real Weierstrass points has not been exploited yet, and so combinatorics is the only accessible approach. In his unpublished thesis [108], Preston proved the real counterpart of Poincaré's uniformization theorem, [106]: S is the quotient of H under the action of a non-euclidean crystallographic (NEC in short) group, that is, a discrete subgroup of the extended modular group. This NEC group can be assumed having no orientation preserving mapping of finite order. This theorem, together with the classification of NEC groups due to Macbeath [86] and Wilkie [126], opened the door to the combinatorial approach to groups of automorphisms presented here.

Since 1975 many papers have appeared studying groups of automorphisms of compact Klein surfaces of genus ≥ 2, analogous to the ones described before on Riemann surfaces. We shall either present or report them along the book.

It is remarkable that no technique is developed to study groups of automorphisms of algebraic curves defined over real closed ground fields distinct to $\mathbb{R}$. Only in some cases, Tarski's Transfer Principle allows us to translate the results, but not the proofs, from the "true reals" to arbitrary real closed fields. As an example, one can see [23].

Contents of this book.

The general problem we analyze can be stated as follows: given a class $\mathscr{G}$ of finite groups and a class $\mathscr{K}$ of compact Klein surfaces of algebraic genus $p\geq 2$, under what conditions do a surface S in $\mathscr{K}$ and a group G in $\mathscr{G}$ exist, such that G acts as a (or the full) group of automorphisms on S? This is essentially the concern of chapters 3 to 6, which constitute the core of the book.

In chapter 3, $\mathscr{G}$ is the class of cyclic groups and we decide, in a computable way, whether a given natural number is the order of an automorphism of some surface of fixed topological type. Moreover, if the surface is orientable, we can decide whether such an automorphism preserves or reverses the given orientation. As a consequence, we determine the minimum genus of surfaces admitting an automorphism of a given order and the maximum order of an automorphism of a surface of a given genus, with all precisions concerning orientability.

In chapter 4 we consider several classes of finite groups: soluble, supersoluble, nilpotent, p-groups, abelian. For each of these classes $\mathscr{G}$ we compute an upper bound $N(p,\mathscr{G})$ for the order of a group of automorphisms G of a surface of genus p, provided $G\in\mathscr{G}$. Moreover, we calculate the topological type of the surfaces attaining this maximal group of automorphisms and the algebraic structure of this group. Most results are original except a few which were proved first by May using different techniques. We present here a unified treatment of the problem. We must remark that, as technical lemmata for the proofs, we obtain several results on abstract group theory, as presentations of supersoluble and nilpotent groups.

In chapter 5 we look at the family $\mathscr{K}$ of surfaces with nonempty connected boundary. Groups acting as groups of automorphisms in them are cyclic or dihedral and we determine for a fixed group G of this type the existence of a surface $S\in\mathscr{K}$ with $\mathrm{Aut}(S)=G$, in terms of the genus and orientability of S.

The same problem is solved in chapter 6 for the family of hyperelliptic surfaces. Here groups of automorphisms are extensions by $\mathbb{Z}_2$ of cyclic and dihedral groups.

As said before, if a surface S is written as H/Γ for a surface NEC group Γ, then $\mathrm{Aut}(S)=\Lambda/\Gamma$ where Λ is the normalizer of Γ in the extended modular group. Hence, the computation of Aut(S) involves the knowledge of this normalizer, which is a rather difficult problem. We avoid it by using Teichmüller spaces theory to prove that maximal signatures correspond to maximal NEC groups. Then we call to the lists of non-maximal signatures of Singerman [115] and Bujalance [13]. The results of chapters 5 and 6 follow from this, together with specific procedures highly depending on the features of the involved surfaces - *e.g.* the existence of a unique central involution in the hyperelliptic case.

The remaining chapters have a different nature. In chapter 0 we describe briefly Klein surfaces, NEC groups and some general properties on Teichmüller spaces that are needed later to compute full groups of automorphisms. For the proofs of some results in this chapter, the reader is refered to the original papers where these fundamental facts were established.

Chapter 1 is a detailed presentation of the fundamental results of Preston quoted above, and a theorem by May saying that groups of automorphisms of H/Γ are quotients Λ/Γ where Λ is another NEC group. This naturally leads to study, in chapter 2, the relation between the presentations of an NEC group Λ and of its normal subgroup Γ, using combinatorial methods based mainly in surgery on a fundamental region of Λ. Most of this is classical, but was scattered in the literature. So we develop it here in a unified and precise fashion for later use in the combinatorial study of automorphism groups.

All in all, the reader familiar with the prerequisites mentioned in chapter 0, will find the book self-contained.

Of course, part of the material belongs to other people, and the proper attributions appear in the historical notes included at the end of each chapter.

We have also included both a subject and a symbol indices at the end of the text.

Interdependence of chapters

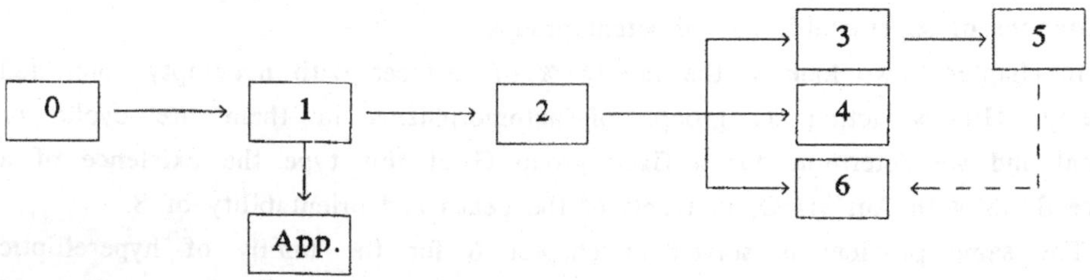

The first three listed authors enjoy a grant of the Comision Interministerial de Ciencia y Tecnologia, whilst G. Gromadzki has obtained a grant of the Spanish Ministerio de Educación y Ciencia to be spent in the Universidad Nacional de Educación a Distancia, and a sabbatical stay at Universidad Complutense de Madrid.

We wish to thank some of our colleagues and students to have suggested improvements in language and proofs, specially in chapter 1 and appendix.

The authors
Madrid, October 1989

Table of Contents

CHAPTER - 0
Preliminary results

We present here basic results concerning Klein surfaces and non-euclidean crystallographic (NEC in short) groups. In section 1 we introduce the category of Klein surfaces, whilst section 2 is devoted to the notion of NEC group, the main tool to study groups of automorphisms of Klein surfaces from the combinatorial point of view. Finally in section 3 we deal with Teichmüller and moduli spaces of NEC groups, that distinguish dianalytic structures that a given topological surface can admit.

0.1. THE CATEGORY OF KLEIN SURFACES

Definition 1. A *surface* is a Hausdorff, connected, topological space S together with a family $\Sigma=\{(U_i,\phi_i)|i\in I\}$ such that $\{U_i|i\in I\}$ is an open covering of S and each map $\phi_i:U_i \longrightarrow A_i$ is a homeomorphism onto an open subset A_i of $\mathbb{C}$ or $\mathbb{C}^+=\{z\in\mathbb{C}:\mathrm{Im}z\geq 0\}$. The family Σ is said to be a *topological atlas* on S. The *boundary of* S is the set

$$\partial S=\{x\in S|\text{there exists } i\in I,\ x\in U_i,\ \phi_i(x)\in\mathbb{R} \text{ and } \phi_i(U_i)\subseteq\mathbb{C}^+\}.$$

Each (U_i,ϕ_i) is said to be a *chart* and we say that this chart is *positive* if $\phi_i(U_i)\subseteq\mathbb{C}^+$. The *transition functions* of Σ are the homeomorphisms

$$\phi_{ij}=\phi_i\phi_j^{-1}:\phi_j(U_i\cap U_j)\longrightarrow\phi_i(U_i\cap U_j).$$

Simple arguments show that $\phi_j(U_i\cap U_j)$ is open in $\mathbb{C}$ (resp $\mathbb{C}^+$) if and only if the same is true for $\phi_i(U_i\cap U_j)$.

The *orientability* of S is defined as for a real 2-manifold (under the identification of $\mathbb{C}$ with $\mathbb{R}^2$).

If k(S), g(S) and χ(S) are the number of connected components of ∂S, the topological genus and the Euler characteristic of S, respectively, then the following formula holds true:

Proposition 0.1.1.

$$g(S)=\begin{cases}(2-\chi(S)-k(S))/2 & \text{if S is orientable,}\\ 2-\chi(S)-k(S) & \text{if S is nonorientable.}\end{cases}$$

Comments. A Klein surface is a surface whose transition functions verify additional conditions. To give the precise definition we introduce the following notations:

Let A be a nonempty open subset of $\mathbb{C}$, $f:A \longrightarrow \mathbb{C}$ a map. We write $z=x+iy \in \mathbb{C}$, $x,y \in \mathbb{R}$, $i=\sqrt{-1}$, $\bar{z}=x-iy$, and $f(z)=u(x,y)+iv(x,y)$, for certain functions $u,v:A \longrightarrow \mathbb{R}$ of the class 2. We define

$$\frac{\partial f}{\partial z} = \frac{\partial u}{\partial z} + i\frac{\partial v}{\partial z} = \frac{\partial u}{\partial x}\frac{\partial x}{\partial z} + \frac{\partial u}{\partial y}\frac{\partial y}{\partial z} + i\left[\frac{\partial v}{\partial x}\frac{\partial x}{\partial z} + \frac{\partial v}{\partial y}\frac{\partial y}{\partial z}\right],$$

$$\frac{\partial f}{\partial \bar{z}} = \frac{\partial u}{\partial \bar{z}} + i\frac{\partial v}{\partial \bar{z}} = \frac{\partial u}{\partial x}\frac{\partial x}{\partial \bar{z}} + \frac{\partial u}{\partial y}\frac{\partial y}{\partial \bar{z}} + i\left[\frac{\partial v}{\partial x}\frac{\partial x}{\partial \bar{z}} + \frac{\partial v}{\partial y}\frac{\partial y}{\partial \bar{z}}\right].$$

Since $x=(z+\bar{z})/2$ and $y=i(\bar{z}-z)/2$ we obtain

$$\frac{\partial x}{\partial z} = \frac{\partial x}{\partial \bar{z}} = 1/2, \quad \frac{\partial y}{\partial z} = -i/2, \quad \frac{\partial y}{\partial \bar{z}} = i/2,$$

and so

$$\frac{\partial f}{\partial z} = \frac{1}{2}\left[\frac{\partial u}{\partial x} - i\frac{\partial u}{\partial y} + i\frac{\partial v}{\partial x} + \frac{\partial v}{\partial y}\right]; \quad \frac{\partial f}{\partial \bar{z}} = \frac{1}{2}\left[\frac{\partial u}{\partial x} + i\frac{\partial u}{\partial y} + i\frac{\partial v}{\partial x} - \frac{\partial v}{\partial y}\right].$$

In particular, if $\bar{f}:A \longrightarrow \mathbb{C}: z=x+iy \longmapsto \bar{f}(z)=u(x,y)-iv(x,y)$ we obtain the fundamental equalities

$$\frac{\partial \bar{f}}{\partial \bar{z}} = \overline{\left[\frac{\partial f}{\partial z}\right]}; \quad \frac{\partial \bar{f}}{\partial z} = \overline{\left[\frac{\partial f}{\partial \bar{z}}\right]}.$$

Definition 2. The map $f:A \longrightarrow \mathbb{C}$ is *analytic on* A if $\frac{\partial f}{\partial \bar{z}}=0$ (these are the Cauchy-Riemann conditions). The map f is *antianalytic on* A if $\frac{\partial f}{\partial z}=0$. The map f is said to be *dianalytic* if its restriction to every connected component of A is either analytic or antianalytic.

The equalities above show that f is analytic if and only if $\bar{f}$ is antianalytic. These formulas allow us also to conclude that for a connected subset A of $\mathbb{C}$, the only analytic and antianalytic maps $f:A \longrightarrow \mathbb{C}$ are constants. It is also clear that the composition $gf:A \longrightarrow \mathbb{C}$ of dianalytic maps $f:A \longrightarrow B \subseteq \mathbb{C}$ and $g:B \longrightarrow \mathbb{C}$ is dianalytic. Moreover if f and g are both analytic or antianalytic, then gf is analytic. In the other case gf is antianalytic.

Given f, it is clear that

$$\det\begin{pmatrix} \frac{\partial u}{\partial x} & \frac{\partial u}{\partial y} \\ \frac{\partial v}{\partial x} & \frac{\partial v}{\partial y} \end{pmatrix} = \varepsilon\left[\left(\frac{\partial u}{\partial x}\right)^2 + \left(\frac{\partial v}{\partial x}\right)^2\right]$$

where $\varepsilon=1$ if f is analytic and $\varepsilon=-1$ if f is antianalytic. Thus analytic functions preserve orientation whilst antianalytic ones reverse it.

Example 0.1.2. Let a,b,c,d be real numbers, $c \neq 0$, $A=\mathbb{C}\setminus\{-d/c\}$. Clearly the map $f:A \longrightarrow \mathbb{C}: z \longmapsto \frac{az+b}{cz+d}$ is analytic. So the map $\bar{f}:A \longrightarrow \mathbb{C}$ given by $\bar{f}(z)=\frac{a\bar{z}+b}{c\bar{z}+d}$ is antianalytic.

To deal with bordered Klein surfaces we need an extension of the notion of dianalyticity to functions having an open subset of $\mathbb{C}^+$ as a domain.

Definition 3. Let A be an open subset of $\mathbb{C}^+$. A continuous function $f:A \longrightarrow \mathbb{C}$ is said to be *analytic* (resp. *antianalytic*) *on* A if there exists an extension of f to an analytic (resp. antianalytic) function f_U on an open subset U of $\mathbb{C}$. *Dianalyticity* of f is defined similarly (using the notion of dianalyticity given in the previous definition).

The modern definition of Klein surfaces that we give now is due to Alling and Greenleaf, [5].

Definition 4. Let S be a surface with atlas Σ. We say that Σ is *dianalytic* if the transition functions of Σ are dianalytic in the sense of definitions 2 and 3. The atlas Σ is *analytic* if its transition functions are analytic.

Remarks and examples 0.1.3. (1) As in differential topology, two dianalytic (resp. analytic) atlas Σ and Σ' on S are said to be *dianalytically* (resp. *analytically*) *equivalent* if $\Sigma \cup \Sigma'$ is a dianalytic (resp. analytic) atlas on S. A *dianalytic* (resp. *analytic*) *structure* on a surface S is the equivalence class of a dianalytic (resp. analytic) atlas on S. The surface S equipped with the dianalytic structure induced by a dianalytic atlas Σ is said to be a *Klein surface*.

(2) It is clear that a classical Riemann surface can be viewed as an orientable Klein surface with empty boundary. In such a case it is easily seen that Σ can be assumed to be analytic.

(3) Let $H=\{z\in\mathbb{C} \mid \mathrm{Im} z>0\}$, $D=\{z\in\mathbb{C} \mid |z|<1\}$ be the upper half plane and the open unit disc in $\mathbb{C}$, respectively. Both of them are Klein surfaces with empty boundary, with the analytic structures induced by the trivial analytic atlas $\{(U_1=H,\phi_1=1_H)\}$, and $\{(U_1=D,\phi_1=1_D)\}$ on H and D respectively. From now on we shall use letters H and D only to denote these surfaces.

(4) The surface $\mathbb{C}^+$ with the structure induced by the analytic atlas $\{(\mathbb{C}^+,1_{\mathbb{C}^+})\}$ is a Klein surface with boundary $\partial\mathbb{C}^+=\mathbb{R}$.

(5) Let $\bar{\mathbb{C}}=\mathbb{C}\cup\{\infty\}$ and $\Delta=\mathbb{C}^+\cup\{\infty\}$ be the Alexandroff compactifications of $\mathbb{C}$ and $\mathbb{C}^+$. They are compact Klein surfaces with the structures induced by the atlas

$$\{(U_1=\mathbb{C},\ \phi_1=1_{\mathbb{C}}),\ (U_2=\bar{\mathbb{C}}\setminus\bar{B}(0,1),\ \phi_2=z^{-1})\}$$

and

$$\{(U_1=\mathbb{C}^+,\ \phi_1=1_{\mathbb{C}^+}),\ (U_2=\Delta\setminus\bar{B}(0,1),\ \phi_2=\bar{z}^{-1})\}$$

respectively. Of course $\partial\bar{\mathbb{C}}$ is empty and $\partial\Delta=\mathbb{R}\cup\{\infty\}$.

In order to define the notion of morphism between Klein surfaces, we need the following

Definition 5. The *folding map* is the continuous map

$$\Phi:\mathbb{C}\longrightarrow\mathbb{C}^{+}:x+iy\longmapsto x+i|y|.$$

For a given subset A of $\mathbb{C}^{+}$ we write $\bar{A}=\{z\in\mathbb{C}\,|\,\bar{z}\in A\}$. It is obvious that $\Phi^{-1}(A)=A\cup\bar{A}$. In particular, $\Phi^{-1}(\mathbb{R})=\mathbb{R}$. Clearly Φ is an open map.

Definition 6. A *morphism* between the Klein surfaces S and S′ is a continuous map $f:S\longrightarrow S'$ such that:

(i) $f(\partial S)\subseteq\partial S'$,

(ii) Given $s\in S$, there exist charts (U,ϕ) and (V,ψ) at s and f(s) respectively, and an analytic function $F:\phi(U)\longrightarrow\mathbb{C}$ such that the following diagram

(0.1.3.1)
$$\begin{array}{ccccc} U & & \xrightarrow{\quad f \quad} & & V \\ \downarrow{\scriptstyle\phi} & & & & \downarrow{\scriptstyle\psi} \\ \phi(U) & \xrightarrow{\;F\;} & \mathbb{C} & \xrightarrow{\;\Phi\;} & \mathbb{C}^{+} \end{array}$$

commutes.

Example 0.1.4. Let H and D be the Klein surfaces without boundary introduced in 0.1.3. The map $\rho:D\longrightarrow H:\ z\longmapsto(z+i)/(iz+1)$ is well defined because if $z=x+iy\in D$, then $x^2+y^2<1$ and so

$$\rho(z)=\frac{2x+i\,[\,1-(x^2+y^2)]}{x^2+(1-y)^2}\in H.$$

Moreover, it is analytic and so in particular continuous. Given $s\in D$ we choose $(U=D,1_D)$ and $(V=H,1_H)$ as the charts at s and $\rho(s)$ respectively. Then $\Phi\rho=\rho$, since $\rho(D)\subseteq H\subseteq\mathbb{C}^{+}$, and we have a commutative diagram

$$\begin{array}{ccccc} U & & \xrightarrow{\quad \rho \quad} & & V \\ \downarrow{\scriptstyle 1_U} & & & & \uparrow \\ U & \xrightarrow{\;F=\rho\;} & \mathbb{C} & \xrightarrow{\;\Phi\;} & \mathbb{C}^{+} \end{array}$$

Consequently, ρ is a morphism between the Klein surfaces D and H.

Remarks 0.1.5. (1) In case we are dealing with orientation preserving morphisms between Riemann surfaces, diagram (0.1.3.1) can be replaced after translations by

$$\begin{array}{ccc} U & \xrightarrow{\quad f \quad} & V \\ \downarrow{\scriptstyle\phi} & & \downarrow{\scriptstyle\psi} \\ \phi(U) & \xrightarrow{\quad F \quad} & \psi(V) \end{array}$$

(2) We can always extend F to an analytic map $\hat{F}:\phi(U)\cup\overline{\phi(U)} \longrightarrow \mathbb{C}$ by Schwartz Reflection Principle, that is, taking

$$\hat{F}(z)=\overline{F(\bar{z})} \text{ if } z\in\overline{\phi(U)},$$

because $F(\phi(U\cap\partial S)) \subseteq \mathbb{R}$. In fact, given $x\in U\cap\partial S$ it is $f(x)\in V\cap\partial S'$ and so $\Phi F\phi(x)= =\psi(f(x))\in\mathbb{R}$, *i.e.* $F\phi(x)\in\mathbb{R}$.

(3) The main difference between morphisms in the categories of Klein surfaces and Riemann surfaces, is the "appearance" of Φ in diagram (0.1.3.1). However, "given two nonconstant analytic maps $F_1,F_2:A\longrightarrow\mathbb{C}$, where A is an open connected subset of $\mathbb{C}$, such that $\Phi F_1=\Phi F_2$, we have $F_1=F_2$."

To see that, we choose an open connected and nonempty subset Y of the inverse image of $\mathbb{C}\setminus\mathbb{R}$ under F_1. Then $M_1=\{F_1=F_2\}\cap Y$ and $M_2=\{F_1=\bar{F}_2\}\cap Y$ are closed and disjoint subsets of Y with $M_1\cup M_2=Y$. Then either $M_1=Y$ or $M_2=Y$. In the second case F_1 should be both analytic and antianalytic on Y, *i.e.* $F_1|Y$ is constant and so, F_1 is constant by Riemann's Identity Principle. Hence $F_1=F_2$ on Y, *i.e.* $F_1=F_2$.

(4) With notations in (2), it is easily seen that $\Phi F\Phi=\Phi\hat{F}$ on $\phi(U)\cup\overline{\phi(U)}$.

(5) At the first sight, the notion of morphism looks "very local" in the sense it only guarantees that for a given point $s\in S$ there exist suitable charts making commutative certain diagrams. However

Proposition 0.1.6. *Let $f:S\longrightarrow S'$ be a nonconstant morphism. Let (U,ϕ), (V,ψ) be any two charts in S and S' such that $f(U)\subseteq V$ and $\psi(V)\subseteq\mathbb{C}^+$. Then there exists a unique analytic map $F:\phi(U)\longrightarrow\mathbb{C}$ such that the diagram*

$$\begin{array}{ccccc} U & & \xrightarrow{\quad f \quad} & & V \\ \downarrow\phi & & & & \downarrow\psi \\ \phi(U) & \xrightarrow{F} & \mathbb{C} & \xrightarrow{\Phi} & \psi(V) \end{array}$$

commutes.

Proof. Uniqueness is obvious from (3) in Remark above. Let us suppose that we can cover U by $(U_j|j\in J)$ such that for some analytic maps $F_j:\phi(U_j)\longrightarrow\mathbb{C}$, we have diagrams

$$\begin{array}{ccccc} U & & \xrightarrow{\quad f \quad} & & V \\ \downarrow\phi & & & & \downarrow\psi \\ \phi(U_j) & \xrightarrow{F_j} & \mathbb{C} & \xrightarrow{\Phi} & \psi(V) \end{array} \tag{0.1.6.1}$$

Using once more (3) in 0.1.5, the functions F_j glue together to produce the function F we are looking for. So, all reduces to produce 0.1.6.1.

For every $x\in U$, $y=f(x)$, there exist charts (U^x,ϕ_x), (V^y,ψ_y) and an analytic map F_x such that $U^x\subseteq U$, $V^y\subseteq V$ and the following diagram commutes:

$$\begin{array}{ccccc} U^x & \xrightarrow{\quad f \quad} & & & V^y \\ \downarrow \phi_x & & & & \downarrow \psi_y \\ \phi_x(U^x) & \xrightarrow{\;F_x\;} & \mathbb{C} & \xrightarrow{\;\Phi\;} & \psi_y(V^y) \end{array}$$

We construct F_x^* such that

$$\begin{array}{ccccc} U^x & \xrightarrow{\quad f \quad} & & & V^y \\ \downarrow \phi & & & & \downarrow \psi \\ \phi(U^x) & \xrightarrow{\;F_x^*\;} & \mathbb{C} & \xrightarrow{\;\Phi\;} & \psi(V^y) \end{array}$$

In fact, given $\xi \in \phi(U^x)$ it is $F_x\phi_x\phi^{-1}(\xi) \in \Phi^{-1}(\mathrm{im}\psi_y) = \psi_y(V^y) \cup \overline{\psi_y(V^y)}$, and we can consider $(\psi\psi_y^{-1})^\wedge : \psi_y(V^y) \cup \overline{\psi_y(V^y)} \longrightarrow \mathbb{C}$ defined as in (2) of 0.1.5. Then, according with $\phi_x\phi^{-1}$ and $\psi\psi_y^{-1}$ were analytic or antianalytic, we take F_x^* or $\overline{F_x^*}$ equal to $(\psi\psi_y^{-1})^\wedge F_x\phi_x\phi^{-1}$. This provides 0.1.6.1.

A fundamental result concerning the behaviour of morphisms under composition is the following one due to Alling-Greenleaf [5] and Cazacu [32]. The proof follows the same lines that the classical ones for Riemann surfaces by using Remarks 0.1.5.

Proposition 0.1.7. *Let* S, S′ *and* S″ *be Klein surfaces and let* $f:S \longrightarrow S'$, $g:S' \longrightarrow S''$ *be continuous maps such that* $f(\partial S) \subseteq \partial S'$, $g(\partial S') \subseteq \partial S''$. *Let us consider the following statements:*

(1) f *is a morphism,*

(2) g *is a morphism,*

(3) gf *is a morphism.*

Then (1) and (2) imply (3). If f *is surjective (1) and (3) imply (2). Moreover, if* f *is open, (2) and (3) imply (1).*

A trivial consequence of the last part in the previous proposition is the following

Corollary 0.1.8. *Let* S *and* S′ *be Klein surfaces and let* $f:S \longrightarrow S'$ *be a morphism. If* f *is a homeomorphism, then* f *is an isomorphism in the category of Klein surfaces.*

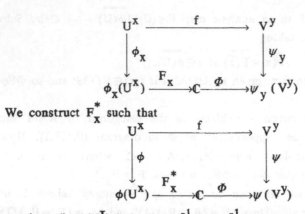

Example 0.1.9. We have proved in 0.1.4 that the map $\rho:D \longrightarrow H$: $z \longmapsto (z+i)/(iz+1)$ is a morphism of Klein surfaces. Moreover if $g:H \longrightarrow \mathbb{C}$ sends z to $(z-i)/(1-iz)$ we have $g\rho = 1_H$. Since ρ is onto, $\mathrm{im}g \subseteq D$. In addition $\rho g = 1_H$. So ρ is an

isomorphism of Klein surfaces.

Remark 0.1.10. In view of the previous example we do not need to distinguish between D and H as Klein surfaces.

Another obvious but useful consequence of 0.1.7 is the following

Corollary 0.1.11. *Let S and S′ be topological surfaces and let* $f:S\longrightarrow S'$ *be a continuous map. Then*

(1) If S′ is a Klein surface, there is at most one structure of Klein surface on S, such that the map f is a morphism.

(2) If f is surjective and S is a Klein surface, then there exists at most one structure of Klein surface on S′ such that f is a morphism.

One of the main tools in the study of Klein surfaces is the existence of certain double covers which are Riemann surfaces. Following **[5]** we construct now the double cover that we shall use in this book.

Construction 0.1.12. Let S be a Klein surface with dianalytic atlas $\Sigma=\{(U_i,\phi_i)|i\in I\}$. Suppose that S is not a Riemann surface and let us define

$$U_i'=U_i\times\{i\}\times\{1\},\ \text{and}\ U_i''=U_i\times\{i\}\times\{-1\}$$

where i runs over I. Now we shall identify some points in $X=\bigcup_{i\in I}U_i'\cup\bigcup_{i\in I}U_i''$.

(1) Given $i\in I$ and $D_i=\partial S\cap U_i$, we identify $D_i\times\{i\}\times\{1\}$ with $D_i\times\{i\}\times\{-1\}$.

(2) Given $(j,k)\in I\times I$ such that U_j meets U_k, and given a connected component W of $U_j\cap U_k$, we identify $W\times\{j\}\times\{\delta\}$ with $W\times\{k\}\times\{\delta\}$, for $\delta=\pm1$, if $\phi_j\phi_k^{-1}:\phi_k(W)\longrightarrow\mathbb{C}$ is analytic, and $W\times\{j\}\times\{\delta\}$ with $W\times\{k\}\times\{-\delta\}$, for $\delta=\pm1$, if $\phi_j\phi_k^{-1}:\phi_k(W)\longrightarrow\mathbb{C}$ is antianalytic.

We put $S_c=X/\{\text{identifications above}\}$. Let us write now, for each $i\in I$

$$\phi_i':U_i'\longrightarrow\mathbb{C}:\ (x,i,1)\longmapsto\phi_i(x);\quad \phi_i'':U_i''\longrightarrow\mathbb{C}:\ (x,i,-1)\longmapsto\overline{\phi_i(x)}$$

Obviously, if $p:X\longrightarrow S_c$ denotes the canonical projection and $\tilde{U}_i=p(U_i'\cup U_i'')$, the family $\{\tilde{U}_i|i\in I\}$ is an open cover of S_c. Even more, each map

$$\tilde{\phi}_i:\tilde{U}_i\longrightarrow\mathbb{C}:\ u\longmapsto\begin{cases}\phi_i'(u) & \text{if } u\in U_i',\\ \phi_i''(u) & \text{if } u\in U_i'',\end{cases}$$

is a homeomorphism onto its image. Thus $\Sigma_c=\{(\tilde{U}_i,\tilde{\phi}_i)|i\in I\}$ is an analytic atlas on S_c. It is also clear that ∂S_c is empty. In such a way S_c is a Riemann surface.

We claim that there exists a morphism $f:S_c\longrightarrow S$ and an antianalytic map

$\sigma:S_c \longrightarrow S_c$ such that $f\sigma=f$ and $\sigma^2=1_{S_c}$. In fact it suffices to take

$$f:S_c \longrightarrow S:\ u=p(v,i,\delta) \longmapsto v,\ v\in U_i,\ \delta=\pm 1.$$

The fibers of f have one or two points, and we define

$$\sigma:S_c \longrightarrow S_c:\ u \longmapsto \begin{cases} u & \text{if } \#f^{-1}(f(u))=1, \\ f^{-1}(f(u))\setminus\{u\} & \text{if } \#f^{-1}(f(u))=2. \end{cases}$$

We shall call the triple (S_c,f,σ) *the double cover of* S.

Remarks 0.1.13. (1) The surface S_c is constructed in such a way that a triangulation T of S can be lifted to a triangulation T′ of S_c. Let β be the number of 0-simplices of T′ in the connected components B of ∂S such that $f|f^{-1}(B):f^{-1}(B) \longrightarrow B$ is a bijection. If n_i (resp. m_i) is the number of i-simplices of T (resp. T′), it is obvious that

$$m_2=2n_2,\ m_1=2n_1-\beta,\ m_0=2n_0-\beta.$$

Thus $\chi(S_c)=2\chi(S)$.

(2) With the notations in 0.1.1, since S_c is orientable without boundary,

$$g(S_c)=[2-\chi(S_c)]/2=1-\chi(S).$$

Thus from 0.1.1 we deduce that

$$g(S_c)=\begin{cases} 2g(S)+k(S)-1 & \text{if S is orientable,} \\ g(S)+k(S)-1 & \text{if S is nonorientable.} \end{cases}$$

The topological genus $g(S_c)$ is also called the *algebraic genus of* S, and is denoted by p(S).

(3) In order to extend isomorphisms between Klein surfaces to their double covers we need the following (cf. [5], Prop 1.6.2).

"Given a morphism g from a Riemann surface S onto a Klein surface S′, with the double cover (S'_c,f',σ'), there exists a unique morphism $g':S \longrightarrow S'_c$ such that $f'g'=g$."

Comment. The purpose of this book is the study of groups of automorphisms of Klein surfaces. We end this section analyzing this question for the Klein surface H introduced in 0.1.3.

Definition 7. An *automorphism* of a Klein surface S is an isomorphism $f:S \longrightarrow S$ in the category of Klein surfaces.

From 0.1.7 follows that the set Aut(S) of automorphisms of S forms a group with respect to the composition.

Example 0.1.14. To each non singular matrix with real entries $A=\begin{pmatrix} a & b \\ c & d \end{pmatrix}$ we associate the map

$$f_A:H \longrightarrow H : z \longmapsto \begin{cases} \dfrac{az+b}{cz+d} & \text{if det } A>0 \\ \dfrac{a\bar{z}+b}{c\bar{z}+d} & \text{if det } A<0. \end{cases}$$

It is obvious that each $f_A \in \text{Aut}(H)$. Of course $f_A = f_{rA}$ for every non zero $r \in \mathbb{R}$. Hence the group $PGL(2,\mathbb{R})=GL(2,\mathbb{R})/\mathbb{R}\setminus\{0\}$ can be embedded into Aut(H) via

$$PGL(2,\mathbb{R}) \longrightarrow \text{Aut}(H): [A] \longmapsto f_A.$$

We are going to prove that this map is surjective. Let $f \in \text{Aut}(H)$. If f preserves the orientation we apply (1) in 0.1.5 and 0.1.6 to deduce that f is analytic. Let $\rho:D \longrightarrow H$ be the isomorphism constructed in 0.1.4. It is analytic, and so, the same holds true for $g=\rho^{-1}f\rho$. Hence, by the maximum principle

$$g(z)=\mu(z-\alpha)(1-\bar{\alpha}z)^{-1}, \; z \in D,$$

for some $\alpha \in D$ and $\mu \in \mathbb{C}$ with $|\mu|=1$. Consequently,

$$f(z)=\frac{az+b}{cz+d}, \text{ for some } a,b,c,d \in \mathbb{C}.$$

Since $f(H)=H$ we get $f(\mathbb{R}\setminus\{-d/c\}) \subseteq \mathbb{R}$, by continuity, and it is easy to see that we can choose a,b,c,d to be real numbers. Moreover, $f(i) \in H$ implies $ad-bc>0$. In case f reverses the orientation of H, the map $h:H \longrightarrow H:z \longmapsto \overline{-f(z)}$ is an automorphism of H and it preserves the orientation. From the above

$$f(z)=\frac{a\bar{z}+b}{c\bar{z}+d} \text{ for some } a,b,c,d \in \mathbb{R}, \; ad-bc<0$$

and we are done.

We shall denote $\text{Aut}^+(H)$ the subgroup of orientation preserving elements in Aut(H). We have obtained:

Theorem 0.1.15. *(1)* $\text{Aut}(H)=PGL(2,\mathbb{R})$.
(2) The group Aut(H) *is a topological group.*
(3) Conjugate subgroups of Aut(H) *are homeomorphic.*
(4) The map $e_v:\text{Aut}(H)\times H \longrightarrow H: (f,x) \longmapsto f(x)$ *is continuous.*

We finish this section introducing certain subgroups of Aut(H) to be studied in the next one.

Definition 8. A subgroup Γ of Aut(H) is said to be *discrete* if it is discrete as a topological subspace of Aut(H).

From (3) above it is evident that given two conjugate subgroups Γ_1 and Γ_2

of Aut(H), Γ_1 is discrete if and only if Γ_2 is discrete. Also, by (2), if there exists $\{f_n:n\in\mathbb{N}\}\subseteq\Gamma$ with $f_n\neq f_m$ for $n\neq m$ and $\{f_n\}\longrightarrow f\in\mathrm{Aut(H)}$, then Γ is not discrete.

0.2. NON-EUCLIDEAN CRYSTALLOGRAPHIC GROUPS

Non-euclidean crystallographic groups were introduced by Wilkie **[126]**. Along this section we shall state some results to be used later.

Definition 9. Let Γ be a discrete subgroup of Aut(H). We say that Γ is a *non-euclidean crystallographic group* (shortly *NEC group*) if the quotient H/Γ is compact.

Definition 10. For a given $f\in\mathrm{Aut(H)}$ there exist just two matrix $A,B\in GL(2,\mathbb{R})$ such that $f_A=f=f_B$ whose determinant has absolute value equal to 1. Moreover $B=-A$ and so $\det A = \det B$, $\mathrm{tr}A = -\mathrm{tr}B$. We define $\det f=\det A$, $|\mathrm{tr}f|=|\mathrm{tr}A|$. Of course $f\in\mathrm{Aut}^+(H)$ if and only if $\det f=1$.

(1) We say that f is

(i) *hyperbolic* if $\det f=1$, $|\mathrm{tr}f|>2$,

(ii) *elliptic* if $\det f=1$, $|\mathrm{tr}f|<2$,

(iii) *parabolic* if $\det f=1$, $|\mathrm{tr}f|=2$,

(iv) *glide reflection* if $\det f=-1$, $|\mathrm{tr}f|\neq 0$,

(v) *reflection* if $\det f=-1$, $\mathrm{tr}f=0$.

(2) An NEC group Γ is said to be a *fuchsian group* if $\Gamma\subseteq\mathrm{Aut}^+(H)$. Otherwise Γ is said to be a *proper* NEC group.

(3) Given a proper NEC group Γ, the *canonical fuchsian subgroup of* Γ is $\Gamma^+= =\Gamma\cap\mathrm{Aut}^+(H)$.

Remarks 0.2.1. Let Γ and Γ' be proper NEC groups.

(1) It is rather obvious that $\det(fg)=(\det f)(\det g)$ for $f,g\in\mathrm{Aut(H)}$. In particular if $f,g\in\Gamma^+$ or $f,g\in\Gamma\setminus\Gamma^+$, then $fg\in\Gamma^+$. Thus

(2) Γ^+ is a normal subgroup of Γ of index two.

(3) If $\Gamma\triangleleft\Gamma'$, then $\Gamma^+\triangleleft\Gamma'^+$. In fact, if $f\in\Gamma'^+$ and $g\in\Gamma^+$, we know that $u=fgf^{-1}\in\Gamma$ because Γ is a normal subgroup of Γ'. Moreover $\det u=\det g$, and so $u\in\Gamma^+$.

(4) If Γ is a subgroup of Γ' then $[\Gamma':\Gamma]=[\Gamma'^+:\Gamma^+]$. In fact

$$[\Gamma':\Gamma] = [\Gamma':\Gamma^+]/[\Gamma:\Gamma^+] = [\Gamma':\Gamma'^+][\Gamma'^+:\Gamma^+]/2 = [\Gamma'^+:\Gamma^+].$$

(5) We shall prove in 1.2.3 that if a Klein surface S which is not a Riemann surface is represented as H/Γ for some proper NEC group Γ, then the double cover S_c constructed in the previous section verifies $S_c=H/\Gamma^+$.

Elementary computations give us the following:

Proposition 0.2.2. *Let us extend each* $f_A \in \mathrm{Aut}(H)$, where $A=\begin{bmatrix} a & b \\ c & d \end{bmatrix}$, *to* $\bar{\mathbb{C}}=\mathbb{C}\cup\{\infty\}$ *in the natural way:*

$$\tilde{f}_A:\bar{\mathbb{C}}\longrightarrow\bar{\mathbb{C}}:\ z\longmapsto\begin{cases} -d/c & \text{if} & & z=\infty, \\ \infty & \text{if} & & z=-d/c, \\ (az+b)/(cz+d) & \text{if} & \det f_A=+1, & z\neq\infty,\ -d/c, \\ (a\bar{z}+b)/(c\bar{z}+d) & \text{if} & \det f_A=-1, & z\neq\infty,\ -d/c. \end{cases}$$

Let $f\in\mathrm{Aut}(H)$ *and let* $\mathrm{Fix}(f)=\{z\in\bar{\mathbb{C}}\mid\tilde{f}(z)=z\}$. *Then*

$$\mathrm{Fix}(f)=\begin{cases} \textit{two points on } \bar{\mathbb{R}}=\mathbb{R}\cup\{\infty\}, \textit{ if f is hyperbolic or glide reflection,} \\ \textit{one point on } \bar{\mathbb{R}} \textit{ if f is parabolic,} \\ \textit{a circle or a line perpendicular to } \mathbb{R} \textit{ if f is a reflection,} \\ \textit{two non real conjugate points if f is elliptic.} \end{cases}$$

In particular, if we look again on each f as on a map $f:H\longrightarrow H$, *it has fixed points (on H) if and only if it is elliptic or a reflection.*

Our next goal is to obtain a presentation of an NEC group by means of generators and defining relations. First we need

Definition 11. Let Γ be an NEC group. A *fundamental region for* Γ is a closed subset F of H satisfying

(i) If $z\in H$, then there exists $g\in\Gamma$ such that $g(z)\in F$.
(ii) If $z\in H$ and $f,g\in\Gamma$ verify $f(z),g(z)\in \mathrm{Int}F$, then $f=g$.
(iii) The non-euclidean area of $F\setminus\mathrm{Int}F$ is zero, *i.e.*

$$\mu(F\setminus\mathrm{Int}F)=\iint_{F\setminus\mathrm{Int}F}\frac{dx\,dy}{y^2}=0.$$

The existence of a fundamental region can be seen as follows [126]:

Construction 0.2.3. Let Γ be an NEC group.
(1) There exists a point $p\in H$ having trivial stabilizer in Γ, *i.e.* $g(p)\neq p$ for every $g\in\Gamma$, $g\neq 1_H$. Suppose, to get a contradiction, that this is not true.

We can assume the existence of an upper half euclidean line l perpendicular to $\mathbb{R}$ such that $l\neq\mathrm{Fix}(\gamma)$ for every $\gamma\in\Gamma$. Otherwise we choose a sequence $\{x_n\mid n\in\mathbb{N}\}$ convergent to a point $a\in H$, lying on an euclidean line parallel to $\mathbb{R}$, and the upper half euclidean line l_n perpendicular to $\mathbb{R}$ and passing through x_n verifies $l_n=\mathrm{Fix}(\gamma_n)$ for some $\gamma_n\in\Gamma$. Consequently $\gamma_n\neq\gamma_m$ if

$n \neq m$ and $\lim\{\gamma_n(a)\} = \lim\{\gamma_n(x_n)\} = \lim\{x_n\} = a$ what contradicts 1.1.3 in Ch. 1.

Now we choose a sequence $\{y_n | n \in \mathbb{N}\}$ of points of H lying on l convergent to some point $b \in H$. By our assumption there exists a sequence of pairwise distinct transformations $\{g_n | n \in \mathbb{N}\} \subseteq \Gamma$ such that $g_n(y_n) = y_n$ for every $n \in \mathbb{N}$. This leads us to a contradiction as before.

(2) Now it is easy to check that

$$F = F_p = \{z \in H | d(z,p) \leq d(g(z),p) \text{ for each } g \in \Gamma\}$$

is a fundamental region for Γ, where $d(u,v)$ is the non-euclidean distance between u and v, *i.e.*

$$d(u,v) = \int_{C_{u,v}} \frac{(dx^2 + dy^2)^{1/2}}{y},$$

$C_{u,v}$ being the geodesic joining u and v, *i.e.* a circle or a line orthogonal to $\mathbb{R}$.

F_p verifies conditions (i), (ii), and (iii) of definition 11:

(i) Let z be a point in H. Then, since Γ is discrete, the orbit O_z of z under Γ is closed. Thus there exists $w \in O_z$ such that $d(w,p) \leq d(w',p)$ for each $w' \in O_z$. If $w = g(z)$, $g \in \Gamma$, then it is obvious that $g(z) = w \in F_p$.

(ii) It is rather obvious that

$$\mathrm{Int}F_p = \{z \in H | d(z,p) < d(g(z),p), \text{ for each } g \in \Gamma \setminus \{1_H\}\}.$$

Then, $z \in H$, $f,g \in \Gamma$ and $f(z), g(z) \in \mathrm{Int}F_p$ imply for $f \neq g$

$$d(f(z),p) < d(gf^{-1}(f(z)),p) = d(g(z),p),$$

$$d(g(z),p) < d(fg^{-1}(g(z)),p) = d(f(z),p),$$

which is false. Thus $f = g$.

(iii) It follows easily from (1) in the proposition stated below.

The fundamental region F_p just constructed is called the *Dirichlet region with center* p. The following proposition collects the main properties of F_p, and it is also due to Wilkie **[126]**:

Proposition 0.2.4. *(1)* F_p *is a bounded convex polygon (in the non-euclidean metric) with a finite number of edges (or sides).*

(2) F_p *is homeomorphic to the closed disc.*

(3) $F_p \setminus \mathrm{Int}F_p$ *is a closed Jordan curve.*

(4) There is a finite number of points (vertices) on $F_p \setminus \mathrm{Int}F_p$ *which divide it into Jordan arcs (edges).*

(5) The edges of F_p *are classified as follows:*

(5.1) $e = F_p \cap gF_p$, *where* $g \in \Gamma$ *is a reflection,*

(5.2) $e = F_p \cap gF_p$, *where* $g \in \Gamma$, $g^2 \neq 1_H$,

(5.3) e *for which there exists an elliptic transformation* $g \in \Gamma$ *of order* 2, *such*

that $e \cup ge = F_p \cap gF_p$.
In the second and third cases ge *is another edge of* F.
(6) If $g \in \Gamma$ *and* F_p, gF_p *do not have an edge in common, then, since* F_p *is convex,* $F_p \cap gF_p$ *has just one vertex.*

Definition 12. A fundamental region F for Γ verifying the properties listed in the previous proposition is called *regular.* In particular the fundamental region F_p for Γ, just constructed, is regular.

Construction 0.2.5. Let F be a regular fundamental region for an NEC group Γ. Given $g \in \Gamma$, gF is said to be a *face.* It is obvious that the map

$$\Gamma \longrightarrow \{\text{faces}\}: g \longmapsto gF$$

is a bijection and $H = \bigcup_{g \in \Gamma} gF$. In fact $\{gF | g \in \Gamma\}$ is a tessellation of H.

(1) Given an edge e of F let g_e be the unique transformation for which g_eF meets F in the edge e, *i.e.* $e = F \cap g_eF$. We claim that

$$\{g_e | e \in \text{edges of } F\}$$

is a set of generators for Γ. In fact given $g \in \Gamma$ there exists a sequence of elements of Γ

$$g_1 = 1_H,\ g_2, \ldots, g_{n+1} = g$$

such that g_iF and $g_{i+1}F$ meet one to another in an edge, say $g_i(e_i)$, where e_i is an edge of F. Now clearly $g_i(g_{e_i}F) = g_{i+1}F$ and so $g_{i+1} = g_ig_{e_i}$, for $i = 1, \ldots, n$. As a result $g = g_{e_1}g_{e_2}\ldots g_{e_n}$ for some edges $e_1, \ldots, e_n$ of F.

(2) Now we label the edges of F as follows. First we label edges of type (5.1). Afterwards, if we label e an edge of type (5.2) or (5.3), the edge ge is labelled e' if $g \in \Gamma^+$, and e^* if $g \in \Gamma \setminus \Gamma^+$. We write down the labels of the edges in counter-clockwise order. We say that (e,e') and (e,e^*) are *paired* (or *congruent)* edges. In this way we obtain the *surface symbol* for Γ, which will determine the presentation of Γ as well as the topological structure of H/Γ.

(3) A regular fundamental region F for Γ can be modified. Let e and $\hat{e}$ be congruent edges and let $g \in \Gamma$ be an element for which $g^{-1}(e) = \hat{e}$. Given a hyperbolic interval f joining two vertices of F and splitting F into two regions A and B containing e and $\hat{e}$ respectively, $A \cup gB$ is a new fundamental region for Γ, having two new congruent edges f and $\hat{f}$ where $\hat{f} = g^{-1}(f)$ instead of $(e, \hat{e})$ and suitably relabelled other edges. Repeating this procedure in suitable way one can arrive to the fundamental region with the following surface symbol

(*) $\xi_1\xi_1' \ldots \xi_r\xi_r'\varepsilon_1\gamma_{10}\ldots\gamma_{1s_1}\varepsilon_1'\ldots\varepsilon_k\gamma_{k0}\ldots\gamma_{ks_k}\varepsilon_k'\alpha_1\beta_1\alpha_1'\beta_1'\ldots\alpha_g\beta_g\alpha_g'\beta_g'$

(**) $\xi_1\xi_1' \ldots \xi_r\xi_r'\varepsilon_1\gamma_{10}\ldots\gamma_{1s_1}\varepsilon_1'\ldots\varepsilon_k\gamma_{k0}\ldots\gamma_{ks_k}\varepsilon_k'\delta_1\delta_1^*\ldots\delta_g\delta_g^*$

according with H/Γ is orientable or not. It is known that this region can be chosen to be convex (see [72],[94],[107]).

(4) Identifying points in the paired edges, we arrive to

$$H/\Gamma = \text{sphere with k discs removed and } \begin{cases} \text{g handles added if (*)} \\ \text{g cross-cups added if (**)} \end{cases}$$

(5) To obtain the set of defining relations for Γ, consider the faces meeting at each vertex of F. Since Γ is discrete, the number of these faces is finite. Let us pick one of the vertices of F and let

$$F=F_o,F_1,\dots,F_n,F_{n+1}=F,$$

be the corresponding chain of faces. Clearly there exists a sequence $g_1,\dots,g_n$ of elements of Γ such that

$$F_1=g_1F,\ F_2=g_2g_1F,\dots,\ F=F_{n+1}=g_n\dots g_1F.$$

So each vertex induce a relation

$$g_n\dots g_1=1_H.$$

It turns out that the relations of this type and the relations $g_e^2=1_H$ coming from the edges of F fixed by a unique nontrivial element g_e of Γ form the set of all defining relations for Γ.

(6) Having a surface symbol (*) or (**) and using procedures described in (1) and (5) one can arrive to the following presentation of Γ:

generators: x_i; $i=1,\dots,r$,
e_i; $i=1,\dots,k$,
c_{ij}; $i=1,\dots,k$, $j=0,\dots,s_i$,
a_i,b_i; $i=1,\dots,g$, in the case (*),
d_i; $i=1,\dots,g$, in the case (**),

(0.2.5.1)

relations: $x_i^{m_i}=1$; $i=1,\dots,r$,
$e_i^{-1}c_{i0}e_ic_{is_i}=1$, $i=1,\dots,k$,
$c_{i,j-1}^2=c_{ij}^2=(c_{i,j-1}c_{ij})^{n_{ij}}=1$,
$x_1\dots x_re_1\dots e_k[a_1,b_1]\dots[a_g,b_g]=1$, in case (*),
$x_1\dots x_re_1\dots e_kd_1^2\dots d_g^2=1$, in case (**),

where a,b,c,d,e,x correspond to the transformations induced by the edges $\alpha,\beta,\gamma,\delta,\varepsilon,\xi$ (see (3)), $[a_i,b_i]=a_ib_ia_i^{-1}b_i^{-1}$, and m_i, n_{ij} are certain positive integers, being the number of faces meeting F at common vertices for the edges, (ξ_i,ξ_i') and $(\gamma_{i,j-1},\gamma_{ij})$ respectively. The generators listed above will be said in this book to be *canonical generators* and we reserve the letters a,b,c,d,e,x only to denote them.

The elements x_i are elliptic (in particular $x_i\neq 1_H$ and $m_i\neq 1$), c_{ij} are reflections (and since $c_{i,j-1}\neq c_{ij}$, we deduce that $n_{ij}\neq 1$), d_i are glide reflections, a_i and b_i are hyperbolic and e_i are hyperbolic or, in a few cases, elliptic.

We concentrate this presentation of Γ by means of generators and relations in the following way:

Definition 13. (1) A *signature* is a collection of nonnegative integers and symbols of the following form:

$$\sigma=(g;\pm;[m_1,\ldots,m_r];\{(n_{11},\ldots,n_{1s_1}),\ldots,(n_{k1},\ldots,n_{ks_k})\}).$$

If the sign "+" appears we write sign(σ)="+", and otherwise sign(σ)="-". The numbers $m_1,\ldots,m_r$ are called the *proper periods of* σ and the numbers $n_{i1},\ldots,n_{is_i}$ are the *periods* of the *period-cycle* $(n_{i1},\ldots,n_{is_i})$. We shall write $g(\sigma)=g$, and $k(\sigma)=k$. An empty set of proper periods, (*i.e.* $r=0$) will be denoted by [-], an empty period-cycle (*i.e.* $s_i=0$) by (-), and finally the fact that σ has no period-cycles (*i.e.* $k=0$) by {-}.

(2) Now given an NEC group Γ with presentation (0.2.5.1), we define the *signature* $\sigma(\Gamma)$ *of* Γ:

$$\sigma(\Gamma)=(g;\pm;[m_1,\ldots,m_r];\{(n_{11},\ldots,n_{1s_1}),\ldots,(n_{k1},\ldots,n_{ks_k})\})$$

the sign of $\sigma(\Gamma)$ being "+" in the case (*) and "-" otherwise. (In the first case we say that Γ is *orientable*, and *nonorientable* in the second one.) The integer g is called the *orbit genus* of Γ.

In the obvious manner, the presentation of Γ can be read off from this signature. Moreover, it describes the topological structure of H/Γ and gives us a procedure to classify NEC groups up to isomorphism, **[86]**, **[126]**:

Remark 0.2.6. Let Γ be an NEC group, $S=H/\Gamma$ and $\sigma=\sigma(\Gamma)$. Then

(1) S is orientable if and only if sign(σ)="+",

(2) $g(\sigma)$ equals the topological genus of S,

(3) $k(\sigma)$ equals the number of connected components of ∂S. In particular, if $g(\sigma)=0$, then sign(σ)="+". Also, if Γ is a fuchsian group, then it has no reflections. So $k(\sigma)=0$ and consequently S has empty boundary.

(4) Let Γ' be another NEC group with signature

$$\sigma'=\sigma(\Gamma')=(g';\pm;[m'_1,\ldots,m'_{r'}];\{(n'_{11},\ldots,n'_{1s'_1}),\ldots,(n'_{k'1},\ldots,n'_{k's'_{k'}})\}).$$

Let us write $C_i=(n_{i1},\ldots,n_{is_i})$ and $C'_i=(n'_{i1},\ldots,n'_{is'_i})$. Then Γ and Γ' are isomorphic as abstract groups if and only if

(i) sign(σ)=sign(σ'),

(ii) $g=g'$; $r=r'$; $k=k'$; $s_i=s'_i$, for $i=1,\ldots,k$,

(iii) $(m'_1,\ldots,m'_r)$ is a permutation of $(m_1,\ldots,m_r)$,

(iv) if $\mathrm{sign}(\sigma)=$"+", then there exists a permutation ϕ of $\{1,\ldots,k\}$ such that for each i, $1\leq i\leq k$, one of the following conditions holds true:
(*) C_i' is a cyclic permutation of $C_{\phi(i)}$.
(**) C_i' is a cyclic permutation of the inverse of $C_{\phi(i)}$.

(v) if $\mathrm{sign}(\sigma)=$"-", then there exists a permutation ϕ of $\{1,\ldots,k\}$ such that either every C_i' is a cyclic permutation of $C_{\phi(i)}$ or every C_i' is a cyclic permutation of the inverse of $C_{\phi(i)}$.

Definition 14. An NEC group Γ having signature

$$\sigma(\Gamma)=(g;\pm;[-];\{(-),\overset{k}{\ldots},(-)\}),\ k\geq 0$$

is said to be a *surface group*. When $k\geq 1$, Γ is a *bordered surface group*.

Remark 0.2.7. (1) It is easy to prove (see *e.g.* **[78]**) that the only elements of finite order in Aut(H) are those that are reflections or elliptic. Moreover, with the precedent notations, an element g of an NEC group Γ has finite order if and only if it is conjugate to either a power of some x_i or a power of $c_{i,j-1}c_{ij}$ or some c_{ij}.

(2) As a consequence, Γ is a surface group if and only if it has no orientation preserving elements of finite order.

Definition 15. (1) Let $\sigma=(g;\pm;[m_1,\ldots,m_r];\{(n_{11},\ldots,n_{1s_1}),\ldots,(n_{k1},\ldots,n_{ks_k})\})$ be a signature and let $\alpha(\sigma)=\alpha=2$ if $\mathrm{sign}(\sigma)=$"+", and $\alpha(\sigma)=\alpha=1$ otherwise. The *area of* σ is defined to be

$$\mu(\sigma)=2\pi[\alpha g+k-2+\sum_{i=1}^{r}(1-1/m_i)+1/2\sum_{i=1}^{k}\sum_{j=1}^{s_i}(1-1/n_{ij})].$$

We denote $\bar{\mu}(\sigma)=\mu(\sigma)/2\pi$.

(2) If Γ is an NEC group, then the *area of* Γ is the hyperbolic area of a fundamental region of Γ. We denote it by $\mu(\Gamma)$.

The following theorem justifies these definitions (see **[130]**).

Theorem 0.2.8. *(1) The hyperbolic area does not depend on the fundamental region for Γ we choose.*
(2) If Γ is an NEC group, then $\mu(\Gamma)=\mu(\sigma(\Gamma))$.
(3) The signature σ is the signature of some NEC group if and only if $\mu(\sigma)>0$, and $\alpha+g(\sigma)\geq 2$.

Remarks 0.2.9. (1) Let Γ be a surface NEC group with signature

$$\sigma=\sigma(\Gamma)=(g;\pm;[-];\{(-),\overset{k}{\ldots},(-)\}),\ k\geq 0.$$

Then $\mu(\Gamma)=2\pi(\alpha g+k-2)$. Thus if $S=H/\Gamma$ we deduce from 0.1.13 (2) and 0.2.6 that $\mu(\Gamma)=2\pi(p(S)-1)$.

(2) Let Γ' be an NEC group and let Γ be a subgroup of Γ' of finite index. We shall see, in 2.1.1, that Γ is an NEC group. Then if $[\Gamma':\Gamma]=N$, $\Gamma'=\Gamma g_1\cup\ldots\cup\Gamma g_N$ for some $g_1,\ldots,g_N$ in Γ'. Thus if F' is a fundamental region for Γ', it is easy to prove that $F=g_1F'\cup\ldots\cup g_NF'$ is a fundamental region for Γ. Consequently

$$\mu(\Gamma)=\text{area}(F)=\sum_{i=1}^{N}\text{area}(g_iF')=N\times\text{area}(F')=N\times\mu(\Gamma'),$$

i.e.

(0.2.9.1) $$\frac{\mu(\Gamma)}{\mu(\Gamma')}=[\Gamma':\Gamma]\ \text{(Hurwitz Riemann formula).}$$

0.3. TEICHMÜLLER SPACES

Along this section Γ is an NEC group and we consider $\Omega=\text{Aut}(H)$ as a topological group (see 0.1.15). For an abstract group G we denote by Aut(G) the group of its automorphisms and by Inn(G) the normal subgroup of Aut(G) consisting of all inner automorphisms of G.

Definition 16. The *Weil space of Γ* (with respect to Ω) is the set $\mathbf{R}(\Gamma)=\{$group monomorphisms $r:\Gamma\longrightarrow\Omega$ such that $\Gamma'=r(\Gamma)$ is an NEC group$\}$, with the topology inherited from Ω^{Γ}.

Remarks 0.3.1. (1) The notion of Weil space was introduced by Weil **[125]** in the context of study of discrete subgroups of a semisimple Lie group. See also **[84]**.

(2) Using the equality $\text{Aut}(\Omega)=\text{Inn}(\Omega)$, we obtain that for $\alpha\in\text{Aut}(\Omega)$, $r=\alpha|\Gamma\in\mathbf{R}(\Gamma)$, and also

(3) The map $\text{Aut}(\Omega)\times\mathbf{R}(\Gamma)\longrightarrow\mathbf{R}(\Gamma)$: $(\alpha,r)\longmapsto\alpha r$ defines an action of $\text{Aut}(\Omega)$ on $\mathbf{R}(\Gamma)$. The orbit $[r]$ of $r\in\mathbf{R}(\Gamma)$ equals

$$\{s\in\mathbf{R}(\Gamma)\mid\text{there exists } g\in\Omega \text{ with } r(\gamma)=gs(\gamma)g^{-1} \text{ for each } \gamma\in\Gamma\}.$$

Definition 17. The *Teichmüller space of Γ* is the orbit space $\mathbf{T}(\Gamma)=\mathbf{R}(\Gamma)/\text{Aut}(\Omega)$ of $\mathbf{R}(\Gamma)$ under the action of $\text{Aut}(\Omega)$, endowed with the quotient topology.

The main properties of $\mathbf{T}(\Gamma)$ are collected in the following

Theorem 0.3.2. *(a) The topology of* $\mathbf{T}(\Gamma)$ *can be derived from the Teichmüller metric, see* **[88]**, *and* $\mathbf{T}(\Gamma)$ *is a complete metric space of finite dimension* $d(\Gamma)$.

(b) *Fricke and Klein proved in* **[50]** *that if* Γ *is a fuchsian group with* $\sigma(\Gamma)=(g;\pm;[m_1,\dots,m_r];\{-\})$ *then* $T(\Gamma)$ *is a cell of dimension* $d(\Gamma)=6(g-1)+2r$.

(c) *Singerman proved in* **[116]** *that if* Γ *is a proper NEC group, then* $T(\Gamma)$ *is a cell of dimension* $d(\Gamma)=d(\Gamma^+)/2$. (See also **[73]**).

(d) *Given two NEC groups* Γ_1 *and* Γ_2, *and a group monomorphism* $\alpha:\Gamma_2\hookrightarrow\Gamma_1$ *the induced map*

$$T(\alpha):T(\Gamma_1)\longrightarrow T(\Gamma_2):\ [r]\longmapsto [r\alpha],$$

is an isometric embedding **[88]**.

Definition 18. *(Moduli space and modular group)* Given $r\in\mathbf{R}(\Gamma)$ and $\beta\in\mathrm{Aut}(\Gamma)$ it is obvious that $r\beta\in\mathbf{R}(\Gamma)$. Even more, if $\alpha\in\mathrm{Aut}(\Omega)$ and $s=\alpha r$, then $s\beta=\alpha(r\beta)$. Thus we obtain an action

$$\mathrm{Aut}(\Gamma)\times T(\Gamma)\longrightarrow T(\Gamma):\ (\beta,[r])\longmapsto [r\beta].$$

The *moduli space of* Γ is the quotient $M(\Gamma)=T(\Gamma)/\mathrm{Aut}(\Gamma)$ endowed with the quotient topology.

The *modular group of* Γ is the quotient $\mathrm{Mod}(\Gamma)=\mathrm{Aut}(\Gamma)/\mathrm{Inn}(\Gamma)$.

Remarks 0.3.3. (a) The map $\mathrm{Mod}(\Gamma)\times T(\Gamma)\longrightarrow T(\Gamma)$: $([\beta],[r])\longmapsto [r\beta]$ defines an action and $M(\Gamma)=T(\Gamma)/\mathrm{Mod}(\Gamma)$. In fact, if $\alpha\in\mathrm{Inn}(\Gamma)$, then α acts as an automorphism of Ω. Thus $[r]=[\alpha r]$ for $r\in\mathbf{R}(\Gamma)$ and so $\mathrm{Inn}(\Gamma)$ acts trivially on $T(\Gamma)$. Hence we are done.

(b) $\mathrm{Mod}(\Gamma)$ acts as a group of isometries of $T(\Gamma)$, see **[88]**.

(c) It is proved in **[88]** that if $\sigma(\Gamma)=(g;+;[-];\{(-)\})$, then the elements of any finite subgroup of $\mathrm{Mod}(\Gamma)$ have a common fixed point.

(d) $\mathrm{Mod}(\Gamma)$ acts as a totally discontinuous group of transformations of $T(\Gamma)$ (see **[75]**,**[88]**,**[92]**).

(e) The action of $\mathrm{Mod}(\Gamma)$ on $T(\Gamma)$ is not necessarily faithful (*i.e.* the group homomorphism

$$\mathrm{Mod}(\Gamma)\longrightarrow\mathrm{Isom}(T(\Gamma)):[\beta]\ \longmapsto\ T(\beta)$$

is not necessarily injective). In fact Macbeath and Singerman **[88]** proved

Theorem 0.3.4. *The following conditions are equivalent:*

(1) $\mathrm{Mod}(\Gamma)$ *fails to act faithfully on* $T(\Gamma)$,

(2) There exist an NEC group Γ' *and a group monomorphism* $\alpha:\Gamma\longrightarrow\Gamma'$ *such that* $d(\Gamma)=d(\Gamma')$ *and* $\alpha(\Gamma)$ *is a normal subgroup of* Γ'.

The full list of pairs (σ,σ') of *fuchsian signatures* (*i.e.* signatures of fuchsian groups) such that $\sigma=\sigma(\Gamma)$, $\sigma'=\sigma(\Gamma')$ for some groups Γ and Γ' verifying (2) in the theorem above was obtained by Singerman in **[115]**. We write $[\sigma':\sigma]=[\Gamma':\alpha(\Gamma)]$.

List of Singerman (0.3.5)

σ′	σ	[σ′:σ]
(0;+;[2,2,2,2t])	(1;+;[t])	2
(0;+;[2,2,2,2,t])	(1;+;[t,t])	2
(0;+;[2,2,2,2,2,2])	(2;+;[-])	2
(0;+;[2,2,2,t]), t ≥ 3	(0;+;[t,t,t,t])	4
(0;+;[2,2,t,u]), t + u ≥ 5	(0;+;[t,t,u,u])	2
(0;+;[3,3,t]), t ≥ 4	(0;+;[t,t,t])	3
(0;+;[2,3,2t]), t ≥ 4	(0;+;[t,t,t])	6
(0;+;[2,t,2u]), t ≥ 4, t + u ≥ 7	(0;+;[t,t,u])	2

We shall obtain the corresponding list of pairs of NEC signatures in 2.4.7.

Comment. In order to decide if a given finite group can be the full group of automorphisms of some compact Klein surface we shall need all pairs of signatures σ and σ' such that $\sigma(\Gamma)=\sigma$, $\sigma(\Gamma')=\sigma'$, for some NEC groups Γ and Γ' such that $\Gamma\subseteq\Gamma'$ and $d(\Gamma)=d(\Gamma')$. (see section 1 of Ch.5) The list above solves this problem for fuchsian groups in the normal case *i.e.* Γ is a normal subgroup of Γ'. In the nonnormal case, the signatures σ and σ' have genus 0 and three proper periods. The full list of them was also obtained in **[115]** and it is the following:

(0.3.6)

σ′	σ	[σ′:σ]
(0;+;[2,3,7])	(0;+;[7,7,7])	24
(0;+;[2,3,7])	(0;+;[2,7,7])	9
(0;+;[2,3,7])	(0;+;[3,3,7])	8
(0;+;[2,3,8])	(0;+;[4,8,8])	12
(0;+;[2,3,8])	(0;+;[3,8,8])	10
(0;+;[2,3,9])	(0;+;[9,9,9])	12
(0;+;[2,4,5])	(0;+;[4,4,5])	6
(0;+;[2,3,4n]), n ≥ 2	(0;+;[n,4n,4n])	6
(0;+;[2,4,2n]), n ≥ 3	(0;+;[n,2n,2n])	4
(0;+;[2,3,3n]), n ≥ 3	(0;+;[3,n,3n])	4
(0;+;[2,3,2n]), n ≥ 4	(0;+;[2,n,2n])	3

Let Λ and Λ' be two NEC groups with $\sigma(\Lambda')=\tau'$. Let $T=T(\Lambda,\Lambda')$ be the set of $[r]\in T(\Lambda)$ such that there exists an NEC group Λ_r with signature τ' containing $r(\Lambda)$, for some representative r of [r]. Then

(0.3.7.1) $$T=\bigcup_{[\alpha]\in M(\Lambda)} T(\alpha)\left[\bigcup_{[j]\in I(\Lambda,\Lambda')} \mathrm{Im}T(j)\right],$$

where $I(\Lambda,\Lambda')$ is the quotient set, modulo $\mathrm{Aut}(\Lambda)$ and $\mathrm{Aut}(\Lambda')$ of the set $I(\Lambda,\Lambda')$ of all group monomorphisms $j:\Lambda \longrightarrow \Lambda'$.

All we need is to prove the equality

(0.3.7.2) $$T = \bigcup_{j\in I(\Lambda,\Lambda')} \mathrm{ImT}(j).$$

Indeed, once this is done, it is clear that

$$\bigcup_{j\in I(\Lambda,\Lambda')} \mathrm{ImT}(j) = \bigcup_{\alpha\in \mathrm{Aut}(\Lambda)} \bigcup_{\beta\in \mathrm{Aut}(\Lambda')} \bigcup_{[j]\in I(\Lambda,\Lambda')} \mathrm{ImT}(\beta j\alpha)$$

and it is easy to see that given $i,j\in I(\Lambda,\Lambda')$ and $\alpha\in \mathrm{Aut}(\Lambda)$, $\beta\in \mathrm{Aut}(\Lambda')$, with $i=\beta j\alpha$, then $\mathrm{ImT}(i)=T(\alpha)(\mathrm{ImT}(j))$. Hence, since $\mathrm{Inn}(\Lambda)$ acts trivially on $T(\Lambda)$, (0.3.7.1) follows from (0.3.7.2).

Given $[r]\in T$, let Λ_r be an NEC group with signature τ' containing $r(\Lambda)$ for some representative r of [r]. Let $\phi:\Lambda_r \longrightarrow \Lambda'$ be an isomorphism and $j=\phi r\in I(\Lambda,\Lambda')$. Clearly $[r]=[\phi^{-1}j]=T(j)([\phi^{-1}])\in \mathrm{ImT}(j)$. Conversely, let $j\in I(\Lambda,\Lambda')$, $[s]\in T(\Lambda')$ and $[r]=T(j)([s])\in T(\Lambda)$. Then, since $\mathrm{Aut}(\Omega)=\mathrm{Inn}(\Omega)$, there exists $\phi\in\Omega$ such that $r(u)=\phi sj(u)\phi^{-1}$ for every $u\in\Lambda$. Thus we see that $r(\Lambda)=\gamma s(j(\Lambda))\phi^{-1}\subseteq\phi s(\Lambda')\phi^{-1}=\Lambda_r$, and Λ_r is an NEC group with signature τ'. This finishes the proof.

The decomposition above for fuchsian groups Λ and Λ' such that Λ is a normal subgroup of Λ' was found by Harvey in [62].

CHAPTER - 1
Klein surfaces as orbit spaces of NEC groups

With natural restrictions, the quotient S/G of a Klein surface S under the action of a subgroup G of Aut(S) can be endowed with a unique structure of Klein surface, in such a way that the canonical projection $S \longrightarrow S/G$ is a morphism. Our first goal here is to prove this result, which in particular applies when S=H and G is an NEC group.

Conversely we shall show that each compact Klein surface of algebraic genus ≥ 2 can be presented as H/Γ for a suitable surface NEC group Γ.

Finally we use that to study the set of isomorphisms between two compact Klein surfaces. In particular we show the finiteness of the group of automorphisms of all compact Klein surfaces of algebraic genus ≥ 2.

1.1. QUOTIENTS OF KLEIN SURFACES

Definition 1. Let S be a Klein surface, and let G be a subgroup of Aut(S). For each point $x \in S$, the *orbit of* x *under the action of* G is the set $O_x = \{f(x) | f \in G\}$. We denote by S/G the space whose points are the orbits of points in S, endowed with the identification topology with respect to the map $\pi_G = \pi : S \longrightarrow S/G$ which sends x to O_x. The space S/G is the *quotient of* S under G. The *stabilizer of* $x \in S$ under G is the subgroup $G_x = \{f \in G | f(x) = x\}$. Given $f \in G$ we shall write $\mathrm{Fix}(f) = \{x \in S | f(x) = x\}$.

It is rather obvious that the map π defined above is an open continuous surjection. In particular, S/G is connected. Since we are interested in endowing S/G with a structure of Klein surface, we must restrict our attention to subgroups G of Aut(S) such that, at least, S/G is Hausdorff. As we shall see later, it is also necessary for each stabilizer to be finite. From now on we fix the following notations:

Given subsets U and V of S, and a subgroup G of Aut(S),

$$G(U,V) = \{f \in G | U \cap f(V) \text{ is not empty}\}.$$

We shall write $G_U = G(U,U)$. In particular $G_x = G_{\{x\}}$.

Definition 2. (a) G acts *discontinuously on* S if each point $x \in S$ possesses a neighbourhood U such that G_U is finite.

(b) G acts *properly discontinuously on* S if the following conditions hold:
(i) G acts discontinuously on S.
(ii) If $x,y \in S$ with $x \notin O_y$, there exist open neighbourhoods U and V of x and y respectively, such that G(U,V) is empty.
(iii) If $x \in S$, $f \in G_x$, $f \neq 1_S$ and the map $\phi_x f \phi_x^{-1}$ suitably restricted is analytic, then x is isolated in Fix(f).

Proposition 1.1.1. *The following conditions are equivalent:*
(1) G *acts discontinuously on* S.
(2) G_x *is finite for every* $x \in S$. *Moreover each neighbourhood* M *of* x *contains a connected neighbourhood* U *of* x *such that* $G_U = G_x$ *and* $f(U) = U$ *if* $f \in G_x$.
Proof. (2)⇒(1) is evident.
(1)⇒(2) Let us take a neighbourhood U_1 of x, $U_1 \subseteq M$ such that

$$G_{U_1} = \{1_S = f_1, \ldots, f_k\} \text{ is finite.}$$

Then $G_x \subseteq G_{U_1}$ and so G_x is also finite. Let us write $G_{U_1} \setminus G_x = \{g_1, \ldots, g_p\}$. Since S is Hausdorff we can choose disjoint open neighbourhoods V^x and $V^{g_j(x)}$, $j = 1, \ldots, p$. From the continuity of g_j we can assume $g_j(V^x) \subseteq V^{g_j(x)}$. Hence $V = U_1 \cap V^x \subseteq M$ and $G_V = G_x$. Now, G_x being a finite subgroup, $U_2 = \bigcap_{f \in G_x} f(V) \subseteq M$ is a neighbourhood of x, $G_{U_2} = G_x$ and $f(U_2) = U_2$ if $f \in G_x$. Finally it is enough to take as U the connected component of U_2 containing x.

Proposition 1.1.2. *If* G *acts properly discontinuously on* S, *the quotient* S/G *is Hausdorff.*
Proof. Let O_x and O_y be distinct points in S/G. Then $x \notin O_y$ and we can choose open neighbourhoods U and V at x and y such that G(U,V) is empty. Then the images $\Omega_1 = \pi(U)$ and $\Omega_2 = \pi(V)$ under the projection $\pi: S \longrightarrow S/G$ are open disjoint neighbourhoods of O_x and O_y. In fact, if $O_z \in \Omega_1 \cap \Omega_2$ we would have $O_u = O_z = O_v$ for some points $u \in U$, $v \in V$. That means $u = f(z)$, $z = g(v)$, $f,g \in G$. Then $h = fg \in G$ and $u = fg(v) \in U \cap h(V)$, *i.e.* $h \in G(U,V)$. This is absurd.

Discrete subgroups of Aut(H) are examples of groups acting properly discontinuously on H. To see that, we first need the following

Lemma 1.1.3. *Let* Γ *be a subgroup of* Aut(H) *and* $\{f_n | n \in \mathbb{N}\} \subseteq \Gamma$, $f_n \neq f_m$ *if* $n \neq m$.
(1) If $\{f_n | n \in \mathbb{N}\}$ *converges to* $f \in \text{Aut(H)}$, *then* $f(z) = \lim\{f_n(z) | n \in \mathbb{N}\}$ *for each* $z \in H$.
(2) If $\lim\{f_n(x) | n \in \mathbb{N}\} = y$ *for some* $x, y \in H$, *then* Γ *is not discrete.*
Proof. (1) This is immediate from the continuity of $e_v: \text{Aut(H)} \times H \longrightarrow H$, (see 0.1.15).

(2) First we look for $g\in Aut(H)$ with $g(i)=x$. Let $x=u+iv$. Then $v>0$ and we can choose real numbers c,d such that $(c^2+d^2)v=1$. Now $a=dv+cu$, $b=du-cv$ are real numbers, $ad-bc=1$ and so $g(z)=(az+b)/(cz+d)$ defines $g\in Aut(H)$ with $g(i)=x$.

Let us consider $h_n=g^{-1}f_ng\in g^{-1}\Gamma g=\Gamma'$. From the continuity of g^{-1}, $\lim\{h_n(i)|n\in\mathbb{N}\}=g^{-1}(y)=w\in H$.

If we write $h_n=h_{A_n}$, $A_n=\begin{pmatrix}a_n & b_n\\ c_n & d_n\end{pmatrix}$ we see that $|Im h_n(i)|=1/(c_n^2+d_n^2)$, and $|h_n(i)|=(a_n^2+b_n^2)/(c_n^2+d_n^2)$. Hence, $\lim\{1/(c_n^2+d_n^2)\}=Im w$ and $\{(a_n^2+b_n^2)/(c_n^2+d_n^2)\}$ is convergent to $|w|$. In particular, the sequences $\{a_n\},\{b_n\},\{c_n\}$ and $\{d_n\}$ are bounded, and they admit convergent subsequences. Thus some subsequence $\{h_{n_k}\}\subseteq\Gamma'$ converges to $h\in Aut(H)$. Hence Γ' is not discrete, and the same holds true for Γ, by 0.1.15.

Example 1.1.4. Each discrete subgroup Γ of Aut(H) acts properly discontinuously on H.

Proof. We shall see first that the stabilizer Γ_x of each $x\in H$ is finite. Otherwise we take $\{f_n|n\in\mathbb{N}\}\subseteq\Gamma_x$ such that $f_n\neq f_m$ for $n\neq m$, and so $x=\lim\{f_n(x)|n\in\mathbb{N}\}$. This contradicts the previous lemma.

Now let N be the set of natural numbers m such that H contains the euclidean ball B_m with center x and radius 1/m. Let us write $\Gamma_m=\Gamma_{B_m}$. We claim

$$\Gamma_x=\bigcap_{m\in N}\Gamma_m.$$

In fact, if $f\notin\Gamma_x$, we take open disjoint neighbourhoods U and V of x and f(x). If m is big enough, $B_m\subseteq U$, $f(B_m)\subseteq V$. Thus $f\notin\Gamma_m$. The other inclusion is obvious.

Now we prove

(1) Γ acts discontinuously on H. Let x be a point in H. Assume to a contrary that each Γ_m is infinite. Then, the finiteness of Γ_x and the equality above imply

$$\Gamma_{m_1}\supsetneq\Gamma_{m_2}\supsetneq\cdots$$

for some sequence $\{m_k|k\in\mathbb{N}\}\subseteq N$. Let us take now $f_k\in\Gamma_{m_k}\setminus\Gamma_{m_{k+1}}$. Clearly $f_k\neq f_l$ when $k\neq l$. However, if we pick $x_k\in B_{m_k}\cap f_k(B_{m_k})$ and $y_k\in B_{m_k}$ with $x_k=f_k(y_k)$ we conclude that

$$\lim\{x_k|k\in\mathbb{N}\}=x=\lim\{y_k|k\in\mathbb{N}\}$$

and so,

$$\lim\{f_k(x)|k\in\mathbb{N}\}=x$$

against the discreteness of Γ.

(2) Given two points $x,y\in H$, $x\notin O_y$ we must find open neighbourhoods U of x and V of y such that $\Gamma(U,V)$ is empty.

Let P be the set of $m\in\mathbb{N}$ such that the balls B_m and B'_m of radius 1/m and centers x and y respectively, are contained in H. We shall prove that $D_m=\Gamma(B_m,B'_m)$ is empty for some $m\in P$. Clearly $\bigcap_{m\in P} D_m$ is empty. Otherwise, for some $f\in\Gamma$ there are points $x_m\in B_m$, $y_m\in B'_m$, $f(y_m)=x_m$, $m\in P$, which implies $f(y)=x$. That means $x\in O_y$, absurd. So if no one of D_m is empty, we have

$$D_{m_1}\supsetneqq D_{m_2}\supsetneqq\cdots$$

for some sequence $\{m_k|k\in\mathbb{N}\}\subseteq P$. As before we pick $f_k\in D_{m_k}\setminus D_{m_{k+1}}$. Clearly $\lim\{f_k(y)|k\in\mathbb{N}\}=x$, $f_k\neq f_l$ if $k\neq l$. This again contradicts lemma 1.1.3.

(3) Finally, given $f\in\Gamma$, $f\neq 1_H$ verifying the hypothesis (iii) in Def. 2, f has the form:

$$f(z)=(az+b)/(cz+d),\text{ where }(b,c,d-a)\neq(0,0,0).$$

Thus $\mathrm{Fix}(f)\setminus\{x\}$ is finite.

In order to prove the main result of this chapter we must first analyze germs of automorphisms in $\mathbb{C}$ and $\mathbb{C}^+$.

Definition 3. Let us denote by $\mathscr{A}$ (resp $\mathscr{A}^+$) the group (under composition) of germs of dianalytic automorphisms of neighbourhoods of 0 in $\mathbb{C}$ (resp. in $\mathbb{C}^+$). We use τ for the germ $z\longmapsto\bar{z}$, and 1_A for the identity germ.

Let $\sigma\in\mathscr{A}$ or $\mathscr{A}^+$. Then there exist connected neighbourhoods U and U′ of 0 in $\mathbb{C}$ or $\mathbb{C}^+$ and a representative $\sigma:U\longrightarrow U'$ which admits an expansion

$$\sigma(z)=\sum_{n=1}^{\infty} a_n z^n \quad\text{or}\quad \sigma(z)=\sum_{n=1}^{\infty} a_n(\bar{z})^n$$

according with the analytic or antianalytic character of σ.

Remarks 1.1.5. (i) Of course this expansion has no term of degree 0 because $\sigma(0)=0$. Moreover, $a_1\neq 0$ since σ is injective.

(ii) If $\sigma\in\mathscr{A}^+$ it is $\sigma(U\cap\mathbb{R})\subseteq U'\cap\mathbb{R}$ and so each $a_n\in\mathbb{R}$.

(iii) In particular $\tau\sigma=\sigma\tau$ for $\sigma\in\mathscr{A}^+$.

(iv) If σ is an involution, then $|a_1|=1$.

Proposition 1.1.6. *Let $\sigma\in\mathscr{A}^+$. Then σ is analytic if and only if $a_1>0$.*

Proof. Let us choose neighbourhoods U and U′ as above. If σ is analytic it preserves the orientation in U, and in particular, on $U\cap\mathbb{R}$. Hence $a_1=\frac{d\sigma}{dz}(0)>0$. The converse uses the same argument.

Lemma 1.1.7. *Let $\sigma\in\mathscr{A}$ be an antianalytic involution. There exists an analytic $\beta\in\mathscr{A}$ such that $\sigma=\beta\tau\beta^{-1}$.*

Proof. We write $\sigma(z)=\sum_{n=1}^{\infty} a_n(\bar{z})^n$ and $\alpha=\sigma\tau+\varepsilon 1_A$ with $\varepsilon=\begin{cases} 1 & \text{if } a_1\neq -1 \\ -1 & \text{if } a_1=-1 \end{cases}$. Clearly $\alpha(z)=(a_1+\varepsilon)z+\sum_{n=2}^{\infty} a_n z^n \in \mathscr{A}$, and it is analytic. Multiplying by τ and σ we get $\alpha\tau=\sigma+\varepsilon\tau$, $\sigma\alpha=\tau+\varepsilon\sigma$ and so $\alpha^{-1}\sigma\alpha=\varepsilon\tau=\gamma^{-1}\tau\gamma$ for $\gamma=\begin{cases} 1_A & \text{if } \varepsilon=1 \\ i\,1_A & \text{if } \varepsilon=-1 \end{cases}$. Hence we can choose $\beta=\alpha\gamma^{-1}$.

Lemma 1.1.8. *Let $\sigma\in\mathscr{A}^+$ be an antianalytic involution. There exists $\beta\in\mathscr{A}^+$ such that $\sigma=\beta(-\tau)\beta^{-1}$.*

Proof. If $\sigma(z)=\sum_{n=1}^{\infty} a_n(\bar{z})^n$, then $z=\sigma^2(z)=a_1^2 z+\dots$ .Hence $a_1^2=1$, $a_1<0$, *i.e.* $a_1=-1$. Now $\beta=\sigma\tau-1_A\in\mathscr{A}^+$ is analytic and $\beta\tau=-\sigma\beta$.

Lemma 1.1.9. *(i) 1_A is the only analytic element in $\mathscr{A}^+$ of finite order.*
(ii) Let $\sigma\in\mathscr{A}^+$ be an element of finite order, $\sigma\neq 1_A$. Then σ is an antianalytic involution.
Proof. (i) Let $f\in\mathscr{A}^+$ be analytic of finite order k. Let us write

$$f(z)=zu(z),\quad u\in\mathbb{R}\{z\},\quad u(0)>0.$$

Then $z=f^k(z)=zu(z)\dots u(f^{k-1}(z))$, *i.e.*, $u(0)^k=1$, and so $u(0)=1$. Hence if $f\neq 1_A$, there exists $h\in\mathbb{R}\{z\}$ of order $r\geq 2$, such that $f(z)=z+h(z)$. Consequently

$$1_A=f^k=h^k+kh^{k-1}+\dots+kh+1_A$$

and so $0=h^k+kh^{k-1}+\dots+kh$. This is false because the order of each h^j, $2\leq j\leq k$, is greater than or equal to jr.
(ii) Of course $\sigma^2\in\mathscr{A}^+$ is analytic of finite order. Thus $\sigma^2=1_A$. Hence σ is an involution and using (i) again it must be antianalytic.

Theorem 1.1.10. *Let* G *be a subgroup of* Aut(S) *which acts properly discontinuously on the Klein surface* S. *Then* $S'=S/G$ *admits a unique structure of Klein surface such that* $\pi:S\longrightarrow S'$ *is a morphism.*
Proof. By 1.1.2 S′ is Hausdorff and connected. We shall first construct a topological atlas on S′. We shall check afterwards that the atlas we construct is dianalytic, and that π is a morphism. Once this is done, the uniqueness follows from 0.1.11.2.

Let $x\in S$. There exists a chart $(U^x=U,\phi_x=\phi)$ on S such that $\phi(x)=0$ and $G_x=\{1_S=f_1,f_2,\dots,f_k\}$ is finite.

Let us write $H_x=\{f\in G_x \mid \phi f\phi^{-1}$ is analytic on $\phi(U)\}$. It is a subgroup of G_x and from 1.1.1 and condition (iii) in Def 2, we can assume that

(1.1.10.1) U is connected.

$U \cap f(U)$ is empty, if $f \in G \setminus G_x$.

$U = f_j(U)$, $1 \le j \le k$.

x is the only fixed point of f_j on U, $2 \le j \le k$, if $f_j \in H_x$.

In what follows we shall write f_j instead of $f_j|U$.

The natural domain for a chart on S′ at O_x is $V = V^x = U^x/G_x$. In fact, if we denote by π' the canonical projection $U \longrightarrow V$ we get a diagram

$$\begin{array}{ccc} & \longrightarrow & S \\ \Big\downarrow & \pi\,|\,U & \downarrow \pi \\ U & \longrightarrow & S' \\ & & \uparrow u \\ & \xrightarrow{\quad \pi' \quad} & V \end{array}$$

where u maps the orbit of $y \in U$ under G_x to its orbit under G. By 1.1.10.1, u is injective, and so V and $\pi(U)$ are homeomorphic.

Claim: $[G_x:H_x] \le 2$. In fact, take $f,g \in G_x \setminus H_x$. Since $f,g: U \longrightarrow U$ are morphisms, we get diagrams

$$\begin{array}{ccc} U & \xrightarrow{\quad f \quad} & U \\ \downarrow \phi & & \downarrow \phi \\ \phi(U) & \xrightarrow{F} M \xrightarrow{\Phi} & \phi(U) \end{array} \qquad \begin{array}{ccc} U & \xrightarrow{\quad g \quad} & U \\ \downarrow \phi & & \downarrow \phi \\ \phi(U) & \xrightarrow{G} N \xrightarrow{\Phi} & \phi(U) \end{array}$$

where F and G are analytic maps and $M = \operatorname{im} F$, $N = \operatorname{im} G$. Of course ΦF and ΦG are bijective and so F and G are. This together with the hypothesis means $M,N \subseteq \{z \in \mathbb{C} \mid \operatorname{Im} z \le 0\}$. As a consequence $\Phi \mid M$ and $\Phi \mid N$ are also injective. Hence $\phi g^{-1} f \phi = G^{-1} F$ is analytic, *i.e.* $g^{-1} f \in H_x$. Thus $[G_x:H_x] \le 2$.

To construct the chart $(V=V^x, \psi=\psi_x)$ at $O_x \in S'$ we distinguish four cases according with $x \in \partial S$ or $x \notin \partial S$ and $[G_x:H_x] = 1$ or 2.

Case 1. $x \notin \partial S$. Let us denote $W^x = U/H_x$ and $\tilde{O}_y$ the orbit of $y \in U$ under the action of H_x. Let us construct the maps

$$\alpha_x: W^x \longrightarrow \mathbb{C} : \tilde{O}_y \longmapsto \prod_{f \in H_x} \phi f(y) \text{ and } g = g_x: \phi(U) \longrightarrow \mathbb{C} : \xi \longmapsto \prod_{f \in H_x} \phi f \phi^{-1}(\xi).$$

Then g is analytic and the following diagram commutes:

$$\begin{array}{ccc} U & \xrightarrow{\quad \pi_1 \quad} & W^x \\ \downarrow \phi & & \downarrow \alpha_x \\ \phi(U) & \xrightarrow{\quad g \quad} & \mathbb{C} \end{array}$$

Hence α_x is an open continuous map and $\alpha_x(W^x)$ is an open subset of $\mathbb{C}$.

We prove now that α_x is injective. After shrinking U if necessary we can assume the fibers of g have at most $l = \# H_x$ points, because each $\phi f \phi^{-1}$ is bijective and $\phi f \phi^{-1}(0) = 0$, *i.e.* g has a zero of order l at 0. On the other hand, from 1.1.10.1, $\pi^{-1}(\tilde{O}_y) = \{f(y) \mid f \in H_x\}$ has l elements, if $\tilde{O}_y \ne \tilde{O}_x$. As a consequence $\alpha_x \mid W^x\backslash\{\tilde{O}_x\}$ is injective. Moreover, since $\alpha_x(\tilde{O}_x)=0$ and $\alpha_x(\tilde{O}_y)\neq 0$ for $\tilde{O}_y\neq\tilde{O}_x$, we deduce that α_x is injective, as desired. Now we are forced to consider two subcases:

Subcase 1.1. $x\notin\partial S$, $[G_x:H_x]=1$. In this case $V^x=W^x$ and so $(V^x, \psi_x=\alpha_x)$ is the searched chart. Notice that $O_x=\tilde{O}_x\notin\partial S'$ in this case, because $\psi_x(V^x)=\alpha_x(W^x)$ is open in $\mathbb{C}$.

Subcase 1.2. $x\notin\partial S$. $[G_x:H_x]=2$. Let us take $\theta\in G_x\backslash H_x$ and the induced map $\tilde{\theta}:W^x\longrightarrow W^x$: $\tilde{O}_y\longmapsto\tilde{O}_{\theta(y)}$. Clearly $G_x/H_x=<\tilde{\theta}>$. First of all we prove that $\sigma=\alpha_x\tilde{\theta}\alpha_x^{-1}:\alpha_x(W^x)\longrightarrow\alpha_x(W^x)$ is an antianalytic involution. By the definition of $\tilde{\theta}$ and α_x we have $\tilde{\theta}\pi_1=\pi\theta$ and $\alpha_x\pi_1=g\phi$. Also $\phi\theta\phi^{-1}$ is antianalytic, because $\theta\notin H_x$.

Now $\alpha_x\tilde{\theta}\pi_1=\alpha_x\pi_1\theta=g\phi\theta$, *i.e.* $\alpha_x\tilde{\theta}\alpha_x^{-1}\alpha_x\pi_1=g\phi\theta$, and so

$$\sigma g=(\alpha_x\tilde{\theta}\alpha_x^{-1})g=g(\phi\theta\phi^{-1}) \text{ is antianalytic.}$$

By Riemann's Extension theorem, σ is antianalytic. Of course $\sigma^2=1_A$. Consequently, if $\tau:z\longmapsto\bar{z}$ we get, from 1.1.7 $\tau=h\sigma h^{-1}$ for a certain analytic automorphism $h:\alpha_x(W^x)\longrightarrow\alpha_x(W^x)$. The chart we are looking for is given by $\psi_x:V^x\longrightarrow\mathbb{C}^+:O_y\longmapsto\Phi h\alpha_x(\tilde{O}_y)$ where Φ is the folding map. To see that it is well defined it suffices to check $\Phi h\alpha_x\tilde{\theta}=\Phi h\alpha_x$, because $G_x/H_x=<\tilde{\theta}>$. But $\tau h=h\sigma$, *i.e.* $\tau h\alpha_x=h\alpha_x\tilde{\theta}$, and this implies $\Phi h\alpha_x\tilde{\theta}=\Phi\tau h\alpha_x=\Phi h\alpha_x$. To prove the injectivity of ψ_x, we put $\pi_2:W^x\longrightarrow V^x:\tilde{O}_y\longmapsto O_y$. We know $\pi=\pi_2\pi_1$ and the following diagram commutes:

$$\begin{array}{ccccc}
W^x & & \xrightarrow{\pi_2} & & V^x \\
\downarrow \alpha_x & & & & \downarrow \psi_x \\
\alpha_x(W^x) & \xrightarrow{h} & \alpha_x(W^x) & \xrightarrow{\Phi} & \mathbb{C}^+
\end{array}$$

The fibers of $\pi_2|W^x\backslash\{\tilde{O}_x\}$ have two points, since $[G_x:H_x]=2$. The fibers of Φ have at most two points. Thus, since h and α_x are bijections and $\psi_x(O_z)\neq 0=\psi_x(O_x)$ for $O_z\neq O_x$, ψ_x is injective.

Also the commutativity of the last diagram shows that ψ_x is an open map onto an open subset of $\mathbb{C}^+$.

Case 2. $x\in\partial S$. In this situation H_x is trivial. In fact, given $f\in H_x$, we get $\sigma=\phi f\phi^{-1}\in\mathcal{A}^+$ and it is analytic of finite order $\leq\#G_x$. Hence, by 1.1.9 $\sigma=1_A$, *i.e.*, f is the identity. Thus

Subcase 2.1. $x\in\partial S$, $[G_x:H_x]=1$. Then G_x is trivial and $\pi:U\longrightarrow V$ is a homeomorphism, with

$$\begin{array}{ccc} U & \xrightarrow{\quad\pi\quad} & V=U \\ \downarrow \phi & & \downarrow \psi_x=\phi \\ \phi(U) & \xrightarrow{\quad 1_{\phi(U)}\quad} & \phi(U)=\psi_x(U) \end{array}$$

So, nothing must be proven here.

Subcase 2.2. $x\in\partial S$, $[G_x:H_x]=2$. Then G_x is generated by an element η of order 2 and $\sigma=\phi\eta\phi^{-1}$ is antianalytic. By 1.1.8 there exists an analytic homeomorphism $h:\phi(U)\longrightarrow\phi(U)$ with $-\tau=h\sigma h^{-1}$, *i.e.*, $-\tau h\phi=h\phi\eta$.

Now, if $s:\mathbb{C}\longrightarrow\mathbb{C}: z\longmapsto z^2$ and Φ is the folding map, we construct the chart $\psi_x:V\longrightarrow\mathbb{C}^+:O_y\longmapsto \Phi sh\phi(y)$. In order to prove it is well defined it suffices, since $G_x=\{1,\eta\}$, to check the equality $\Phi sh\phi=\Phi sh\phi\eta$. But $\tau s=s(-\tau)$ and $\Phi\tau=\Phi$. Hence $\Phi sh\phi\eta=\Phi s(-\tau)h\phi=\Phi\tau sh\phi=\Phi sh\phi$. Moreover, we get a diagram

$$\begin{array}{ccccc} U & & \xrightarrow{\quad\pi\quad} & & V \\ \downarrow \phi & & & & \downarrow \psi_x \\ \phi(U) & \xrightarrow{\ sh\ } & \mathbb{C} & \xrightarrow{\ \Phi\ } & \psi_x(V)\subseteq\mathbb{C}^+ \end{array}$$

and this proves that ψ_x is injective and open onto an open subset of $\mathbb{C}^+$. In fact, since $\phi(U)\subseteq\mathbb{C}^+$, $sh|\phi(U)\backslash\mathbb{R}$ is injective, and so the fibers of Φsh have at most two points. Also, the fibers of $\pi|U\backslash\{x\}$ have two points, because $G_x=\{1,\eta\}$. This together with $\psi_x(O_y)\neq 0=\psi_x(O_x)$ for $O_y\neq O_x$, proves the injectivity of ψ_x. Now openness is obvious.

We have constructed a topological atlas $\Sigma'=\{(V^x,\psi_x)|O_x\in S'\}$ on S', and we are going to prove it is dianalytic.

Let us denote, as we did in case 1, $g_x=\prod_{f\in H_x}\phi f\phi^{-1}$. Of course it is the identity when $x\in\partial S$. We have defined ψ_x in such a way that for certain analytic homeomorphism h, and $s(z)=z^2$, the map

$$F_x=\begin{cases}\Phi g_x & \text{if} \qquad\qquad G_x=H_x\\ \Phi hg_x & \text{if} \quad x\notin\partial S,\quad G_x\neq H_x\\ \Phi sh & \text{if} \quad x\in\partial S,\quad G_x\neq H_x\end{cases}$$

makes commutative the diagram

(1.1.10.2)
$$\begin{array}{ccc} U^x & \xrightarrow{\quad\pi\quad} & V^x \\ \downarrow \phi_x & & \downarrow \psi_x \\ \phi_x(U^x) & \xrightarrow{\quad F_x\quad} & \psi_x(V^x) \end{array}$$

All we need is to prove the dianalyticity of $\psi_x\psi_y^{-1}:\psi_y(V^x\cap V^y)\longrightarrow\psi_x(V^x\cap V^y)$ for $x,y\in S$ with $O_x\neq O_y$. As it is easily seen, we can restrict ourselves to the inner points. By 1.1.10.1, the stabilizer G_z of a point $z\in U^x$, $z\neq x$, is trivial. This implies G_z is trivial for $O_z\in V^x\cap V^y$ and so, given an inner point $O_z\in V^x\cap V^y$ we can find an open neighbourhood $U^z\subseteq U^x\cap U^y$ such that $\pi|U^z$ is injective. Hence

$$\begin{array}{ccccc}
\phi_x(U^z) & \xleftarrow{\quad\phi_x\quad} & U^z & \xrightarrow{\quad\phi_y\quad} & \phi_y(U^z) \\
\Big\downarrow F_x & & \Big\downarrow \pi & & \Big\downarrow F_y \\
\psi_x(V^x) & \xrightarrow{\quad\psi_x\quad} & \pi(U^z) & \xrightarrow{\quad\psi_y\quad} & \psi_y(V^y)
\end{array}$$

where F_x and F_y are analytic and bijective. As a consequence $\psi_x\psi_y^{-1}=F_x\phi_x\phi_y^{-1}F_y^{-1}$ is dianalytic. Note also that diagram 1.1.10.2 proves that π is a morphism. This finishes the proof.

Remark 1.1.11. We have shown that $\pi(x)\notin\partial S'$ when $x\notin\partial S$ and $G_x=H_x$.

Now from 1.1.4 and the theorem just proved we derive

Corollary 1.1.12. *Given a discrete subgroup* Γ *of* Aut(H), *the quotient* H/Γ *admits a unique structure of Klein surface, such that the canonical projection* $H\longrightarrow H/\Gamma$ *is a morphism of Klein surfaces. In particular this holds true when* Γ *is an NEC group. Moreover, if* Γ *is a fuchsian group, from the last remark we conclude that* H/Γ *has empty boundary.*

1.2. COMPACT KLEIN SURFACES AS ORBIT SPACES

From now on we restrict our attention to compact Klein surfaces. We are going to establish some kind of converse of the last corollary in 1.1. This can be seen as the starting point for the combinatorial approach to study automorphism groups of compact Klein surfaces that we shall develop in this book. First we prove

Proposition 1.2.1. *Let* S *be a compact orientable Klein surface without boundary which admits the upper half plane as the universal covering. Let* $p:H\longrightarrow S$ *be the covering projection. Let us assume* p *is a morphism of Klein surfaces. Put*

$$\Gamma=\{f\in \mathrm{Aut}(H)\,|\,pf=p\}.$$

Given $x\in H$ *we denote by* O_x *the orbit of* x *under* Γ. *Then:*

(1) The map $h:H/\Gamma\longrightarrow S\colon O_x\longmapsto p(x)$ *is a homeomorphism.*

(2) The stabilizer Γ_x *of each point* $x\in H$ *is trivial.*

(3) Γ *is a surface NEC group.*

Proof. (1) First we observe that h is well defined. If $x,y\in H$ and $O_x=O_y$, then there exists $f\in\Gamma$ such that $y=f(x)$. Hence, $p(y)=pf(x)=p(x)$. By the definition, $h\pi=p$, where $\pi:H\longrightarrow H/\Gamma$ is the canonical projection. Thus since p is continuous, the same holds true for h. Even more h is an open map. For, if

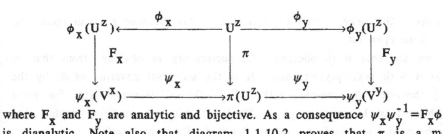

$A \subseteq H/\Gamma$ is open, the same is true for $h(A)=p(\pi^{-1}(A))$, because p is an open map and $\pi^{-1}(A)$ is an open set.

Now we see that h is bijective; the surjectivity is obvious, from that of p. Let $x,y \in H$ with $p(x)=p(y)=z$. Since H is the universal covering of S, by the Monodromy theorem, the covering $p:H \longrightarrow S$ is Galois. Hence $y=f(x)$ for some homeomorphism f of H verifying $pf=p$. By 0.1.7 this last condition implies that f is a morphism of Klein surfaces, and by 0.1.8, $f \in \Gamma$. Thus $O_x = O_y$.

(2) Let us take $f \in \Gamma_x$ and $M=\text{Fix}(f)$. We must see that $M=H$. Since H is connected it is enough to prove that the closed and nonempty set M is also open. Let us take $z \in M$, and open neighbourhoods U, V with $z \in U \subseteq H$ and $p(z) \in V \subseteq S$ such that $p|U:U \longrightarrow V$ is a homeomorphism. Since f is continuous and $f(z)=z$ we have $f(U') \subseteq U$ for some neighbourhood $z \in U' \subseteq U$. Clearly $pf|U'=p|U'$. We claim that $U' \subseteq M$. In fact, if $u \in U'$, we know that $pf(u)=p(u)$, and $u,f(u) \in U$. Since $p|U$ is injective, $f(u)=u$, *i.e.* $u \in M$.

(3) All reduces to check that Γ is discrete. In that case, $H/\Gamma \approx S$ being compact, Γ is an NEC group. From part (2) Γ has not elliptic elements, *i.e.* it is a surface group by (0.2.7).

Let us assume, by the way of contradiction, that Γ is not discrete. Then $\{g_n | n \in \mathbb{N}\} \to 1_H \in \Gamma$, for some sequence $\{g_n | n \in \mathbb{N}\} \subseteq \Gamma$, $g_n \neq g_m$ if $n \neq m$.

Let us take a point $x \in H$. From (1) in 1.1.3 $x=\lim\{g_n(x) | n \in \mathbb{N}\}$. Now if we write $y=p(x)$, the restriction $p|U:U \longrightarrow V$ is injective for suitable neighbourhoods U and V at x and y respectively. Consequently, $g_n \in \Gamma$ implies that $pg_n(x)=p(x)$, and $p|U$ is injective. Thus $g_n(x) \notin U \setminus \{x\}$. This together with $\lim\{g_n(x) | n \in \mathbb{N}\}=x$ means that $g_n(x)=x$ for n big enough. Using (2), $g_n = 1_H$ if $n \gg 0$, which is false.

Proposition 1.2.2. *Let* S *be an orientable compact Klein surface, without boundary and* $g(S) \geq 2$. *Then* $S \approx H/\Gamma$ *as Klein surfaces, for some surface NEC group* Γ. *In fact* Γ *is a fuchsian group.*

Proof. S is a Riemann surface. From the uniformization theorem, H is the universal covering for S. Let $p:H \longrightarrow S$ be the corresponding projection, which is a morphism of Klein surfaces. By the previous proposition, $\Gamma=\{f \in \text{Aut}(H) | pf=p\}$ is a surface NEC group, and there is a homeomorphism $h:H/\Gamma \longrightarrow S$ for which $h\pi=p$, where $\pi:H \longrightarrow H/\Gamma$ is the canonical projection. By (0.1.8) all reduces to check that h is a morphism of Klein surfaces. But p is a morphism and π is a surjective morphism, and so h is a morphism by 0.1.7. Finally, it is clear that Γ is a fuchsian group because, if $f \in \text{Aut}(H)$ reverses the orientation of H, the equality $pf=p$ cannot hold, since p is compatible with the orientations in H and S. Thus $f \notin \Gamma$.

Our next goal is to extend this result to a much bigger class of Klein surfaces. To do that we shall use the double cover we constructed in 0.1.12.

Theorem 1.2.3. *Let* S *be a compact Klein surface,* g=g(S) *and* k=k(S). *Let us assume that*

$2g+k\geq 3$ *if S is orientable and*

$g+k\geq 3$ *if S is nonorientable.*

Then there exists a surface NEC group Γ such that S *and* H/Γ *are isomorphic as Klein surfaces. Moreover if* $\pi':H\longrightarrow H/\Gamma$ *is the canonical projection, then* $\Gamma=\{f\in Aut(H)\,|\,\pi'f=\pi'\}$. *Notice that the conditions above just mean that the algebraic genus of* S *is bigger than or equal to* 2. *This number is also called the algebraic genus of* Γ.

Remark 1.2.4. *The only compact topological surfaces which do not satisfy these conditions are:*

orientable without boundary:	$g=0$	*sphere*
	$g=1$	*torus*
orientable with nonempty boundary:	$g=0,\ k=1$,	*closed disc*
	$g=0,\ k=2$,	*closed annulus*
nonorientable:	$g=1,\ k=0$,	*projective plane*
	$g=2,\ k=0$,	*Klein bottle*
	$g=1,\ k=1$,	*Möbius strip.*

Proof of theorem 1.2.3. The case of orientable S without boundary has been solved in 1.2.2. Thus assume that S fails to be a Riemann surface and let $f:S_c\longrightarrow S$ be the double cover constructed in 0.1.12. By 0.1.13.2, $g(S_c)\geq 2$. So using 1.2.2 there exists a surface NEC group Γ_c such that $S_c=H/\Gamma_c$. Moreover, if we denote by π the canonical projection $H\longrightarrow S_c$, then from 1.2.2 it follows that

$$\Gamma_c=\{h\in Aut(H)\,|\,\pi h=\pi\}.$$

Let $\sigma:S_c\longrightarrow S_c$ be the involution constructed in 0.1.12. The group $<\sigma>$ is finite - it has order 2 - and so $S_c/<\sigma>$ is a Klein surface.

Claim 1. $\tilde{f}:S_c/<\sigma>\longrightarrow S:O_x\longmapsto f(x)$ is an isomorphism of Klein surfaces.

This is an immediate consequence of 0.1.7, 0.1.8 and the equality $f=\tilde{f}\pi_1$, where $\pi_1:S_c\longrightarrow S_c/<\sigma>$ is the obvious projection, using that given $x,y\in S_c$, $f(x)=f(y)$ if and only if $x\in\{y,\sigma(y)\}$.

Consequently $S\approx\dfrac{H/\Gamma_c}{<\sigma>}$, and we look for an NEC group Γ such that

$\langle\sigma\rangle=\Gamma/\Gamma_c$. In such a case $S\approx\dfrac{H/\Gamma_c}{\Gamma/\Gamma_c}$ and "clearing denominators", $S\approx H/\Gamma$. We do that in detail. Using 0.1.7 and 0.1.8 it is enough to work into the topological category.

First notice that both $\pi:H\longrightarrow S_c$ and $\sigma\pi:H\longrightarrow S_c$ are universal coverings and so there exists $g\in\mathrm{Aut}(H)$ such that

$$\begin{array}{ccc} H & \xrightarrow{\quad g\quad} & H \\ \downarrow{\scriptstyle\pi} & & \downarrow{\scriptstyle\pi} \\ S_c & \xrightarrow{\quad\sigma\quad} & S_c \end{array}$$

Claim 2. $\Gamma=\langle\Gamma_c,g\rangle$ is a discrete subgroup of Aut(H) and $[\Gamma:\Gamma_c]=2$. Once we show the last, the discreteness of Γ will be obvious from that of Γ_c. Thus it suffices to show $g\notin\Gamma_c$, $g^2\in\Gamma_c$, $g\Gamma_cg^{-1}\subseteq\Gamma_c$. The first is clear because $\Gamma_c\subseteq\mathrm{Aut}^+(H)$ and $\pi g=\sigma\pi$ implies that g reverses the orientation of H, as σ reverses the one of S_c. Moreover $\pi g^2=\sigma\pi g=\sigma^2\pi=\pi$, *i.e.* $g^2\in\Gamma_c$. Finally, if $h\in\Gamma_c$, it is $\pi ghg^{-1}=\sigma\pi hg^{-1}=\sigma\pi g^{-1}=\pi gg^{-1}=\pi$, and so $g\Gamma_cg^{-1}\subseteq\Gamma_c$.
Claim 3. We construct an isomorphism $h:S_c/\langle\sigma\rangle\longrightarrow H/\Gamma$ as follows. Let us call π_2 the canonical projection $\pi_2:H\longrightarrow H/\Gamma$. Now

$$\begin{array}{ccc} H & \xrightarrow{\quad\pi_2\quad} & H/\Gamma \\ \downarrow{\scriptstyle\pi} & & \uparrow{\scriptstyle h} \\ S_c & \xrightarrow{\quad\pi_1\quad} & S_c/\langle\sigma\rangle \end{array}$$

where h: $a=\pi_1\pi(u)\longmapsto\pi_2(u)$. First we see that h is a well defined map. If $a=\pi_1\pi(u)=\pi_1\pi(v)$, $u,v\in H$, it is either $\pi(u)=\pi(v)$ or $\pi(u)=\sigma(\pi(v))$. In the first case, since $\Gamma_c\subseteq\Gamma$, we get $\pi_2(u)=\pi_2(v)$. In the second one $\pi(u)=\pi(g(v))$ and so $u=hg(v)$ for some $h\in\Gamma_c\subseteq\Gamma$. Since also $g\in\Gamma$ we obtain $\pi_2(u)=\pi_2(v)$.

Similarly we check that h is injective: given $a=\pi_1\pi(u)$ and $b=\pi_1\pi(v)$ with $h(a)=h(b)$ we deduce $v=l(u)$ for some $l\in\Gamma$. If $l\in\Gamma_c$ we get $\pi(u)=\pi(v)$ and so $a=b$. If $l\notin\Gamma_c$ it is $l=gh$, $h\in\Gamma_c$ and so, if $w=h(u)$, we get $\pi(u)=\pi(v)$, $v=g(w)$. Thus, since $\sigma\pi(w)=\pi g(w)=\pi(v)$ it is $b=\pi_1\pi(v)=\pi_1\pi(w)=\pi_1\pi(u)=a$, and we are done.

Now it is immediate, from the commutativity of the last diagram, that h is an isomorphism of Klein surfaces.
Claim 4. As a consequence $S\approx S_c/\langle\sigma\rangle\approx H/\Gamma$. Moreover S is compact, and Γ is discrete. So Γ is an NEC group. It remains only to prove that Γ is a surface group and to study its relation with the covering map $\pi_2:H\longrightarrow H/\Gamma$.
Claim 5. $\Gamma=\{l\in\mathrm{Aut}(H)\,|\,\pi_2l=\pi_2\}$.
The inclusion "$\subseteq$" is evident. To see the converse, notice that

$$\begin{array}{ccccc} H & \xrightarrow{\quad\pi_2\quad} & & & H/\Gamma \\ \downarrow \pi & & & & \uparrow h \\ S_c & \xrightarrow{\ f\ } & S & \xrightarrow{(\tilde{f})^{-1}} & S_c/\langle\sigma\rangle \end{array}$$

Hence, if $\pi_2 l=\pi_2$ we get $h(\tilde{f})^{-1}f\pi l=h(\tilde{f})^{-1}f\pi$ and so $f\pi l=f\pi$. Since the orbits of f have at most 2 points, it is

$$A_l=\{x\in H|\pi(x)\neq\pi l(x)\}\subseteq B_l=\{x\in H|\pi l(x)=\sigma\pi(x)\}.$$

Since σ reverses the orientation of S_c, we deduce that A_l is empty when l is analytic. Thus, in this case $\pi l=\pi$, *i.e.* $l\in\Gamma_c\subseteq\Gamma$. If l is not analytic we consider $l_1=gl$. It is analytic and also $f\pi l_1=f\pi gl=f\sigma\pi l=f\pi l=f\pi$. Thus, as before, $l_1\in\Gamma_c$ and so $l\in g^{-1}\Gamma=\Gamma$. Finally

Claim 6. Γ is a surface NEC group.

Otherwise there would exist an elliptic element $e\in\Gamma$. Since Γ_c is a surface group and $[\Gamma:\Gamma_c]=2$, we get $g\in\Gamma=\Gamma_c\cup e\Gamma_c\subseteq\mathrm{Aut}^+(H)$, a contradiction. The proof is finished.

Remark 1.2.5. Given S we have represented its double cover S_c as the quotient H/Γ_c in such a way that S and H/Γ are isomorphic as Klein surfaces, for some NEC group Γ containing Γ_c as a subgroup of index 2. By the definition, $\Gamma_c\subseteq\Gamma^+$. Thus $\Gamma_c=\Gamma^+$ and so $S_c=H/\Gamma^+$, as announced in 0.2.1. The group Γ we constructed turns out to be the group of deck-transformations of H with respect to the canonical projection $\pi':H\longrightarrow H/\Gamma=S$, *i.e.* $\Gamma=\{f\in\mathrm{Aut}(H)|\pi'f=\pi'\}$.

1.3. GENERAL RESULTS ON AUTOMORPHISMS OF KLEIN SURFACES

Let S and S′ be compact Klein surfaces satisfying the assumption of theorem 1.2.3. We devote this section to study the set Isom(S′,S) of isomorphisms from S′ to S. Then we specify the obtained results to the case S=S′. By theorem 1.2.3, S and S′ can be represented as H/Γ and H/Γ' for some surface NEC groups Γ and Γ' respectively. Moreover, let $\pi:H\longrightarrow S$ and $\pi':H\longrightarrow S'$ be the canonical projections. As we remarked in 1.2.5, Γ and Γ' can be chosen in such a way that they coincide with the groups of deck-transformations of H with respect to π and π' respectively. Finally let

$$A(\Gamma,\Gamma')=\{g\in\mathrm{Aut}(H)|\pi'(x)=\pi'(y)\ \Leftrightarrow\ \pi g(x)=\pi g(y)\}.$$

All these assumptions and notations are fixed along the section.

First we state the following easy but essential result.

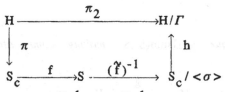

Proposition 1.3.1. *Let* g *be an automorphism of* H. *The following statements are*

equivalent:

(1) $g \in A(\Gamma,\Gamma')$.

(2) There exists a unique $\bar{g} \in \mathrm{Isom}(S',S)$ *making commutative the following diagram:*

(1.3.1.2)
$$\begin{array}{ccc} H & \xrightarrow{\;g\;} & H \\ \downarrow{\scriptstyle \pi'} & & \downarrow{\scriptstyle \pi} \\ S' & \xrightarrow{\;\bar{g}\;} & S \end{array}$$

(3) $\Gamma' = g^{-1}\Gamma g$.

Proof. (1)⇒(2) Given $x' = \pi'(x) \in S'$, 1.3.1.2 forces us to define $\bar{g}(x') = \bar{g}\pi'(x) = \pi g(x)$. Using 0.1.7 and 0.1.8, it is enough to see that $\bar{g}$ is a homeomorphism. This is evident by the definition of $A(\Gamma,\Gamma')$.

(2)⇒(3) As we remarked at the beginning, Γ and Γ' coincide with the groups of deck-transformations of H with respect to the canonical projections π and π'. Thus if $f \in \Gamma'$ and $h = gfg^{-1}$ we obtain $\pi h = \pi gfg^{-1} = \bar{g}\pi' fg^{-1} = \bar{g}\pi' g^{-1} = \pi g g^{-1} = \pi$ *i.e.* $h \in \Gamma$ and so, $\Gamma' \subseteq g^{-1}\Gamma g$.

Conversely, if $h \in g^{-1}\Gamma g$, then $ghg^{-1} \in \Gamma$, *i.e.* $\pi ghg^{-1} = \pi$ and so $\bar{g}\pi' h = \bar{g}\pi'$. Since $\bar{g}$ is bijective, $\pi' h = \pi'$. Thus $h \in \Gamma'$.

(3)⇒(1) Let us take $x,y \in H$ with $\pi'(x) = \pi'(y)$ and write $y = f(x)$ for some $f \in \Gamma' = g^{-1}\Gamma g$. Now, $h = gfg^{-1} \in \Gamma$. Since $hg = gf$ and $\pi h = \pi$, we deduce that
$$\pi(g(y)) = \pi(g(f(x))) = \pi(h(g(x))) = \pi(g(x)).$$
Similarly we prove the converse.

Theorem 1.3.2. *With the same notations,*

(1) The map $A(\Gamma,\Gamma') \longrightarrow \mathrm{Isom}(S',S)$: $g \longmapsto \bar{g}$ *is surjective.*

(2) S *and* S′ *are isomorphic if and only if* Γ *and* Γ' *are conjugate subgroups in* $\Omega = \mathrm{Aut}(H)$.

(3) $\mathrm{Aut}(S) \cong N_{\Omega}(\Gamma)/\Gamma$, *where* $N_{\Omega}(\Gamma)$ *is the normalizer of* Γ *in* Ω.

Proof. (1) Assume first that S and S′ are Riemann surfaces. Let us take $\phi \in \mathrm{Isom}(S',S)$. Since (H,π) and (H,π') are universal coverings of S and S′ respectively, by the Monodromy theorem and 0.1.7 there exists $g \in \mathrm{Aut}(H)$ making the following diagram commutative

(1.3.2.1)
$$\begin{array}{ccc} H & \xrightarrow{\;g\;} & H \\ \downarrow{\scriptstyle \pi'} & & \downarrow{\scriptstyle \pi} \\ S' & \xrightarrow{\;\phi\;} & S \end{array}$$

It is clear that $g \in A(\Gamma,\Gamma')$, and so $\phi = \bar{g}$ by 1.3.1.

In the general case we consider the double covers $f: S_c \longrightarrow S$ and $f': S'_c \longrightarrow S'$ with the corresponding antianalytic involutions $\sigma: S_c \longrightarrow S_c$ and $\sigma': S'_c \longrightarrow S'_c$. By 0.1.13, (3), there exists $\psi \in \mathrm{Isom}(S'_c,S_c)$ such that the square

(1.3.2.2)
$$\begin{array}{ccc} S'_c & \xrightarrow{\psi} & S_c \\ \downarrow f' & & \downarrow f \\ S' & \xrightarrow{\phi} & S \end{array}$$

commutes.

Let $p:H \longrightarrow S_c$ and $p':H \longrightarrow S'_c$ be the canonical projections. By 1.3.2.1 applied to the Riemann surfaces S_c and S'_c there exists $g \in Aut(H)$ making commutative the following diagram

(1.3.2.3)
$$\begin{array}{ccc} H & \xrightarrow{g} & H \\ \downarrow p' & & \downarrow p \\ S'_c & \xrightarrow{\psi} & S_c \end{array}$$

Now, up to the identifications of S with H/Γ and S′ with H/Γ', the maps $\pi'=f'p':H \longrightarrow S'$ and $\pi=fp:H \longrightarrow S$ are the canonical projections (see the proof of 1.2.3). Gluing together 1.3.2.2 and 1.3.2.3 we obtain the commutative diagram

$$\begin{array}{ccc} H & \xrightarrow{g} & H \\ \downarrow \pi' & & \downarrow \pi \\ S' & \xrightarrow{\phi} & S \end{array}$$

Using once more 1.3.1, $g \in A(\Gamma,\Gamma')$ and $\phi=\bar{g}$.

(2) It is obvious that S and S′ are isomorphic if and only if $A(\Gamma,\Gamma')$ is nonempty. Hence the assertion follows from 1.3.1.

(3) Using 1.3.1 with S=S′ it follows that $A(\Gamma,\Gamma)=N_\Omega(\Gamma)$. Thus from part (1)
$$\mu:N_\Omega(\Gamma) \longrightarrow Aut(S):\ g \longmapsto \bar{g}$$
is a surjective map. Even more, given $g_1,g_2 \in A(\Gamma,\Gamma)$, $\bar{g}_1$ and $\bar{g}_2$ verify $\pi g_1=\bar{g}_1\pi$ and $\pi g_2=\bar{g}_2\pi$. As a result $\pi(g_1g_2)=(\bar{g}_1\bar{g}_2)\pi$. On the other hand, $\pi(g_1g_2)=(\overline{g_1g_2})\pi$. By 1.3.1, $\bar{g}_1\bar{g}_2=\overline{g_1g_2}$ and so μ is a group epimorphism. Finally we must only check that $\ker\mu=\Gamma$. Clearly, if $g \in \Gamma$ we have $\pi g=\pi$ *i.e.*

$$\begin{array}{ccc} H & \xrightarrow{g} & H \\ \downarrow \pi & & \downarrow \pi \\ S & \xrightarrow{1_S} & S \end{array}$$

Thus using 1.3.1, $\bar{g}=1_S$ and so $g \in \ker\mu$. Conversely, $\bar{g}=1_S$ means that $\pi g=\pi$ and so $g \in \Gamma$.

Remark 1.3.3. (1) Let us fix a surface NEC group Γ and $S=H/\Gamma$, $\Omega=Aut(H)$. Then the moduli space $M(\Gamma)$ parametrizes the isomorphism classes of compact Klein surfaces homeomorphic to S as a topological space. In fact given such a surface, say S′, we can represent it as H/Γ' for some surface NEC group Γ' isomorphic to Γ. Let $\sigma:\Gamma \longrightarrow \Gamma'$ be an isomorphism. Now if $\iota:\Gamma' \subseteq \Omega$ denotes the set theoretical inclusion and $r=\iota\sigma:\Gamma \longrightarrow \Omega$ is the corresponding element in the

Weil space (Ch. 0, Def 16) of Γ, then $S'=H/r(\Gamma)$.

Moreover if r_1 and r_2 belong to $R(\Gamma)$ and $S_1=H/r_1(\Gamma)$, $S_2=H/r_2(\Gamma)$ we claim that their classes $[r_1]$ and $[r_2]$ in $M(\Gamma)$ coincide if and only if S_1 and S_2 are isomorphic. In fact, if S_1 and S_2 are isomorphic, there exists $g\in\Omega$ with $r_2(\Gamma)=gr_1(\Gamma)g^{-1}$, by 1.3.2. Thus if ϕ_g denotes the conjugation by g, we obtain a commutative diagram

$$\begin{array}{ccc} \Gamma & \xrightarrow{\alpha} & \Gamma \\ \downarrow r_1 & & \uparrow r_2^{-1} \\ r_1(\Gamma) & \xrightarrow{\phi_g} & r_2(\Gamma) \end{array}$$

and consequently $\alpha\in\mathrm{Aut}(\Gamma)$.

Hence the classes in $T(\Gamma)$ of $r_2\alpha$ and r_1 are the same, *i.e.* $[r_1]=[r_2]$ as elements in $M(\Gamma)$.

Conversely if $[r_1]=[r_2]$, then there exist $\alpha\in\mathrm{Aut}(\Gamma)$ and $\beta\in\mathrm{Aut}(\Omega)$ such that $\beta r_1\alpha=r_2$. Thus $r_2(\Gamma)=\beta(r_1(\Gamma))$. Since β is an inner automorphism of Ω, the groups $r_1(\Gamma)$ and $r_2(\Gamma)$ are conjugate and so by 1.3.2, S_1 and S_2 are isomorphic.

(2) In order to simplify the proof of 1.3.4 we state now:

"Given $f,g\in\mathrm{Aut}^+(H)\backslash\{1_H\}$ with $fg=gf$, then $\mathrm{Fix}(f)=\mathrm{Fix}(g)$ where f,g are seen as functions $\mathbb{C}\cup\{\infty\}\longrightarrow\mathbb{C}\cup\{\infty\}$."

We can assume $1\leq\#\mathrm{Fix}(f)\leq\#\mathrm{Fix}(g)\leq 2$. From $fg=gf$ we conclude that $g(\mathrm{Fix}(f))=\mathrm{Fix}(f)$, $f(\mathrm{Fix}(g))=\mathrm{Fix}(g)$.

(i) If $\mathrm{Fix}(f)=\{x_0\}$ it is $g(x_0)=x_0$, and if $g(y)=y$ we know $f(y)=y$, *i.e.* $y=x_0$. Thus $\mathrm{Fix}(f)=\{x_0\}=\mathrm{Fix}(g)$.

(ii) If $\mathrm{Fix}(f)=\{x_0,y_0\}$ we deduce $\{g(x_0),g(y_0)\}=\{x_0,y_0\}$. Then, either $\mathrm{Fix}(f)=\mathrm{Fix}(g)$ or $\mathrm{Fix}(f)\neq\mathrm{Fix}(g)$ and $g(x_0)=y_0$, $g(y_0)=x_0$. In this last case we choose $z_0\in\mathrm{Fix}(g)\backslash\mathrm{Fix}(f)$. Since x_0, y_0 and z_0 are distinct fixed points of g^2 we derive $g^2=1_H$. Let us take a matrix $A\in GL(2,\mathbb{R})$ with $\det A=1$ such that, with the notations in 0.1.15, $g=f_A$. From $g^2=1_H$ we get $A^2=\pm I$ and so the minimal polynomial of $A\neq\pm I$ is T^2+1. Hence $g(z)=-1/z$ and $\mathrm{Fix}(g)=\{\pm i\}$. Since $f(H)=H$ and $f(\mathrm{Fix}(g))=\mathrm{Fix}(g)$ we get $f(i)=i$, and so $f(-i)=-i$, *i.e.* $\mathrm{Fix}(f)=\mathrm{Fix}(g)$.

Proposition 1.3.4. *Let Γ be an NEC group, $\Omega=\mathrm{Aut}(H)$. Then $\Lambda=N_\Omega(\Gamma)$, the normalizer of Γ in Ω is an NEC group.*

Proof. The compactness of H/Λ is obvious, since it is the image under the canonical projection of the compact space H/Γ. Since Ω is a topological group, it is now enough to check that the identity $\{1_H\}$ is an open subset of Λ.

We claim that there exist $h_1,h_2\in\Gamma^+$ such that $\mathrm{Fix}(h_1)\neq\mathrm{Fix}(h_2)$, $h_i\neq 1_H$. Let us take $h_1\in\Gamma^+$ defined by $h_1(z)=r_0z$ for some $r_0\in\mathbb{R}^+$. Then, $\mathrm{Fix}(h_1)=\{0,\infty\}$. Let us assume by the way of contradiction that $\mathrm{Fix}(h)=\{0,\infty\}$ for every $h\in\Gamma^+$. Then

$\Gamma^+ \subseteq_{top} A=\{f_r:H\longrightarrow H:z\longmapsto rz,\ r\in\mathbb{R}^+\}$. Since H/Γ^+ is compact, the same holds true for $H/A \approx (0,1)$, absurd.

As a consequence, $C_\Omega(h_1,h_2)=\{h\in\Omega|hh_i=h_ih,\ i=1,2\}$ is trivial. In fact, if $1_H\neq h\in C_\Omega(h_1,h_2)\cap \mathrm{Aut}^+(H)$, then from part (3) in the remark above, $\mathrm{Fix}(h_1)=\mathrm{Fix}(h)=\mathrm{Fix}(h_2)$, absurd. On the other hand, if $h\in C_\Omega(h_1,h_2)\setminus \mathrm{Aut}^+(H)$ we get $h^2=1_H$, and so $h(z)=-\bar{z}$. Now $hh_i=h_ih$ implies $h_i(z)=-1/z$, $i=1,2$, absurd. Also the maps $\zeta_i:\Lambda\longrightarrow\Gamma$: $g\longmapsto gh_ig^{-1}$ are well defined and continuous by 0.1.15. Moreover $\zeta_i(1_H)=h_i$.

Since Γ is discrete, we can find in Λ open neighbourhoods V_1 and V_2 of 1_H such that $\zeta_i(V_i)\subseteq\{h_i\}$, $i=1,2$, *i.e.* $gh_ig^{-1}=h_i$ for each $g\in V=V_1\cap V_2$. In other words, $V\subseteq C_\Omega(h_1,h_2)=\{1_H\}$, and so $\{1_H\}=V$ is open in Λ.

Corollary 1.3.5. *Let* S *be a compact Klein surface satisfying 1.2.3. Then* Aut(S) *is finite.*

Proof. We write $S=H/\Gamma$ as in 1.2.3 and $\Lambda=N_\Omega(\Gamma)$. Since Λ is an NEC group, the index $[\Lambda:\Gamma]$ is finite by the Hurwitz Riemann formula. Thus $|\mathrm{Aut}(S)|=[\Lambda:\Gamma]$ is finite.

Remark 1.3.6. *A group of automorphisms* of a Klein surface S is a subgroup of the group Aut(S). Assume S satisfies 1.2.3. Then S can be written as $S=H/\Gamma$ for some NEC group Γ where, if $\pi:H\longrightarrow S$ is the canonical projection, then $\Gamma=\{f\in \mathrm{Aut}(H)|\pi f=\pi\}$. In such a case, each group G of automorphisms of S is a subgroup of $N_\Omega(\Gamma)/\Gamma$, and so $G=\Gamma'/\Gamma$ for some subgroup Γ' of $N_\Omega(\Gamma)$ containing Γ. Since Γ' contains Γ, the quotient H/Γ' is compact, and since Γ' is contained in $N_\Omega(\Gamma)$, it is discrete. Hence Γ' is also an NEC group.

Thus a group G is a group of automorphisms of a Klein surface represented as $S=H/\Gamma$ if and only if $G\cong\Gamma'/\Gamma$, for some NEC group Γ' containing Γ as a normal subgroup.

1.4. NOTES

The main theorem 1.1.10 was proved by Alling-Greenleaf in **[5]**. Its counterpart 1.2.3 was proved by Hall **[60]**, Preston **[108]** and Singerman **[114]**. The descriptions 1.3.2, 1.3.4, and 1.3.6 of the groups of automorphisms of compact Klein surfaces as quotients of NEC groups are due to May **[98]**. The uniformization theorem of Riemann surfaces we have used in the proof of 1.2.3 was established by Poincaré in **[106]**; a more comprehensive proof can be seen in the books Ahlfors-Sario **[3]** and Beardon **[6]**. The elementary results concerning commuting elements of Aut(H) we have employed in the proof of 1.3.4 appear, for instance, in Lehner **[78]**.

CHAPTER - 2
Normal NEC subgroups of NEC groups

By 1.2.3, a compact Klein surface S of algebraic genus $p \geq 2$ can be represented as $S=H/\Gamma$ for some surface NEC group Γ. Moreover, having a surface so represented, a finite group G acts as a group of automorphisms on S if and only if there exists an NEC group Γ' such that $G=\Gamma'/\Gamma$. For this reason we devote this chapter to analyze the relation between the signatures $\sigma(\Gamma)$ and $\sigma(\Gamma')$, since they determine the algebraic structure of both groups. The relation between the signs is determined in section 1, while proper periods and period-cycles are studied in section 2 and section 3, respectively.

These results are applied in section 4 to solve partially one of the most significant questions: given a surface NEC group Γ, how the signature of an NEC group Γ' containing Γ as a normal subgroup must look like?. This is completely solved in Thm 2.4.2 for odd index $[\Gamma':\Gamma]$ and partially in the even index case, in Thm 2.4.4. We finish obtaining in Thm 2.4.7, the full list of *normal pairs of signatures* (σ,σ') such that there exist NEC groups Γ and Γ' whose Teichmüller spaces have the same dimension, and Γ is a normal subgroup of Γ' with $\sigma(\Gamma)=\sigma$, $\sigma(\Gamma')=\sigma'$. This will be used in Ch.5.

2.1. THE SIGN IN THE SIGNATURE

We first characterize NEC subgroups of a given NEC group.

Proposition 2.1.1. *Let Γ be a subgroup of an NEC group Γ'. Then the following assertions are equivalent:*
(1) Γ is an NEC group.
(2) The index $[\Gamma':\Gamma]$ is finite.
Proof. (1) $\Rightarrow$ (2). It is an immediate consequence of Hurwitz Riemann formula.

Now, for (2) $\Rightarrow$ (1) notice first that Γ is discrete, as a subgroup of the discrete group Γ'. So, all reduces to see that the topological space H/Γ is compact.

Since H/Γ' is compact, Γ' has a compact fundamental region F'. Let $f_1,\ldots,f_k \in \Gamma'$ be the cosets representatives of Γ', that is,

$$\Gamma'=\Gamma f_1 \cup \ldots \cup \Gamma f_k,\ k=[\Gamma':\Gamma].$$

Now $F=f_1(F')\cup\ldots\cup f_k(F')$ is compact and

$$\Gamma F = \bigcup_{i=1}^{k} \Gamma f_i(F') = \Gamma' F' = H.$$

Since H/Γ is the image of F under the canonical continuous projection from H onto H/Γ, the compactness of H/Γ follows.

Comment and notations. From now on, along the chapter, we fix the following notations: Γ and Γ' are NEC groups, Γ is a *normal subgroup* of Γ' of index N. The relation between $\sigma(\Gamma)$ and $\sigma(\Gamma')$ depends strongly on the parity of N. For instance, we shall see that the signs of $\sigma(\Gamma)$ and $\sigma(\Gamma')$ coincide for odd N, but this is not necessarily the case if N is even.

Theorem 2.1.2. *If N is odd, the signatures $\sigma(\Gamma)$ and $\sigma(\Gamma')$ have the same sign.*

Proof. Assume first that Γ' has no reflection. Then, either Γ' is a fuchsian group, and so does Γ, or Γ' contains a glide reflection d. In the first case $\text{sign}\Gamma' = \text{sign}\Gamma = "+"$ whilst in the second one, $d' = d^N \in \Gamma$ and it is a glide reflection, since N is odd. In particular $\text{sign}\Gamma' = \text{sign}\Gamma = "-"$.

In what follows we suppose that some reflection c belongs to Γ'. Since N is odd $c \in \Gamma$ and so Γ' can be written as a disjoint union of cosets, $\Gamma' = \Gamma g_1 \cup \ldots \cup \Gamma g_N$, where, after replacing g_i by cg_i if necessary, we can assume that each g_i preserves the orientation, and $g_1 = 1_H$.

Let us fix a fundamental region F' for Γ'. Then $F = g_1 F' \cup \ldots \cup g_N F'$ is a fundamental region for Γ and all reduces to prove the following:

(2.1.2.1) There exist two congruent sides α and β of F and some orientation reversing $h \in \Gamma$ such that $h(\alpha) = \beta$ if and only if sign $\Gamma' = "-"$.

In fact, if $h(\alpha) = \beta$, $h \in \Gamma$, α and β sides in F, we can write $\alpha = g_i(\bar{\alpha})$, $\beta = g_j(\bar{\beta})$ where $\bar{\alpha}$ and $\bar{\beta}$ are different sides in F' since, in case $\bar{\alpha} = \bar{\beta}$ we would get $g_j g_i^{-1}(\alpha) = \beta$, *i.e.*, $g_j g_i^{-1} \in \Gamma \setminus \{1_H\}$. That means $\Gamma g_i = \Gamma g_j$, $i \neq j$, absurd. Assume $\text{sign}\Gamma' = "+"$. Then, since $g_j^{-1} h g_i(\alpha) = \beta$ we conclude that $g_j^{-1} h g_i$ preserves the orientation, and the same holds true for h.

Conversely, let us suppose that $\text{sign}\Gamma' = "-"$. Then, there exist two different sides γ_1, γ_2 in F' and some orientation reversing element $f \in \Gamma'$ such that $f(\gamma_1) = \gamma_2$. There exist $i \in \{1, \ldots, N\}$ and $h \in \Gamma$ such that $f = hg_i$. Then $\alpha = g_i(\gamma_1)$ and $\beta = \gamma_2 = g_1(\gamma_2)$ are sides in F and $h \in \Gamma$ is an orientation reversing element verifying $h(\alpha) = \beta$.

In order to obtain an analogous result for even N we introduce the notion of *word*. We shall see in next chapters that it is easy in a lot of cases to decide whether a word belongs or not to a given normal subgroup of an NEC group.

Definition 1. Let Γ be a normal subgroup of an NEC group Γ'.

(i) A canonical generator of Γ' is *proper* (with respect to Γ) if it does not belong to Γ.

(ii) The elements of Γ' expressable as composition of proper generators of Γ' are the *words* of Γ' (with respect to Γ).

(iii) A given word is *orientable* if it preserves the orientation of H. Otherwise it is *nonorientable.*

Theorem 2.1.3. *(1) Let us suppose that* N *is even and* Γ' *is orientable. Then* Γ *is orientable if and only if no nonorientable word belongs to* Γ.

(2) Let us suppose that N *is even and* Γ' *is nonorientable. Then* Γ *is nonorientable if and only if either a glide reflection of the canonical generators of* Γ' *or a nonorientable word belongs to* Γ.

Proof. (1) Let $w \in \Gamma$ be a nonorientable word. Then, there exists a canonical reflection $c \in \Gamma' \setminus \Gamma$. (Since these reflections are the only orientation reversing canonical generators)

We shall prove now that w identifies two different sides of a fundamental region of Γ, which implies that w is a canonical glide reflection and therefore the nonorientability of Γ.

Let us define $x=wc$. Clearly $\Gamma x = \Gamma c \neq \Gamma$ and so we can write

$$\Gamma'/\Gamma=\{\Gamma f_1,\Gamma x f_1,\dots,\Gamma f_k,\Gamma x f_k\}, \quad k=N/2$$

for some $f_1=1_H, f_2,\dots,f_k \in \Gamma'$.

Consequently, if F' is a fundamental region for Γ', the set

$$F = F' \cup x(F') \cup f_2(F') \cup f_2x(F') \cup \dots \cup f_k(F') \cup f_kx(F')$$

is a fundamental region for Γ.

If γ' is a side of F' fixed by c, γ' and $\gamma=x(\gamma')$ are different sides of F with $w(\gamma')=x(c(\gamma'))=x(\gamma')=\gamma$, and so w is a glide reflection.

Conversely, if Γ is nonorientable, we look for a nonorientable word in Γ. By the normality of Γ, we can describe the quotient

$$\Gamma'/\Gamma= \{\Gamma g_1,\Gamma g_2,\dots,\Gamma g_N\}, \quad g_1=1_H, g_2,\dots,g_N \in \Gamma'$$

in such a way that each g_i, $i=2,\dots,N$, is a word.

Now, for a fundamental region F' of Γ' we take the corresponding fundamental region of Γ,

$$F= F' \cup g_2(F') \cup \dots \cup g_N(F').$$

Since Γ is nonorientable, some glide reflection $w \in \Gamma$ identifies two different sides α and β of F. We are going to prove that w is the word we are looking for.

Let us write $\alpha=g_i(\bar{\alpha})$, $\beta=g_j(\bar{\beta})$ for some $1 \le i,j \le N$ and some sides $\bar{\alpha}$ and $\bar{\beta}$ of F'. Obviously $g_j^{-1}wg_i(\bar{\alpha})=\bar{\beta}$. Two cases are possible:

(i) $c=g_j^{-1}wg_i$ is nonorientable.

Then it cannot be a glide reflection since Γ' is orientable, $c\in\Gamma'$ and $c(\bar{\alpha})=\bar{\beta}$. Thus c is a reflection in the set of canonical generators of Γ' and to prove that $w=g_j c g_i^{-1}$ is a word, it suffices to check that $c\notin\Gamma$. But if this were not the case,

$$\Gamma g_j = g_j\Gamma = g_j c\Gamma = wg_i\Gamma = w\Gamma g_i = \Gamma g_i$$

and so $i=j$. But c is a reflection and $c(\bar{\alpha})=\bar{\beta}$. Thus $\bar{\alpha}=\bar{\beta}$, *i.e.*

$$\alpha=g_i(\bar{\alpha})=g_j(\bar{\beta})=\beta, \text{ which is false.}$$

(ii) $f=g_j^{-1}wg_i$ is orientable.

Then f is a hyperbolic generator of Γ'. All reduces to show that $f\notin\Gamma$. But in case $f\in\Gamma$ we deduce as before $g_i=g_j$ and so f would be conjugate in Γ' to a glide reflection. This is absurd.

Now, we prove (2). Let $w\in\Gamma$ be a glide reflection of the set of canonical generators of Γ', or a nonorientable word. In the first case it is obvious that Γ is nonorientable. In the second one we can write $w=xc$ or $w=xd$, where c is a reflection and d is a glide reflection. For $w=xc$ we repeat the argument used in (1). If $w=xd$ we shall see that, in fact, w is a glide reflection, and so Γ is nonorientable.

If x belongs to Γ, then so does d and we are done. So assume $x\notin\Gamma$. Then we can write

$$\Gamma'/\Gamma= \{\Gamma g_1,\Gamma g_2,\ldots,\Gamma g_N\},\quad g_1=1_H,\ g_2=x,\ g_3,\ldots,g_N\in\Gamma',$$

and we take fundamental regions F′ and $F=F'\cup g_2(F')\cup\ldots\cup g_N(F')$ for Γ' and Γ respectively. Then, if α,β are different congruent sides of F′ with $d(\alpha)=\beta$, we have $\alpha\in F$, $\gamma=x(\beta)=g_2(\beta)\in F$, $w(\alpha)=\gamma$, and so w is a glide reflection.

For the converse, we repeat the arguments used in (1). The details are left to the reader.

2.2. PROPER PERIODS OF NORMAL SUBGROUPS

We compute here the proper periods in the signature of Γ. As in the first paragraph, the cases of odd or even N are studied separately. The main difference is that in case of even N, proper periods in Γ may proceed from period-cycles of Γ', whilst this is not the case for odd N.

We first introduce the notion of *elliptic complete system* (e.c.s. in short), which will be useful in the sequel.

Definition 2. Let Γ be an NEC group. A finite subset of elliptic elements $E=\{x_1,\ldots,x_r\}\subseteq\Gamma$, none of which is a product of two reflections in Γ, is an *e.c.s.* of Γ if the following conditions hold:

(i) Each elliptic element of Γ which is not a product of two reflections in Γ, is conjugate in Γ to a power of some $x_i \in E$.
(ii) Non trivial powers of different elements of E are not conjugate.

Proposition 2.2.1. *Let* A,B *be two e.c.s. of* Γ. *There exists a bijection* $f: A \longrightarrow B$ *preserving the orders.*

Proof. Let us write $A=\{x_1,\dots,x_r\}$, $B=\{y_1,\dots,y_s\}$.

For each $1 \le i \le r$ we can write

$$x_i = a_i y_{\sigma(i)}^{l_i} a_i^{-1} \text{ for some } 1 \le \sigma(i) \le s,\ l_i \in \mathbb{Z},\ a_i \in \Gamma.$$

The number $\sigma(i)$ is unique with this property, since B is an e.c.s.. Even more, if we use condition (ii) in the previous definition, applied to A, we deduce the injectivity of the map

$$\sigma: \{1,\dots,r\} \longrightarrow \{1,\dots,s\}.$$

Thus $r \le s$ and by symmetry $r=s$. Consequently σ is bijective and so

$$f: A \longrightarrow B : x_i \longmapsto y_{\sigma(i)}$$

is a bijection too. Finally, notice that the equalities

$$x_i = a_i y_{\sigma(i)}^{l_i} a_i^{-1},\quad y_{\sigma(i)} = b_j x_j^{k_j} b_j^{-1},\ \text{for some } 1 \le j \le r,\ k_j \in \mathbb{Z},\ b_j \in \Gamma$$

imply that x_i and $x_j^{l_i k_j}$ are conjugate. That means $i=j$ and so

$$\#(x_i) = \#(y_{\sigma(i)}^{l_i}) \le \#(y_{\sigma(i)}) = \#(x_i^{k_i}) \le \#(x_i).$$

Thus $\#f(x_i) = \#(y_{\sigma(i)}) = \#(x_i)$.

Remarks and notations 2.2.2. (1) From 0.2.7, the canonical elliptic generators of Γ form an e.c.s.

(2) The proper periods of $\sigma(\Gamma)$ are the orders of the members of an arbitrary e.c.s.. This follows immediately from (1) and the last proposition.

(3) From now on, given positive integers u_i, k_i, $i=1,\dots,s$ we shall write $[(u_i)^{k_i} | i=1,\dots,s]$ instead of $[u_1,\overset{k_1}{\dots},u_1,u_2,\overset{k_2}{\dots},u_2,\dots,u_s,\overset{k_s}{\dots},u_s]$.

(4) If the number 1 appears among the proper periods or among periods in a period-cycle, we omit it.

Theorem 2.2.3. *Let* $E' = \{x_1,\dots,x_r\}$ *be the set of canonical elliptic generators of* Γ', $[m_1,\dots,m_r]$ *the proper periods of* Γ'. *Let us denote by* p_i *the order of* $\Gamma x_i \in \Gamma'/\Gamma$. *Then, if* N *is odd, the proper periods in* $\sigma(\Gamma)$ *are*

$$[(m_i/p_i)^{N/p_i} | 1 \le i \le r],$$

where, according to the convention adopted in (4), we omit those periods which are equal to 1 *(i.e., actually* i *runs over* $I=\{i | 1 \le i \le r,\ m_i \ne p_i\}$*).*

Proof. Write $G=\Gamma'/\Gamma$ and $H_i = <\Gamma x_i>$, for each $i \in I$. Since $[G:H_i]=N/p_i$ we can write

$$(1)\qquad G=\bigcup_{j=1}^{N/p_i} g_{ij}H_i,\ \text{where } g_{ij}=\Gamma s_{ij} \text{ for some } s_{ij}\in\Gamma'.$$

Let $$E=\{y_{ij}=s_{ij}x_i^{p_i}s_{ij}^{-1} \mid i\in I,\ 1\le j\le N/p_i\}.$$

By the definition, $x_i^{p_i}\in\Gamma$. Then $E\subseteq\Gamma$, since Γ is normal in Γ'. It is also obvious that $\#(y_{ij})=\#(x_i^{p_i})=m_i/p_i$.

So, using the second remark above, all reduces to check that E is an e.c.s. of Γ.

It is evident that each y_{ij} is elliptic. Now we prove that $y_{ij}=y$ is not a product of two reflections. In fact assume that $y=cd$ for some reflections c,d.

Now, if we put $s=s_{ij}$, $x=x_i$, $p=p_i$ and $c_1=s^{-1}cs$, $d_1=s^{-1}ds$, we obtain $x^p=c_1d_1$, where obviously, c_1 and d_1 are different reflections. The equality $x^p=c_1d_1$ implies that $Fix(c_1)$ meats $Fix(d_1)$ in one point, say o. Let R be a fundamental region for Γ' such that o is a vertex of R, and let $R'=c_1(R)$. Since $x(o)=o$, we deduce the existence of $q\in\mathbb{N}$ with $R'=x^q(R)$. Consequently $c_1=x^q$, a contradiction.

Now we must show that E verifies both conditions (i) and (ii) in Def. 2.

(i) Let $e\in\Gamma$ be an elliptic element which is not a product of two reflections in Γ. Then e is not a product of two reflections in Γ', because N is odd. Since E' is an e.c.s. of Γ', $e=tx_i^qt^{-1}$ for some $t\in\Gamma'$, $q\in\mathbb{N}$, $1\le i\le r$.

From (1) we can write $t=ws_{ij}vx_i^k$ for some $1\le j\le N/p_i$, $k\in\mathbb{N}$, and $v,w\in\Gamma$. Since Γ is normal, $u=ws_{ij}vs_{ij}^{-1}\in\Gamma$ and $t=us_{ij}x_i^k$. Therefore $e=us_{ij}x_i^qs_{ij}^{-1}u^{-1}$.

We claim that p_i divides q. In fact $s_{ij}x_i^qs_{ij}^{-1}=u^{-1}eu\in\Gamma$ and Γ being normal, $x_i^q\in\Gamma$. Thus we can write $q=\rho p_i$ for some $\rho\in\mathbb{Z}$, *i.e.*, $e=uy_{ij}^{\rho}u^{-1}$. But $e\ne1$ implies $y_{ij}\ne1$, *i.e.* $x_i^{p_i}\ne1$ or equivalently $i\in I$. We have seen that e is conjugate in Γ to $y_{ij}\in E$.

(ii) Let us suppose that y_{ij}^k and y_{eh}^l are conjugate in Γ. Then $x_i^{kp_i}$ and $x_e^{lp_e}$ are conjugate in Γ'. This implies i=e, and we put

$$s_{ij}x_i^{kp_i}s_{ij}^{-1}=ts_{ih}x_i^{lp_i}s_{ih}^{-1}t^{-1},\ t\in\Gamma.$$

The element $v=s_{ij}^{-1}ts_{ih}\in\Gamma'$ verifies $x_i^{kp_i}v=vx_i^{lp_i}$.

But $Fix(x_i)=Fix(x_i^{kp_i})=Fix(x_i^{lp_i})=\{o\}$. So, the equality $x_i^{kp_i}v=vx_i^{lp_i}$ implies $Fix(x_i)=Fix(v)$ and so v is a power of x_i, say $x_i^g=v$. Consequently, $s_{ij}x_i^g=ts_{ih}$, $t\in\Gamma$. In particular, $\Gamma s_{ij}<\Gamma x_i>=\Gamma s_{ih}<\Gamma x_i>$, *i.e.* $j=h$ and $y_{ij}=y_{eh}$. The proof is finished.

We study the case of even N. As was said before, the situation is slightly more involved.

Theorem 2.2.4. *Let us suppose that* N *is even. Let us write*

$$\sigma(\Gamma')=(g;\pm;[m_1,\dots,m_r];\{(n_{i1},\dots,n_{is_i})|1\le i\le k\}).$$

Consider the sets:

(1) $E'=\{x_1,\dots,x_r\}$ = *canonical elliptic generators of* Γ'.

(2) $C=\{(c_{i,j-1},c_{ij})|i\in I,\ j\in J_i\}$ *consisting of those pairs of consecutive reflections in* $\Gamma'\setminus\Gamma$.

Let us denote by p_l *the order of* $\Gamma x_l\in\Gamma'/\Gamma$ *for* $1\le l\le r$ *and by* q_{ij} *the order of* $\Gamma c_{i,j-1}c_{ij}$. *Then, the proper periods of* Γ *are*

$$[(m_l/p_l)^{N/p_l},(n_{ij}/q_{ij})^{N/2q_{ij}}\ |\ 1\le l\le r,\ 1\le i\le k,\ 0\le j\le s_i].$$

Note that actually l *runs on* $X=\{l|1\le l\le r,\ p_l\ne m_l\}$ *and* (i,j) *runs on* $Y=\{(i,j)|q_{ij}\ne n_{ij}\}$.

Proof. First we shall see that $2q_{ij}$ divides N. In fact, if $A=\Gamma c_{i,j-1}$ and $B=\Gamma c_{ij}$, we get $A^2=B^2=(AB)^{q_{ij}}=1$. Hence they generate the dihedral group of order $2q_{ij}$. Since this is a subgroup of Γ'/Γ, which has order N, the claim is proved.

Let us construct an elliptic complete system E for Γ.

Write $G=\Gamma'/\Gamma$, $G_l=\langle\Gamma x_l\rangle$, $l\in X$ and $H_{ij}=\langle\Gamma c_{i,j-1}c_{ij}\rangle$, $(i,j)\in Y$. Then we get equalities

$$G=\bigcup_{n=1}^{N/p_l}\Gamma s_{ln}G_l;\ G=\bigcup_{m=1}^{N/q_{ij}}\Gamma t_{i,j,m}H_{ij}$$

for each $l\in X$, $(i,j)\in Y$, where s_{ln}, $t_{i,j,m}$ are elements in Γ'. Even more, each $c_{i,j-1}\notin\Gamma c_{i,j-1}c_{ij}$, since $c_{ij}\notin\Gamma$, and so we can assume $t_{i,j,1}=1$, $t_{i,j,2}=c_{i,j-1}$, $t_{i,j,2m}=t_{i,j,2m-1}c_{i,j-1}$, for $1\le m\le N/2q_{ij}$. Denote

$$y_{ln}=s_{ln}(x_l^{p_l})s_{ln}^{-1};\ z_{i,j,m}=t_{i,j,m}(c_{i,j-1}c_{ij})^{q_{ij}}t_{i,j,m}^{-1}$$

and $F_1=\{y_{ln}|l\in X,\ 1\le n\le N/p_l\}$, $F_2=\{z_{i,j,m}|(i,j)\in Y,\ 1\le m\le N/q_{ij},\ m \text{ odd}\}$. Obviously, all reduces to prove that $E=F_1\cup F_2$ is an e.c.s. for Γ. Clearly, the elements in E are elliptic and the argument in 2.2.3 shows that none of them is a product of two reflections in Γ. Moreover, by using the obvious equalities $z_{i,j,2m}^{-1}=z_{i,j,2m-1}$, from our choice of t_{ij}'s, and repeating when necessary the arguments in 2.2.3, the only non trivial fact to prove is the following:

(2.2.4.1) For different triples (i,j,m) and (i′,j′,m′), with (i,j), (i′,j′)∈Y and $1\le m\le N/q_{ij}$, $1\le m'\le N/q_{i'j'}$, m and m′ odd numbers, no powers of $z_{i,j,m}$ and $z_{i',j',m'}$ are conjugate in Γ.

This is trivial if $(i,j)\ne(i',j')$ because the powers of $c_{i,j-1}c_{ij}$ and $c_{i',j'-1}c_{i'j'}$ are not conjugate, since they are distinct elliptic elements. Now, for $(i,j)=(i',j')$ suppose that $z_{i,j,m}^a=gz_{i,j,m'}^bg^{-1}$ for some $g\in\Gamma$, $a,b\in\mathbb{N}$.

Then, if $f=t^{-1}_{i,j,m}gt_{i,j,m'}$, we get $((c_{i,j-1}c_{ij})^{aq_{ij}})f=f(c_{i,j-1}c_{ij})^{bq_{ij}}$ and if $p\in H$ is the fixed point of $c_{i,j-1}c_{ij}$, we deduce $f(p)=p$. Since the stabilizer of p is the dihedral group generated by $c_{i,j-1}$ and c_{ij}, f has necessarily one of the following forms:

$$f=(c_{i,j-1}c_{ij})^{\mu} \text{ or } f=c_{i,j-1}(c_{i,j-1}c_{ij})^{\mu},\ \mu\in\mathbb{N}.$$

In the first case, $t_{i,j,m}(c_{i,j-1}c_{ij})^{\mu}=t_{i,j,m}f=gt_{i,j,m'}$, and so $\Gamma t_{i,j,m}H_{ij}=$ $=\Gamma gt_{i,j,m'}H_{ij}=\Gamma t_{i,j,m'}H_{ij}$, *i.e.*, $m=m'$ and $(i,j,m)=(i',j',m')$. In the second one $\Gamma t_{i,j,m+1}H_{ij}=\Gamma t_{i,j,m}c_{ij}H_{ij}=\Gamma t_{i,j,m}fH_{ij}=\Gamma gt_{i,j,m'}H_{ij}=\Gamma t_{i,j,m'}H_{ij}$, *i.e.*, $m'=m+1$. This is impossible since both m and m′ are odd. So 2.2.4.1 is proved.

Our last result in this section gives $\sigma(\Gamma^+)$ in terms of $\sigma(\Gamma)$. This can be seen as a particular case of 2.2.4.

Corollary 2.2.5. *Let Γ be an NEC group with signature*

$$\sigma(\Gamma)=(g;\pm;[m_1,...,m_r];\{(n_{i1},...,n_{is_i})\mid i=1,...,k\}).$$

Let us denote $g^+=\alpha g+k-1$, where $\alpha=1$ if sign $\sigma(\Gamma)=$"-" and $\alpha=2$ if sign $\sigma(\Gamma)=$"+". Then, the signature of Γ^+ is

$$\sigma(\Gamma^+)=(g^+;+;[m_1,m_1,...,m_r,m_r,n_{11},...,n_{1s_1},...,n_{k1},...,n_{ks_k}];\{-\}).$$

Proof. Obviously sign $\sigma(\Gamma^+)=$"+" and $\sigma(\Gamma^+)$ has not period-cycles, because each element in Γ^+ preserves orientation. Each elliptic canonical generator $x_i\in\Gamma$ preserves the orientation of H. Thus $x_i\in\Gamma^+$. With the notations of 2.2.4, $p_i=1$, $N=2$, and so $N/p_i=2$, $m_i/p_i=m_i$, $i=1,...,r$.

Each reflection $c_{ij}\in\Gamma\backslash\Gamma^+$, but $c_{i,j-1}c_{ij}\in\Gamma^+$, $i=1,...,k$, $1\le j\le s_i$. Consequently $q_{ij}=1$, $N/2q_{ij}=1$, $n_{ij}/q_{ij}=n_{ij}$, $i=1,...,k$, $1\le j\le s_i$. Thus, if q is the genus of Γ^+, we deduce from 2.2.4,

$$\sigma(\Gamma^+)=(q;+;[m_1,m_1,...,m_r,m_r,n_{11},...,n_{1s_1},...,n_{k1},...,n_{ks_k}];\{-\}).$$

We must finally prove that $q=g^+$. The areas are

$$\mu(\Gamma)=2\pi[\alpha g+k-2+\sum_{i=1}^{r}(1-1/m_i)+1/2\sum_{i=1}^{k}\sum_{j=1}^{s_i}(1-1/n_{ij})]$$

$$\text{and } \mu(\Gamma^+)=2\pi[2q-2+2\sum_{i=1}^{r}(1-1/m_i)+\sum_{i=1}^{k}\sum_{j=1}^{s_i}(1-1/n_{ij})]$$

But $\mu(\Gamma^+)=2\mu(\Gamma)$, by the Hurwitz Riemann formula. So

$$2[\alpha g+k-2]=2q-2,\ \textit{i.e. } q=\alpha g+k-1=g^+.$$

2.3. PERIOD-CYCLES

Our first task in this section is to show that for odd N, the period-cycles in $\sigma(\Gamma')$ and the orders modΓ of the hyperbolic canonical generators of Γ' determine the period-cycles in the signature of Γ. This case is complicated enough, but the situation is still much more involved for even N.

Theorem 2.3.1. *Let* N *be odd and let* $H=\{e_1,\dots,e_k\}$ *be the set of hyperbolic canonical generators of* Γ'. *For each* $i=1,\dots,k$ *we denote by* l_i *the order of* $\Gamma e_i \in \Gamma'/\Gamma$. *If* $\{(n_{i1},\dots,n_{is_i}) \mid i=1,\dots,k\}$ *are the period-cycles of* $\sigma(\Gamma')$, *then*

$$\{(\overset{l_i}{n_{i1},\dots,n_{is_i},\dots,n_{i1},\dots,n_{is_i}})^{N/l_i} \mid 1\le i\le k\}$$ *are the period-cycles of* $\sigma(\Gamma)$.

Proof. Let us fix $i\in\{1,\dots,k\}$. From 0.2.5 we can find a fundamental region F' for Γ', whose perimeter is labelled as follows:

$$\varepsilon_i\gamma_{i0}\gamma_{i1}\cdots\gamma_{is_i}\varepsilon_i'\Delta, \text{ where}$$

(i) Δ represents the other sides of the perimeter.

(ii) The reflection $c_{ij}\in\Gamma'$ fixes the side γ_{ij}, $j=0,\dots,s_i$.

(iii) The stabilizer of the common vertex N_{ij} to the sides $\gamma_{i,j-1}$ and γ_{ij} is the dihedral group generated by $c_{i,j-1}$ and c_{ij}, and so it has order $2n_{ij}$.

(iv) For each $i=1,\dots,k$, $e_i(\varepsilon_i')=\varepsilon_i$.

Now we construct a fundamental region F for Γ in the following way:
By the definition of l_i, there exist $\beta_1=1_H, \beta_2,\dots,\beta_{N/l_i}\in\Gamma'$ such that

(1) $\quad \Gamma'/\Gamma=\{\Gamma\beta_1,\Gamma\beta_1e_i,\dots,\Gamma\beta_1e_i^{l_i-1},\dots,\Gamma\beta_{N/l_i},\Gamma\beta_{N/l_i}e_i,\dots,\Gamma\beta_{N/l_i}e_i^{l_i-1}\}$.

Consequently,

$$F=\bigcup_{\substack{r\in\{1,\dots,N/l_i\}\\ h\in\{0,\dots,l_i-1\}}}(\beta_r e_i^h)F'$$

is a fundamental region for Γ.

For each $r=1,\dots,N/l_i$ we focus our attention on the following segment C_r of the perimeter of F:

$$C_r:\quad \beta_r e_i^{l_i-1}(\varepsilon_i),\beta_r e_i^{l_i-1}(\gamma_{i0}),\dots,\beta_r e_i^{l_i-1}(\gamma_{is_i}),\beta_r e_i^{l_i-2}(\gamma_{i0}),\dots,$$
$$\beta_r e_i^{l_i-2}(\gamma_{is_i}),\dots,\beta_r(\gamma_{i0}),\dots,\beta_r(\gamma_{is_i}),\beta_r(\varepsilon_i').$$

Our task now is to check that each C_r generates a hole (if we identify the points in F which are equivalent under the action of Γ). Once this will be proved, $\sigma(\Gamma)$ will consist of N/l_i period-cycles $P_1,\dots,P_{N/l_i}$, whose elements are

P_r: $\qquad \{1/2(\text{order of the stabilizer in } \Gamma \text{ of } \beta_r e_i^h(N_{ij}))\}$

with $j=0,\dots,s_i$ and $h=0,\dots,l_i-1$.

From the description (1) of the quotient Γ'/Γ, the map

$$\eta \longmapsto (\beta_r e_i^h)^{-1}\eta(\beta_r e_i^h)$$

is a bijection between $\mathrm{St}_\Gamma(\beta_r e_i^h(N_{ij}))$ and $\mathrm{St}_{\Gamma'}(N_{ij})$. Since the last one has order $2n_{ij}$, we deduce that

$$P_r:(\overbrace{n_{i1},\dots,n_{is_i},\dots\dots,n_{i1},\dots,n_{is_i}}^{l_i})$$

for each $r=1,\dots,N/l_i$ or with the periods in reverse order. This last possibility can be eliminated: this is obvious by 0.2.6, if $\mathrm{sign}\sigma(\Gamma)=$"-", whilst, for $\mathrm{sign}\sigma(\Gamma)=$"+" it is enough to observe that since N is odd, the elements $\beta_1,\dots,\beta_{N/l_i}$ can be chosen orientation preserving.

Repeating this process for $i=1,\dots,k$, we get the announced period-cycles.

So, all reduces to see that each C_r generates a hole, and for the sake of simplicity, we handle the case $r=1$. Of course, it is enough to prove:

(a) There exists $\alpha\in\Gamma$ with $\alpha(\varepsilon_i')=e_i^{l_i-1}(\varepsilon_i)$.

(b) There is no $\alpha\in\Gamma\backslash\{1_H\}$ which sends a point p in $e_i^h(\gamma_{ij})$ to a distinct point in the perimeter of F, $h=0,\dots,l_i-1$, $j=0,\dots,s_i$.

The first is clear: $\alpha=e_i^{l_i}\in\Gamma$ and $\alpha(\varepsilon_i')=e_i^{l_i-1}(e_i(\varepsilon_i'))=e_i^{l_i-1}(\varepsilon_i)$. To prove (b) let us suppose there exist a pair (j,h), a point $p\in e_i^h(\gamma_{ij})$ and $\alpha\in\Gamma$ such that $q=\alpha(p)$ belongs to the perimeter of F, $p\neq q$. We distinguish:

(b.1) If $q\in e_i^{h_1}(\gamma_{ij_1})$ for some $0\le h_1\le l_i-1$ and $0\le j_1\le s_i$, then $\alpha'=e_i^{-h_1}\alpha e_i^h\in\Gamma'$, $p'=e_i^{-h}(p)\in\gamma_{ij}$, $q'=e_i^{-h_1}(q)\in\gamma_{ij_1}$, and we have $\alpha'(p')=q'$. This implies $p'=q'$, *i.e.* $e_i^{h_1-h}(p)=q$. Then necessarily $h_1=h$ and $p=q$, which is false.

(b.2) Now assume $q\in\beta_r e_i^{h_1}(\gamma_{ij})$ for some $2\le r\le N/l_i$, $0\le h_1\le l_i-1$. We take

$$p'=e_i^{-h}(p)\in\gamma_{ij};\ q'=e_i^{-h_1}\beta_r^{-1}(q)\in\gamma_{ij}$$

and so, for $\alpha'=e_i^{-h_1}\beta_r^{-1}\alpha e_i^h\in\Gamma'$, $\alpha'(p')=q'$. Then $p'=q'\in\gamma_{ij}$ is a fixed point of α'. Consequently α belongs to the group generated by $c_{i,j-1}c_{ij}$ and $c_{i,j+1}$. All these elements, having order two, are in Γ because N is odd. Thus $\alpha'\in\Gamma$, and so $\Gamma\beta_r e_i^{h_1}=\Gamma e_i^h$. Since $r>1$, this contradicts the description (1) of Γ'/Γ.

(b.3) Finally, if q belongs to a side in the perimeter of F different from the ones studied in (b.1) and (b.2), by arguing in the same way we get a point $p'\in\gamma_{ij}$ and another one, q' in a side of the perimeter of F', distinct from γ_{ij}, which are related by a transformation in Γ'. This is impossible and so the theorem is proved.

When the index is even we separate the analysis in two parts.

Theorem 2.3.2. *Let* N *be even and let* $(n_1,\dots,n_s)$ *be a period-cycle in* $\sigma(\Gamma')$ *whose associated generators are denoted* $\{e,c_0,\dots,c_s\}$. *Let* n *be the order of* $\Gamma e\in\Gamma'/\Gamma$. *If we assume that each* $c_j\in\Gamma$, $j=0,\dots,s$, *the signature of* Γ *has* N/n *period-cycles of the form*

$$(1)\qquad (n_1,\dots,n_s,\overset{n}{\dots},n_1,\dots,n_s)$$

if sign $\sigma(\Gamma)=$"-". *If this sign is* "+", *it has* N_1 *period-cycles of type (1) and* $(N/n)-N_1$ *period-cycles of type*

$$(2)\qquad (n_s,\dots,n_1,\overset{n}{\dots},n_s,\dots,n_1)$$

for some nonnegative integer $N_1\le N/n$.

Proof. The proof is very similar to the one of the previous theorem, and we leave the details to the reader. However, we must remark that, possibly, some reflections in Γ' associated to the other period-cycles do not belong to Γ. This produces "inverse" period-cycles of type (2), when the sign "+" occurs. This has not influence if the sign is "-" because, from 0.2.6, two such groups with "inverse" period-cycles are isomorphic.

Notice that this theorem covers the case $(n_1,\dots,n_s)=(-)$.

Our last result in this paragraph deals with the remaining case, *i.e.* not every $c_0,\dots,c_s$ belong to Γ.

Theorem 2.3.3. *Let* N *be even and let* $(n_1,\dots,n_s)$ *be a nonempty period-cycle in* $\sigma(\Gamma')$, *whose associated reflections are denoted* $\{c_0,\dots,c_s\}$. *We assume that the set*

$$J=\{(i,j)\in\{1,\dots,s\}\times\{0,\dots,s-1\}\mid i\le j,\ c_{i-1},c_{j+1}\notin\Gamma,\ c_i,c_{i+1},\dots,c_j\in\Gamma\}$$

is not empty. Let us call n(i,j) *the order of* $\Gamma(c_{i-1}c_{j+1})\in\Gamma'/\Gamma$.

Then, for each pair $(i,j)\in J$, *we have:*

(1) The numbers n_i *and* n_{j+1} *are even.*

(2) The signature of Γ *has* N/2n(i,j) *period-cycles, each consisting of* n(i,j) *copies of periods*

$$n_{j+1}/2,n_j,n_{j-1},\dots,n_{i+1},n_i/2,n_{i+1},\dots,n_j$$

where the quotients $n_i/2$ *and* $n_{j+1}/2$ *are omitted if they are equal to* 1.

Proof. (1) We shall prove that n_i is even. The same argument holds true for n_{j+1}. In fact, if n_i is odd, we put $k=(n_i-1)/2$ and from $(c_ic_{i-1})^{n_i}=1$ we deduce $(c_ic_{i-1})^kc_i(c_{i-1}c_i)^k=c_{i-1}^{-1}=c_{i-1}$. For $f=(c_ic_{i-1})^k$, we have $c_{i-1}=fc_if^{-1}\in\Gamma$, since $c_i\in\Gamma$ and Γ is a normal subgroup of Γ'. A contradiction.

(2) Let us fix $(i,j)\in J$ and $n=n(i,j)$. We can choose a fundamental region F′ for Γ' whose perimeter is labelled

(*) $$\gamma_{j+1},\gamma_j,\gamma_{j-1},\ldots,\gamma_{i+1},\gamma_i,\gamma_{i-1},\Delta$$

where Δ denotes the remaining sides, each γ_h is the side fixed by c_h, $i-1\leq h\leq j+1$, and the stabilizer of the vertex N_h common to the sides γ_{h-1} and γ_h is the dihedral group generated by c_{h-1} and c_h.

We look for the suitable fundamental region of Γ in order to compute the period-cycles in $\sigma(\Gamma)$.

Since $c_{i-1}\notin\Gamma$ and $\Gamma c_{i-1}c_{j+1}\in\Gamma'/\Gamma$ has order $n(i,j)=n$, there exist $\beta_1=1_H$, $\beta_2,\ldots,\beta_{N/2n}\in\Gamma'$, such that

$$\Gamma'/\Gamma=\bigcup_{r=1}^{N/2}\{\Gamma\beta_r,\Gamma\beta_r c_{i-1},\Gamma\beta_r c_{i-1}c_{j+1},\ldots,\Gamma\beta_r(c_{i-1}c_{j+1})^{n-1},\Gamma\beta_r(c_{i-1}c_{j+1})^{n-1}c_{i-1}\}.$$

Consequently, if we denote

$$F_0=F'\cup c_{i-1}F'\cup(c_{i-1}c_{j+1})F'\cup\ldots\cup(c_{i-1}c_{j+1})^{n-1}F'\cup(c_{i-1}c_{j+1})^{n-1}c_{i-1}F',$$

a fundamental region for Γ is given by

$$F=\beta_1(F_0)\cup\beta_2(F_0)\cup\ldots\cup\beta_{N/2n}(F_0).$$

Now we consider the following segment C of the perimeter of F_0, obtained from (*):

$$\gamma_j,\gamma_{j-1},\ldots,\gamma_i,c_{i-1}(\gamma_i),\ldots,c_{i-1}(\gamma_j),(c_{i-1}c_{j+1})(\gamma_j),\ldots,$$
$$(c_{i-1}c_{j+1})(\gamma_i),\ldots,(c_{i-1}c_{j+1})^{n-1}(\gamma_j),\ldots,(c_{i-1}c_{j+1})^{n-1}(\gamma_i),$$
$$(c_{i-1}c_{j+1})^{n-1}c_{i-1}(\gamma_i),\ldots,(c_{i-1}c_{j+1})^{n-1}c_{i-1}(\gamma_j).$$

Repeating arguments used in (a) and (b) in the proof of 2.3.1, we conclude that C produces a hole, which gives rise to a period-cycle in $\sigma(\Gamma)$. Of course, the same holds true for each $C_r=\beta_r(C)$, $1\leq r\leq N/2n$, and all these are different holes, *i.e.*, they provide us N/2n period-cycles, all of them with the same elements. These elements are, precisely, the orders of the products of each pair of reflections in the sides of C meeting at the vertices, arranged in cyclic ordering. Thus, to finish, we only need to compute these orders.

For $h=i+1,\ldots,j$, the reflections c_h and c_{h-1} fix γ_h and γ_{h-1} respectively, and c_hc_{h-1} has order n_h.

However, the side $c_{i-1}(\gamma_i)$ is fixed by $f_i=c_{i-1}c_ic_{i-1}$ and the order of $c_if_i=(c_ic_{i-1})^2$ is $n_i/2$, because n_i is even.

Since the side $c_{i-1}(\gamma_h)$, $h=1,\ldots,j$ is fixed by $f_h=c_{i-1}c_hc_{i-1}$, we must compute the order of f_hf_{h-1}. But

$$\#(f_hf_{h-1})=\#(c_{i-1}c_hc_{h-1}c_{i-1})=\#(c_hc_{h-1})=n_h.$$

Finally, running along C, the next pair of sides is $c_{i-1}(\gamma_j)$, $(c_{i-1}c_{j+1})(\gamma_j)$ which are respectively fixed by

$$f_j=c_{i-1}c_jc_{i-1}\text{ and }g_j=c_{i-1}c_{j+1}c_jc_{j+1}c_{i-1}$$

whose product

$$f_jg_j=c_{i-1}(c_jc_{j+1})^2c_{i-1}$$

has the same order as $(c_jc_{j+1})^2$, *i.e.*, $n_{j+1}/2$. So we have obtained the first block

$$(n_j,\dots,n_{i+1},n_i/2,n_{i+1},\dots,n_j,n_{j+1}/2)$$

of the period-cycle.

Since we can reorder, by 0.2.6, in a cyclic way the elements of a period-cycle, it is enough to repeat the process above with each pair of consecutive sides in C to reach the promised expression for the period-cycles in $\sigma(\Gamma)$. Notice that the case $n_i/2=1$ means $c_ic_{i-1}c_ic_{i-1}=1$, and so the elements $f_i=c_{i-1}c_ic_{i-1}$ and c_i, which fix the sides $c_{i-1}(\gamma_i)$ and γ_i, respectively, are the same. Thus we can delete $n_i/2=1$, and analogously $n_{j+1}/2=1$, in the period-cycle if this situation occurs.

Keeping in mind that, by 0.2.6, the periods in a period-cycle can be cyclically reordered by choosing a different fundamental region, we deduce:

Corollary 2.3.4. *Let us suppose that* N *is even, and let* $(n_1,\dots,n_s)$ *be a nonempty period-cycle of* $\sigma(\Gamma')$. *Let us denote* $\{e,c_0,\dots,c_s\}$ *the generators corresponding to this period-cycle. Then, there is at most one pair of indices* (h,k) *such that*

$$h+1\le k,\ c_0,\dots,c_h,c_{k+1},\dots,c_s\in\Gamma,\ c_{h+1},c_k\notin\Gamma.$$

If such a pair exists, then the signature of Γ *has* N/2n *period-cycles, each consisting of* n *copies of the periods*

$$n_{h+1}/2,n_h,n_{h-1},\dots,n_1,n_s,n_{s-1},\dots,n_{k+2},n_{k+1}/2,n_{k+2},\dots,n_s,n_1,\dots,n_h$$

where n *is the order of* $\Gamma e^{-1}c_{h+1}ec_k\in\Gamma'/\Gamma$.

To finish this paragraph we develop carefully an example, in order to see how the results obtained until now work.

Examples 2.3.5. Let Γ' be an NEC group with signature

$$\sigma(\Gamma')=(1;+;[2];\{(2,2,2,2)\}).$$

From 0.2.5, a presentation of Γ' is given by

Generators: a_1, b_1, x_1, e_1, c_0, c_1, c_2, c_3, c_4.

Relations: $x_1^2=e_1^{-1}c_0e_1c_4=x_1e_1a_1b_1a_1^{-1}b_1^{-1}=c_{j-1}^2=c_j^2=(c_{j-1}c_j)^2=1,\ 1\le j\le 4.$

It is easy to see that the assignment

$$\theta(a_1)=1;\ \theta(b_1)=0;\ \theta(x_1)=2;\ \theta(e_1)=2$$
$$\theta(c_0)=\theta(c_1)=\theta(c_4)=0;\ \theta(c_2)=\theta(c_3)=2$$

induces an epimorphism θ from Γ' onto $\mathbb{Z}_4$. We are going to find the signature of $\Gamma=\ker\theta$.

(i) Since $w=x_1c_2\in\Gamma$ is a nonorientable word, part (1) in 2.1.3 allows us to deduce that sign $\sigma(\Gamma)=$"-".

(ii) With the notations in 2.2.4, N=4, $E'=\{x_1\}$, $C=\{(c_2,c_3)\}$, $c_j=c_{1j}$, and since

$c_2c_3 \in \Gamma$, $x_1 \notin \Gamma$, we obtain

$$m_1=p_1=2;\ n_1=2;\ q_1=1;\ N/2q_1=2, \text{ where } n_j=n_{1j},\ q_j=q_{1j},$$

and thus the proper periods in $\sigma(\Gamma)$ are [2,2].

(iii) To calculate the period-cycles we employ 2.3.4, because N=4 is even, $c_0,c_1,c_4 \in \Gamma$ and $c_2,c_3 \notin \Gamma$.

Here h=1, k=3 and n=1, because $e_1^{-1}c_2e_1c_3 \in \Gamma$. Thus $\sigma(\Gamma)$ has N/2n=2 period-cycles of the form

$$(n_2/2,n_1,n_4/2,n_1)=(2,2).$$

So, $\sigma(\Gamma)=(g;-;[2,2];\{(2,2),(2,2)\})$.

Finally, by the Hurwitz Riemann formula,

$$\mu(\Gamma')=5\pi,\ \mu(\Gamma)=2\pi(g+2) \text{ and } \mu(\Gamma)=4\mu(\Gamma'),\ i.e.\ g=8.$$

Consider another example. We take now the epimorphism ϕ from Γ' onto Z_4 induced by $\phi(a_1)=1$; $\phi(b_1)=\phi(x_1)=\phi(e_1)=\phi(c_j)=0$, $0 \le j \le 4$, and denote by Γ its kernel. From 2.1.3, the sign in the signature of Γ is "+". Now the set C of 2.2.4, is empty and $x_1 \in \Gamma$. Consequently, $m_1=2$, $p_1=1$, $N/p_1=4$, and the proper periods in $\sigma(\Gamma)$ are [2,2,2,2].

Apply 2.3.2 to compute the period-cycles, because each $c_j \in \Gamma$, $0 \le j \le 4$. But also $e_1 \in \Gamma$ and so $\sigma(\Gamma)$ has four period-cycles of the form (2,2,2,2). Finally, by the Hurwitz Riemann formula, we obtain that the genus g of Γ is equal to 1 and so Γ has signature

$$(1;+;[2,2,2,2];\{(2,2,2,2),(2,2,2,2),(2,2,2,2),(2,2,2,2)\}).$$

Remark 2.3.6. (1) Period-cycles obtained in this section are all that a normal subgroup Γ of an NEC group Γ' has. In fact assume that C is a period-cycle of Γ and let c be one of the canonical reflections of Γ corresponding to C. Then c is a reflection of Γ' of course. Although it may not be a canonical one for Γ', by 0.2.7 it is conjugate in Γ' to such a reflection, say c'. But since Γ is normal in Γ', $c' \in \Gamma$ and we are in the starting point for constructing a period-cycle in the proofs of the results in question.

(2) Of course in section 2 we have calculated all proper periods of Γ.

(3) Combining facts above with results of section 1 concerning the sign of $\sigma(\Gamma)$ and with the Hurwitz Riemann formula we are able to find the signature of Γ (as we showed, as a function of the signature σ' of Γ' and the orders of the images in Γ'/Γ of certain canonical generators of Γ' and some of its products).

Remark 2.3.7. Although the results of this chapter are quite general (they provide a way to find the signature of a normal subgroup Γ of an NEC group Γ' given by the signature as was remarked in (3) above), we shall use them in rather particular form (with the exception of chapter 6). We shall be mainly concerned with the conditions for the signature σ' of an NEC group Γ' and for the quotient Γ'/Γ, for Γ to be a surface group.

In such a case we shall use results of section 2 just to deduce that Γ has no proper periods if and only if for any canonical elliptic generator x_i of Γ' and for any two canonical reflections c,c' belonging to a period in a period-cycle of Γ' the elements $x_i \in \Gamma'$ and $\Gamma x_i \in \Gamma'/\Gamma$ have the same orders and the same holds true for $cc' \in \Gamma'$ and $\Gamma cc' \in \Gamma'/\Gamma$. If two consecutive canonical reflections of Γ' belong to Γ, then Γ has a nonempty period-cycle. Thus if Γ is a surface group then in particular no two consecutive reflections of Γ' belong simultaneously to Γ. So, for example, theorem 2.3.3 will be used for the set $J=\{(i,i)|c_i \in \Gamma\}$ just to find the number of empty period-cycles of Γ (every $(i,i) \in J$ produces $N/2n(i,i)$ empty period-cycles, where $n(i,i)$ is the order of $\Gamma c_{i-1}c_{i+1} \in \Gamma'/\Gamma$).

2.4. NORMAL SUBGROUPS

In this section we study the inverse problem to the one we have analyzed in previous ones. Namely, how the signature of a group Γ' containing a bordered surface group Γ with prescribed signature as a normal subgroup looks like.

As we shall see, in certain cases the results are fairly complete. They will be used in coming chapters.

Theorem 2.4.1. *Let Γ' be an NEC group with signature*

$$\sigma(\Gamma')=(g';\pm;[m_1,\dots,m_r];\{(n_{i1},\dots,n_{is_i})|\ i=1,\dots,k\})$$

and $k \geq 1$. Let us suppose that there exists a normal bordered surface NEC subgroup Γ of Γ'. Then $\sigma(\Gamma')$ has either an empty period-cycle or a period-cycle with two consecutive periods equal to 2.

Proof. From 2.1.1, $N=[\Gamma':\Gamma]$ is finite. For odd N, every period-cycle in $\sigma(\Gamma')$ is empty by 2.3.1, and we are done. So, we assume that N is even and each period-cycle in $\sigma(\Gamma')$ is nonempty. By assumption, Γ has a period-cycle. Thus, a canonical reflection c_{ij} of Γ' belongs to Γ. Since this period-cycle is empty we deduce from 2.3.3 that $c_{i,j-1} \notin \Gamma$, $c_{i,j+1} \notin \Gamma$, since otherwise n_{ij} or $n_{i,j+1}$ would appear as periods of a period-cycle in $\sigma(\Gamma)$. Therefore, using 2.3.3 once more, we see that $n_{ij}/2$ and $n_{i,j+1}/2$ actually appear as periods in a period-cycle of $\sigma(\Gamma)$, unless they both are equal to one. Since Γ is a surface group, the last is the case, *i.e.*, $n_{ij}=n_{i,j+1}=2$.

If N is odd we can deduce much more precise information about $\sigma(\Gamma')$.

Theorem 2.4.2. *Let Γ be an NEC group with signature*

$$\sigma(\Gamma)=(g;\pm;[-];\{(-),\overset{k}{\ldots},(-)\})$$

and let Γ' be an NEC group containing Γ as a normal subgroup of odd index N.

Then the signature of Γ' has the form

$$\sigma(\Gamma')=(g';\pm;[m_1,\ldots,m_r];\{(-),\overset{k'}{\ldots},(-)\}),$$

where:

(i) sign $\sigma(\Gamma)$=sign $\sigma(\Gamma')$.

(ii) Each m_i *divides* N, $i=1,\ldots,r$.

(iii) $k=N\sum_{i=1}^{k'} 1/l_i$, *for* l_i*=order of* $\Gamma e_i\in\Gamma'/\Gamma$, $H=\{e_1,\ldots,e_{k'}\}$ *the set of canonical hyperbolic generators of* Γ'.

(iv) $\alpha g-2+k=N[\alpha g'+k'-2+\sum_{i=1}^{r}(1-1/m_i)]$, $\alpha=\begin{cases}1 \text{ if sign } \sigma(\Gamma)=\text{"}-\text{"}\\ 2 \text{ if sign } \sigma(\Gamma)=\text{"}+\text{"}.\end{cases}$

Proof. Condition (i) follows from 2.1.2, whilst (ii) is an obvious consequence of 2.2.3.

Applying 2.3.1 it is pretty obvious that the period-cycles in $\sigma(\Gamma')$ are empty, and also

$$k=\sum_{i=1}^{k'} N/l_i=N\sum_{i=1}^{k'} 1/l_i$$

Finally, (iv) is nothing else but the Hurwitz Riemann formula.

For even N we have a result of the same kind provided Γ'/Γ is cyclic. First we need an easy but useful observation.

Proposition 2.4.3. *Let* N *be an even integer. Let Γ and Γ' be NEC groups and $\theta:\Gamma'\longrightarrow Z_N$ be a group epimorphism with* $\ker\theta=\Gamma$. *Let us suppose that Γ is a bordered surface NEC group. Then, if $(n_1,\ldots,n_s)$ is a nonempty period-cycle in $\sigma(\Gamma')$ with associated reflections $\{c_0,\ldots,c_s\}$, the following conditions hold:*

(1) $\theta(c_0)=\theta(c_{2l})$; $\theta(c_{2l-1})=\theta(c_0)+N/2$, *for* $1\le l\le[s/2]$, $\theta(c_0)=0$ or $\theta(c_0)=N/2$.

(2) Each $n_j=2$, $1\le j\le s$.

(3) The integer s *is even.*

Proof. Let us take two consecutive reflections $c=c_j$, $c'=c_{j+1}$. Then $\#(c)=\#(c')=2$ and $\#(cc')=n=n_{j+1}$. Consequently, for $u=\theta(c)$ and $v=\theta(c')$, $2u=2v=n(u+v)=0$ modN. Using 2.3.3 we discard the possibility $u=v=0$, because in such a case the number n would be a period in some period-cycle of $\sigma(\Gamma)$. From 2.2.4 however, it follows that $u=0$ or $v=0$. Otherwise u would be a proper period in $\sigma(\Gamma)$. So, we can assume that $u=0$, $v=N/2$ or $u=N/2$, $v=0$. This proves (1). Even more, we deduce that

$$0=n(u+v)=nN/2 \mod N$$

and so $n=2n'$ for some $n'\in\mathbb{N}$. Now by 2.3.3 n' is a period in some period-cycle of the signature of the surface group Γ unless $n'=1$. Consequently $n'=1$, *i.e.*, $n=2$ and condition (2) is proved. Finally, let e be the canonical hyperbolic generator associated with $(n_1,\ldots,n_s)$. Since $c_0e^{-1}c_se=1$, we deduce $\theta(c_0)+\theta(c_s)=0$ and so by part (1), s is even.

Now we can state

Theorem 2.4.4. *Let Γ be a surface NEC group and let* N *be an even integer. If* $\sigma(\Gamma)=(g;\pm;[-];\{(-),\overset{k}{\ldots},(-)\})$ *and Γ' is another NEC group containing Γ as a normal subgroup with cyclic factor of order* N, *then for some $0\le s'\le k'$ and some positive even integers $r'_{s'+1},\ldots,r'_{k'}$,*

$$\sigma(\Gamma')=(g';\pm;[m_1,\ldots,m_r];\{(-),\overset{s'}{\ldots},(-),(2,\overset{r'_i}{\ldots},2)\mid s'+1\le i\le k'\}).$$

Moreover, assume that $c_{i0}\in\Gamma$ *for* $1\le i\le p'$ *and* $c_{i0}\notin\Gamma$ *for* $p'+1\le i\le s'$, *for some* p' *in range* $0\le p'\le s'$. *Let $e_i\in\Gamma'$ be the corresponding hyperbolic generators and let l_i be the order of $\Gamma e_i\in\Gamma'/\Gamma$, $1\le i\le p'$. Then:*

(1) *Each* m_j *divides* N, $j=1,\ldots,r$.

(2) $k=N\sum_{i=1}^{p'}1/l_i+N/2\sum_{i=s'+1}^{k'}r'_i/2.$

(3) $\alpha g\text{-}2 = N[\alpha' g'\text{-}2+k'+\sum_{j=1}^{r}(1\text{-}1/m_j)\text{-}\sum_{i=1}^{p'}1/l_i]$, *with α and α' equal to one or two according with* sign $\sigma(\Gamma)$ *and* sign $\sigma(\Gamma')$.

Proof. With the notations of 2.2.4, $m_i=p_i$ is the order of $\Gamma x_i\in\Gamma'/\Gamma$. In particular this proves (1), and by the last proposition, $\sigma(\Gamma')$ has the required form. To prove condition (2) we shall employ 2.3.2 and 2.3.3.

Let us fix one of the period-cycles $(2,\overset{r'_i}{\ldots},2)$. Since Γ'/Γ is isomorphic to $\mathbb{Z}_N$, there exists an epimorphism $\theta:\Gamma'\longrightarrow\mathbb{Z}_N$ with kernel Γ. Using part (1) in 2.4.3, we can assume without loss of generality that $\theta(c_0)=0$, and so

$$c_0\in\Gamma,\ c_1\notin\Gamma,\ c_2\in\Gamma,\ldots,c_{r'_i-1}\notin\Gamma,\ c_{r'_i}\in\Gamma,\ c_{2l-1}c_{2l+1}\in\Gamma$$

where $\{c_0,\ldots,c_{r'_i}\}$ are the reflections associated with $(2,\overset{r'_i}{\ldots},2)$.

For each $1\le l\le r'_i/2$, $c_{2l}\in\Gamma$, $c_{2l-1},c_{2l+1}\notin\Gamma$ but the product $c_{2l-1}c_{2l+1}$ belongs to Γ. Thus by 2.3.3, for each l we obtain N/2 period-cycles in $\sigma(\Gamma)$. Consequently, the total number of period-cycles in $\sigma(\Gamma)$ "coming from" the nonempty period-cycles of $\sigma(\Gamma')$ is

$$N/2 \sum_{i=s'+1}^{k'} r_i'/2.$$

The other period-cycles in $\sigma(\Gamma)$ arise from those in $\sigma(\Gamma')$ whose associated reflections are in Γ. Using 2.3.2 the number of them is $\sum_{i=1}^{p'} N/l_i$, and so, by adding, $k= \sum_{i=1}^{p'} N/l_i+ N/2 \sum_{i=s'+1}^{k'} r_i'/2$, which proves (2).

Finally (3) is a trivial consequence of the relation of areas $\mu(\Gamma)=N\mu(\Gamma')$ after substituting the value of k obtained in (2).

We finish this chapter introducing the notion of normal pair of signatures and obtaining the full list of such pairs. This will be later useful to decide whether a given finite group is the full group of automorphisms of some Klein surface with prescribed properties.

Definition 3. Let σ and σ' be the signatures of two NEC groups. We say that (σ,σ') is a *normal pair*, and we write $\sigma \triangleleft \sigma'$ if there exist an NEC group Γ' with signature σ' containing an NEC group Γ with signature σ as a normal subgroup and the dimensions $d(\Gamma)$ and $d(\Gamma')$ of the Teichmüller spaces of Γ and Γ' are equal.

In such a case we say that (Γ,Γ') is associated to (σ,σ'). Notice that if another pair (Γ_1,Γ_1') is associated to (σ,σ'), then $\mu(\Gamma)=\mu(\Gamma_1)$ and $\mu(\Gamma')=\mu(\Gamma_1')$, since the signature of an NEC group determines its area. In particular

$$[\Gamma':\Gamma]=\mu(\Gamma)/\mu(\Gamma')=\mu(\Gamma_1)/\mu(\Gamma_1')=[\Gamma_1':\Gamma_1]$$

and we denote $[\sigma':\sigma]=[\Gamma':\Gamma]$ for any associated pair (Γ,Γ').

The pair (σ,σ') is *proper* if σ' has period-cycles.

Remarks 2.4.5. (1) Let (σ,σ') be a normal pair and let Γ be an arbitrary NEC group with signature σ. Then there exists an NEC group Γ' with signature σ' containing Γ as a normal subgroup. In fact, let (Γ_1,Γ_2) be a pair associated to (σ,σ') and let $r:\Gamma \hookrightarrow \Omega$ and $i:\Gamma_1 \hookrightarrow \Gamma_2$ be the canonical embeddings. Let $\beta:\Gamma \longrightarrow \Gamma_1$ be an isomorphism and consider the monomorphism $\alpha=i\beta$ which induces an isometric embedding

$$T(\alpha):T(\Gamma_2) \hookrightarrow T(\Gamma).$$

Since $d(\Gamma_2)=d(\Gamma_1)=d(\Gamma)$, $T(\alpha)$ is onto. Therefore there exists an embedding $s:\Gamma \hookrightarrow \Omega$ such that $[r]=T(\alpha)([s])=[s\alpha]$. By the definition of $T(\Gamma)$, there exist $\phi \in Aut(\Gamma)$ and $\psi \in Aut(\Omega)$ for which $r=\psi s\alpha\phi$. Then

$$\Gamma=r(\Gamma)=\psi s\alpha\phi(\Gamma)=\psi s\alpha(\Gamma)\subseteq\psi s(\Gamma_2).$$

Hereby $\Gamma' = \psi s(\Gamma_2)$ is the group we have looked for.

The table 0.3.5 which has its own interest, will be crucial to obtain all normal pairs in view of the following observation:

(2) For a given NEC group Γ with $\sigma(\Gamma) = \sigma$ we denote $\sigma^+ = \sigma(\Gamma^+)$. Then, $\sigma \lhd \sigma'$ implies $\sigma^+ \lhd \sigma'^+$.

In fact, let $\Gamma \lhd \Gamma'$ for some NEC groups Γ and Γ' with $\sigma(\Gamma) = \sigma$, $\sigma(\Gamma') = \sigma'$ and $d(\Gamma) = d(\Gamma')$. By 0.2.1, $\Gamma^+ \lhd \Gamma'^+$. Moreover by 0.3.2, $d(\Gamma^+) = 2d(\Gamma) = 2d(\Gamma') = d(\Gamma'^+)$ and so $\sigma^+ \lhd \sigma'^+$.

(3) *Caution*: The converse is not true. See *e.g.* 2.4.6 below.

(4) The strategy to obtain the full list 2.4.7 of normal pairs is the following. Let us fix a pair (τ, τ') in 0.3.2. We construct the *finite* families $(\sigma_i | i \in I)$ and $(\sigma'_j | j \in J)$ of NEC signatures such that $\sigma_i^+ = \tau$ and $\sigma_j'^+ = \tau'$. In view of our first remark, when (τ, τ') runs along 0.3.5, (σ_i, σ'_j) runs along possible candidates for normal pairs. As was remarked in (3), we must decide in each case whether $\sigma_i \lhd \sigma'_j$ or not. But this is a *finite* (and accessible) task.

Example 2.4.6. We look for all proper pairs of NEC signatures (σ, σ') such that $\sigma'^+ = \tau' = (0;+;[2,2,2,2t])$. From 0.3.5 and from (2), $\sigma^+ = \tau = (1;+;[t])$.

Let us denote by g (resp. g') the genus of σ (resp. σ') and by k (resp. k') the number of period-cycles in σ (resp. σ'). From 2.2.5 it follows that the proper periods 2t in τ' and t in τ proceed from a period-cycle in σ' and σ respectively. In particular $k>0$ and $k'>0$. We start by finding possible candidates for σ. From 2.2.5 we know that

$$1 = \alpha g + k - 1, \qquad \alpha = \begin{cases} 1 & \text{if sign } \sigma = \text{"} - \text{"} \\ 2 & \text{if sign } \sigma = \text{"} + \text{"} \end{cases}$$

and σ has a period-cycle (t).

Now, for $\alpha = 1$ we have $g \neq 0$ and $2 = g + k$. Thus $g = k = 1$ and

$$\sigma = \sigma_1 = (1;-;[-];\{(t)\}).$$

For $\alpha = 2$, $2 = 2g + k$. Since $k > 0$, we deduce that $g = 0$, $k = 2$ and so

$$\sigma = \sigma_2 = (0;+;[-];\{(-),(t)\}).$$

Hence, $\{\sigma_1, \sigma_2\}$ is the family of NEC signatures with $\sigma_i^+ = \tau$.

Now we "determine" σ'. Clearly signσ' = "+", since otherwise $g' + k' = 1$, which is not possible, as $g' > 0$, and $k' > 0$. So signσ' = "+" and $2g' + k' = 1$, *i.e.* $g' = 0$, $k' = 1$.

Clearly $t > 1$ and using 2.2.5 we obtain that either

$$\sigma' = \sigma'_1 = (0;+;[2];\{(2,2t)\})$$

or

$$\sigma' = \sigma'_2 = (0;+;[-];\{(2,2,2,2t)\}.$$

Now we must decide for which pairs $\sigma_i \lhd \sigma'_j$, $i,j = 1,2$. We shall prove that

$$\text{"}\sigma_i \lhd \sigma'_j \text{ if and only if } i = j\text{"}.$$

Let us take an NEC group Γ'_1 with $\sigma(\Gamma'_1) = \sigma'_1$. Consider the epimorphism

$\theta:\Gamma_1' \longrightarrow \mathbb{Z}_2$ induced by the assignment

$$\theta(x_1)=\theta(e_1)=\theta(c_0)=\theta(c_2)=1;\ \theta(c_1)=0.$$

Denote $\ker\theta$ by Γ_1. To prove that $\sigma_1 \triangleleft \sigma_1'$ it is enough to check that $\sigma(\Gamma_1)=\sigma_1$ since, in such a case,

$$d(\Gamma_1)=1/2(d(\Gamma_1^+))=1/2(d(\Gamma_1'^+))=d(\Gamma_1'),$$

the second equality because $\sigma_1^+=\tau \triangleleft \tau'=\sigma'^+$.

Since $w=x_1c_2$ is a nonorientable word in Γ_1, sign $\sigma(\Gamma_1)=$"-" by 2.1.3. In the notations of 2.2.4, $r=1$, $m_1=2$, $p_1=2$, and there are not two consecutive reflections in Γ_1' out of Γ_1 whose product belongs to Γ_1. Thus $\sigma(\Gamma_1)$ has not proper periods.

Now we calculate the period-cycles of Γ_1. We have $c_0,c_2 \notin \Gamma_1$, $c_1 \in \Gamma_1$ and $c_0c_2 \in \Gamma_1$. Thus in terms of 2.3.3, $J=\{(1,1)\}$, $n=n(1,1)=1$. So $N/2n=1$, $n_{11}/2=1$, $n_{12}/2=t$ and $\sigma(\Gamma_1)=(g_1;-;[-];\{(t)\})$. From the Hurwitz Riemann formula $g_1=1$ and so $\sigma(\Gamma_1)=\sigma_1$, as desired.

Analogously, let Γ_2' be an NEC group, $\sigma(\Gamma_2')=\sigma_2'$. The kernel Γ_2 of the epimorphism $\theta:\Gamma_2' \longrightarrow \mathbb{Z}_2$ induced by the assignment

$$\theta(e_1)=\theta(c_1)=\theta(c_3)=0;\ \theta(c_0)=\theta(c_2)=\theta(c_4)=1$$

is a normal subgroup of Γ_2' of index two, and $\sigma(\Gamma_2)=\sigma_2$.

In fact Γ_2 is orientable, in view of 2.1.3, and it is obvious from 2.2.4 that $\sigma(\Gamma_2)$ has not proper periods. To compute its period-cycles we use 2.3.3. Here $J=\{(1,1),(3,3)\}$ and $n(1,1)=n(3,3)=1$ because c_0c_2 and c_2c_4 belong to Γ_2. Thus $N/2n(1,1)=N/2n(3,3)=2/2=1$. Moreover, since $n_{11}=n_{12}=n_{13}=2$, $n_{14}=2t$, we deduce that

$$(n_{12}/2,n_{11}/2)=(1,1)=(-);\ (n_{14}/2,n_{13}/2)=(t,1)=(t).$$

Hence, $\sigma(\Gamma_2)=(g_2;+;[-];\{(t),(-)\})$.

Finally by Hurwitz Riemann formula $g_2=0$ and so $\sigma(\Gamma_2)=\sigma_2$.

Consequently we have already seen that $\sigma_i \triangleleft \sigma_i'$, $i=1,2$. However, although $\sigma_2^+=\tau \triangleleft \tau'=\sigma_1'^+$, we are going to see that (σ_2,σ_1') is not a normal pair. In a similar way the reader can check that (σ_1,σ_2') is not a normal pair.

Let us suppose, by the way of contradiction, the existence of two NEC groups Γ and Γ', such that Γ is a normal subgroup of Γ', and

$$\sigma(\Gamma)=\sigma_2=(0;+;[-];\{(-),(t)\}) \text{ and}$$
$$\sigma(\Gamma')=\sigma_1'=(0;+;[2];\{(2,2t)\}).$$

Then

$$[\Gamma':\Gamma]=[\Gamma'^+:\Gamma^+]=[\sigma_1'^+:\sigma_2^+]=[\tau':\tau]=2.$$

Let x_1,e,c_0,c_1,c_2 be the canonical generators of Γ'. If $x_1 \in \Gamma$ then, by 2.2.4, σ_2 would have 2 as a proper period, which is false. Hence $x_1 \notin \Gamma$.

Now, since sign $\sigma_2=$"+", $x_1c_i \notin \Gamma$, for $i=0,1,2$, by 2.1.3, $c_0,c_1,c_2 \in \Gamma$ and by 2.3.2 $\sigma(\Gamma)$ has 2 and 2t as periods in a period-cycle, a contradiction.

Consequently, the first entry in 0.3.5 provides us exactly two normal pairs (σ_1,σ_1') and (σ_2,σ_2'). After easy but tedious calculations, similar to the ones in this example, we arrive by running all pairs (τ,τ') in 0.3.5 to the following

Theorem 2.4.7. *The list of normal (proper) pairs is the following*

σ'	σ	$[\sigma':\sigma]$
(0;+;[2];{(2,2t)})	(1;-;[-];{(t)})	2
(0;+;[-];{(2,2,2,2t)})	(0;+;[-];{(t)(-)})	2
(0;+;[2,2];{(t)})	(0;+;[-];{(t),(t)})	2
(0;+;[2,2];{(t)})	(2;-;[t];{-})	2
(0;+;[2];{(2,2,t)})	(1;-;[t];{(-)}	2
(0;+;[2];{(2,2,t)})	(1;-;[-];{(t,t)})	2
(0;+;[-];{(2,2,2,2,t)})	(0;+;[t];{(-),(-)})	2
(0;+;[-];{(2,2,2,2,t)})	(0;+;[-];{(t,t),(-)})	2
(0;+;[-];{(2,2,2,2,2,)})	(0;+;[-];{(-),(-),(-)})	2
(0;+;[2];{(2,2,2,2)})	(1;-;[-];{(-),(-)})	2
(0;+;[2,2];{(2,2),(2,2)})	(2;-;[-];{(-)})	2
(0;+;[2,2,2];{(-)})	(1;+;[-];{(-)})	2
(0;+;[2,2,2];{(-)})	(3;-;[-];{-})	2
(0;+;[-];{(2,2,2,t)}),t≥3	(0;+;[t,t];{(-)})	4
(0;+;[-];{(2,2,2,t)}),t≥3	(0;+;[-];{(t,t,t,t)})	4
(0;+;[-];{(2,2,2,t)}),t≥3	(1;-;[t,t];{-})	4
(0;+;[2];{(t,u)}), and max(t,u)≥3	(0;+;[-];{(t,u,t,u)})	2
(0;+;[2];{(t,u)}), and max(t,u)≥3	(1;-;[t,u];{-})	2
(0;+;[-];{(2,2,t,u)}), and max(t,u)≥3	(0;+;[t,u];{(-)})	2
(0;+;[-];{(2,2,t,u)}), and max(t,u)≥3	(0;+;[-];{(t,t,u,u)})	2
(0;+;[-];{(2,t,2,u)}), and max(t,u)≥3	(0;+;[t];{(u,u)})	2
(0;+;[t];{(2,2)}), t≥3	(0;+;[t,t];{(-)})	2
(0;+;[t,2];{(-)}), t≥3	(0;+;[t,t];{(-)})	2
(0;+;[t,2];{(-)}), t≥3	(1;-;[t,t];{-})	2
(0;+;[3];{(t)}), t≥4	(0;+;[-];{(t,t,t)})	3
(0;+;[-];{(2,3,2t)}), t≥4	(0;+;[-];{(t,t,t)})	6
(0;+;[-];{(2,t,2u)}) and t≥3, t+u≥7	(0;+;[t];{(u)})	2
(0;+;[-];{(2,t,2u)}) and t≥3, t+u≥7	(0;+;[-];{(t,t,u)})	2
(0;+;[-];{(t,t,u)}) and t≥3, t+u≥7	(0;+;[t,t,u];{-})	2
(0;+;[t];{(u)}) and t≥3, t+u≥7	(0;+;[t,t,u];{-})	2
(0;+;[3];{(t)}), t≥4	(0;+;[t,t,t];{-})	6
(0;+;[-];{(3,3,t)}), t≥4	(0;+;[t,t,t];{-})	6
(0;+;[-];{(2,t,2u)}) and t≥3, t+u≥7	(0;+;[t,t,u];{-})	4
(0;+;[-];{(2,3,2t)}), t≥4	(0;+;[t,t,t];{-})	12
(0;+;[-];{(m,t,u)}) m≠t≠u≠m t,u≥4 or t≥7	(0;+;[m,t,u];{-})	2

2.5. NOTES

Theorem 2.1.2 was independently obtained in [11] and Hall [60]. Afterwards, a general result concerning the orientability of subgroups of plane groups appeared in Hoare-Singerman [67]. The effective computation of proper periods was established in [11] and [12]. The first proof of 2.2.5 is due to Singerman, [117]. The study of period-cycles is done, according with the parity of the index, in [11] and [30].

Recently it has been proved, in [28], that conditions in Theorem 2.4.1 are not only necessary but also sufficient for the existence of normal bordered surface subgroups. Theorem 2.4.4 was obtained in [30]. The list 2.4.7 of proper normal pairs appeared in [13]. There it is developed a different example to our case in 2.4.6. It is interesting to worth that now, having 2.1.3 and 2.3.3, the proof of 2.4.7 is easier than the given in the original paper.

Particular results on normal subgroups have been stated in [11], [30], [44] and [45]. Finally, let us indicate that a general approach to subgroups of NEC groups has been done in Hoare [64], Hoare-Karras-Solitar [66], Hoare-Singerman [67], Uzzell [124] and Fennessey-Pride [48], whilst similar results to the ones in this chapter, relative to fuchsian groups, have been obtained in Hoare-Karras-Solitar [65], Maclachlan [90], and Singerman [115], [118].

CHAPTER - 3
Cyclic groups of automorphisms of compact Klein surfaces

For fixed data N,g,k we obtain a criterion which allows us to decide whether there exists or not a compact Klein surface of topological genus g with k boundary components, having an automorphism of order N. This is a preliminary in studying necessary and sufficient conditions for the existence of a Klein surface S admitting a given finite group as a group of automorphisms since it leads to limitations of the orders (as a function of the topological type of S). We apply this criterion in Chapter 5 to determine those groups that appear as the automorphism group of surfaces with one boundary component.

A Klein surface of algebraic genus ≥ 2 can be represented as a quotient H/Γ for some surface NEC group Γ, and a group of automorphisms of H/Γ has the form Λ/Γ for another NEC group Λ. So our problem is equivalent to study conditions for Γ to be a normal subgroup of another Λ such that the quotient is cyclic, and to determine the order of Λ/Γ. Roughly speaking, this is the content of the first section.

All this is applied in the second one to give a complete solution of two classical problems:

(i) What is the minimum (algebraic) genus $p=p(N)$ of surfaces admitting an automorphism of order N? Actually we obtain a much more precise information, involving the orientability of the surface and the orientation preserving character of such an automorphism. (Cor. 3.2.16). We also compute the number of boundary components of a surface realizing the minimum genus.

Even more, for prime N we obtain in Cor. 3.2.17, the minimum genus p(N,k) of surfaces having k boundary components and admitting an automorphism of order N. As far as we know, the values of p(N,k) are unknown for non-primes N (although they can be calculated with the aid of a computer and the results of 3.1, for all concrete values of N and k).

(ii) In theorem 3.2.18 we determine, with the precisions above, the maximum order $N=N(p)$ of automorphisms of surfaces of given algebraic genus $p\geq 2$.

3.1. CYCLIC QUOTIENTS

We fix along this chapter an integer $N\geq 2$. The symbol σ will always denote an NEC signature of the following type:

$$\sigma = (g;\pm;[-];\{(-),\overset{k}{\ldots},(-)\}).$$

Thus σ is a function of a sign and the integers g and k, which will always have this meaning. Given an NEC signature τ we say that (σ,τ) is an N-*pair* if there exist an NEC group Λ with signature τ and an epimorphism $\theta:\Lambda \longrightarrow Z_N$ whose kernel is an NEC group Γ with signature σ.

In what follows, the symbols Λ and θ will appear only in this context. The symbols a,b,x,e,c and d will always mean the canonical generators of Λ. This paragraph has rather technical character, and our main task is to obtain necessary and sufficient conditions for a pair of signatures (σ,τ) to be an N-pair. Obviously $\ker\theta$ is a normal subgroup of Λ of index N and so from 2.4.2 and 2.4.4, we deduce that for an N-pair (σ,τ) the signature τ has the form

(3.1.0.1) $\tau=(g';\pm;[m_1,\ldots,m_r];\{(-),\overset{k'}{\ldots},(-)\})$, $\mathrm{sign}\sigma=\mathrm{sign}\tau$ if N is odd, and

(3.1.0.2) $\tau=(g';\pm;[m_1,\ldots,m_r];\{(-),\overset{s'}{\ldots},(-),(2,\overset{r'_i}{\ldots},2)|s'+1\le i\le k'\})$, N even,

for some non-negative integers $g',r,m_1,\ldots,m_r,k',s',r'_{s'+1},\ldots,r'_{k'}$ such that

$$r'_{s'+1},\ldots,r'_{k'} \text{ are even numbers, } s'\le k' \text{ and } m_i\ge 2,\ i=1,\ldots,r.$$

Whenever we use σ and τ we assume they have this form. We shall be able to transform the condition for a pair (σ,τ) to be an N-pair in terms of some relations between these numbers, g and k.

To do that, we first need a technical lemma concerning cyclic groups. If there is no danger of confusion, along the whole paragraph, we shall not distinguish between the elements of Z_N and their representatives.

Lemma 3.1.1. *Let* $m_1,\ldots,m_k$ *be positive integers and* $M=l.c.m.(m_1,\ldots,m_k)$. *Let us write* $M=2^sA$ *for some odd integer* A. *The following statements are equivalent:*

(1) For each multiple N *of* M, *there exist* $\zeta_1,\ldots,\zeta_k\in Z_N$ *such that*

$$\#(\zeta_i)=m_i \text{ and } \zeta_1+\ldots+\zeta_k=0.$$

(2) For each $i=1,\ldots,k$, $M_i=l.c.m.(m_1,\ldots,\hat{m}_i,\ldots,m_k)=M$, *and, if* $s\neq 0$, *the number of* m_i *which are multiple of* 2^s *is even.*

Proof. (1)⇒(2). We take N=M. Let H_i be the subgroup of Z_N generated by $\{\zeta_1,\ldots,\hat{\zeta}_i,\ldots,\zeta_k\}$. It has order M_i and so, since $\zeta_i=-\sum_{j=1,j\neq i}^{k}\zeta_j$, m_i divides M_i. This proves the first part in (2).

Moreover, if M is even we suppose that 2^s does not divide $m_1,\ldots,m_{k-t}$ but it divides $m_{k-t+1},\ldots,m_k$. Since $\#(\zeta_i)=m_i$, we can write

$$\zeta_i=\alpha_iM/m_i,\ g.c.d.(\alpha_i,m_i)=1,\ i=1,\ldots,k.$$

For $i=1,\ldots,k\text{-}t$, M/m_i is even. For $i=k\text{-}t+1$, M/m_i is odd and m_i is even. Thus

α_i and also ζ_i are odd. From the equality

$$0 = \zeta_1+...+\zeta_k = \sum_{i=1}^{k-t} \zeta_i + \sum_{i=k-t+1}^{k} \zeta_i$$

we deduce that the second summand, being an even sum of odd numbers, consists of an even number of summands, as desired.

(2)⇒(1) Z_M can be viewed as a subgroup of Z_N. Thus all reduces to study the case N=M.

Of course, it suffices to find $\alpha_1,...,\alpha_k$, with $g.c.d.(\alpha_i,m_i)=1$ such that

(3.1.1.1) $$\zeta_i = \alpha_i M/m_i,\ i=1,...,k;\ \zeta_1+...+\zeta_k = 0.$$

First we deal with the case $M=p^r$ for some prime number p. Then we can write $m_i=p^{\mu_i}$, with $r=\mu_i$ exactly for $i=k\text{-}t+1,...,k$, $t\in\{1,...,k\}$. Of course t is even if p=2.

We choose $\alpha_1=...=\alpha_{k-2}=1$ and we look for α_{k-1} and α_k. By the definition of M, m_i and t, condition 3.1.1.1 is equivalent to

$$\zeta_{k-1}+\zeta_k=-\sum_{i=1}^{k-2} p^{r-\mu_i}=Rp-\sum_{i=k-t+1}^{k-2} p^{r-\mu_i}=Rp+(2-t)$$

for some integer R.

Now we distinguish:

(a) $t\equiv 1 \bmod p$. Then $p\neq 2$ and we take $\alpha_{k-1}=-t$, $\alpha_k=2+Rp$. It is clear that $g.c.d.(-t,p^r)=g.c.d.(2+Rp,p^r)=1$, and also $\zeta_{k-1}+\zeta_k=\alpha_{k-1}+\alpha_k=(2-t)+Rp$.

(b) $t\not\equiv 1 \bmod p$. Then $g.c.d.(1-t,p^r)=1$ and so, if we choose $\alpha_{k-1}=1$, $\alpha_k=1-t+Rp$ both are coprime with p^r and $\zeta_{k-1}+\zeta_k=2-t+Rp$.

In the general case we factorize $M=p_1^{r_1}...p_n^{r_n}$. Then, for some non-negative integers μ_{ij}, $i=1,...,k$, $j=1,...,n$, we can write $m_i=p_1^{\mu_{i1}}...p_n^{\mu_{in}}$.

Let us define $m_{ij}=p_j^{\mu_{ij}}$, $i=1,...,k$, $j=1,...,n$. We claim that, for a fixed j, the numbers $m_{1j},...,m_{kj}$ verify the hypothesis (2). In fact, given $i=1,...,k$, m_{ij} divides m_i and m_i divides M_i. Thus the factor p_j appears at least μ_{ij} times in M_i and so m_{ij} divides $l.c.m.(m_{1j},...,\hat{m}_{ij},...,m_{kj})$.

The second condition is obvious if $p_j\neq 2$ and, for $p_j=2$, it is $r_j=s$, and 2^s divides m_{ij} if and only if 2^s divides m_i. In particular, $\#\{m_{ij}|2^s \text{ divides } m_{ij}\}$ is even. Consequently, there exist $\{\alpha_{ij}|1\leq i\leq k,\ 1\leq j\leq n\}\subseteq Z$ such that $g.c.d.(\alpha_{ij},m_{ij})=1$ and if $\zeta_{ij}=\alpha_{ij}p_j^{r_j}/p_j^{\mu_{ij}}$ belongs to $Z_{p_j^{r_j}}$, $\xi_{1j}+...+\zeta_{kj}=0$. Now, since the numbers $p_1^{r_1},...,p_n^{r_n}$ are coprime, we can apply the Chinese Remainder Theorem to deduce the existence of integers $\eta_1,...,\eta_k$ such that

$$\eta_i+p_j^{r_j}Z=\zeta_{ij},\ j=1,...,n.$$

Now, if $\zeta_i=\eta_i+MZ\in Z_M$, the elements $\zeta_1,...,\zeta_k$ verify the required conditions. In fact, let a be an integer. Then $a\zeta_i=0$ means $a\eta_i \in MZ$, *i.e.*, $a\eta_i \in p_j^{r_j}Z$ for each j. But $\eta_i+p_j^{r_j}Z=\zeta_{ij}$ and so, $a\zeta_i=0$ if and only if a is a multiple of $\#(\zeta_{ij})=p_j^{\mu_{ij}}$. Consequently, $\#(\zeta_i) = l.c.m.(p_1^{\mu_{i1}},\ldots,p_n^{\mu_{in}}) = m_i$. Secondly, from $\sum_{i=1}^{k} \zeta_{ij}=0$, for $j=1,\ldots,n$, and $\eta_i+p_j^{r_j}Z=\zeta_{ij}$ we deduce that $\eta_1+\ldots+\eta_k$ is a multiple of each $p_j^{r_j}$, $j=1,\ldots,n$. Hence $\zeta_1+\ldots+\zeta_k=0$.

In order to make notations easier, we introduce the following:

Definition 1. The set of positive integers $\{m_1,\ldots,m_k\}$ verifies the *elimination property* if

$$l.c.m.(m_1,\ldots,\hat{m}_i,\ldots,m_k) = l.c.m.(m_1,\ldots,m_k) \text{ for each } i=1,\ldots,k.$$

We adopt the convention: *l.c.m.* of the empty set is 1. Thus $\{m_1\}$ has the elimination property if and only if $m_1=1$.

In the next two theorems we characterize N-pairs (σ,τ) for odd N, according to signσ is "+" or "-", respectively.

Theorem 3.1.2. *Let us suppose that* N *is odd, and* (σ,τ) *is a pair of signatures with* signτ="+". *Then* (σ,τ) *is an* N-*pair if and only if:*

(1) For each $i=1,\ldots,r$, m_i *divides* N, *and* signσ="+".

(2) $\mu(\sigma)=N\mu(\tau)$.

(3) There exist positive divisors $l_1,\ldots,l_{k'}$ *of* N *such that*

(3.1) $k=\sum_{j=1}^{k'} N/l_j$.

(3.2) The set $\{m_1,\ldots,m_r,l_1,\ldots,l_{k'}\}$ *has the elimination property.*

(4) If $g'=0$, *then* $N=l.c.m.(m_1,\ldots,m_r,l_1,\ldots,l_{k'})$.

Proof. If (σ,τ) is an N-pair we write $\Gamma=\ker\theta$, and let l_i be the order of $\theta(e_i)$, $i=1,\ldots,k'$. Then l_i divides N and (1), (2) and (3.1) are obvious consequences of 2.1.2 and 2.4.2.

Since $a_1b_1a_1^{-1}b_1^{-1}\ldots a_{g'}b_{g'}a_{g'}^{-1}b_{g'}^{-1}x_1\ldots x_re_1\ldots e_{k'} = 1$ and Z_N is abelian, we deduce that

$$\sum_{i=1}^{r} \theta(x_i)+\sum_{j=1}^{k'} \theta(e_j)=0.$$

The order of $\theta(x_i)$ is m_i for $i=1,\ldots,r$, since otherwise $\sigma=\sigma(\Gamma)$ would contain proper periods, by 2.2.3. Thus (3.2) follows from lemma 3.1.1.

Since N is odd, Γ contains all reflections of Λ. So, if $g'=0$, then

$$\Lambda/\Gamma = \langle \Gamma x_1,\ldots,\Gamma x_r,\Gamma e_1,\ldots,\Gamma e_{k'}\rangle.$$

Thus $N=|\Lambda/\Gamma|=l.c.m.(m_1,\ldots,m_r,l_1,\ldots,l_{k'})$.

To prove the sufficiency we work as follows: since τ is an NEC signature there exists an NEC group Λ with $\sigma(\Lambda) = \tau$. Let

$$M = l.c.m.(m_1,\ldots,m_r,l_1,\ldots,l_{k'}).$$

From (1) and (3) M divides N. Using 3.1.1 once again there exist $\zeta_1,\ldots,\zeta_r$, $\eta_1,\ldots,\eta_{k'} \in Z_N$ such that

$$\zeta_1+\ldots+\zeta_r+\eta_1+\ldots+\eta_{k'}=0,\ \#(\zeta_i)=m_i,\ \#(\eta_j)=l_j.$$

Now the assignment

$$\theta(x_i)=\zeta_i \text{ for } i=1,\ldots,r;\quad \theta(e_j)=\eta_j,\ \theta(c_j)=0,\ j=1,\ldots,k' \text{ and}$$

$$\theta(a_n)=\theta(b_n)=\begin{cases} 1 & \text{for } n=1, \\ 0 & \text{otherwise,} \end{cases}$$

induces a homomorphism $\theta:\Lambda \longrightarrow Z_N$ because it preserves the relations in Λ: relations $x_i^{m_i}=1$ and $a_1b_1a_1^{-1}b_1^{-1}\ldots a_{g'}b_{g'}a_{g'}^{-1}b_{g'}^{-1}x_1\ldots x_re_1\ldots e_{k'}=1$, become in Z_N:

$$m_i\zeta_i=0 \text{ and } \zeta_1+\ldots+\zeta_r+\eta_1+\ldots+\eta_{k'}=0.$$

If $g'\neq 0$, a_1 appears among the generators of Λ, and $\theta(a_1)=1$. So θ is surjective. Thus, let $g'=0$. Clearly,

$$|\mathrm{im}\theta| = l.c.m.(m_1,\ldots,m_r,l_1,\ldots,l_{k'}).$$

But by (4), the right hand side of the last equality is N, and so θ is also surjective in this case.

To finish, we must only check that $\Gamma=\ker\theta$ has signature σ. First notice that the order of x_i equals the order of its image $\theta(x_i)$. Hence $\sigma(\Gamma)$ has no proper periods, by 2.2.3. Moreover, the period-cycles in $\sigma(\Gamma)$ are empty by 2.3.1 and $\mathrm{sign}\sigma(\Gamma)=\mathrm{sign}\tau=$"+"$=\mathrm{sign}\sigma$. Thus it remains to show that $\sigma(\Gamma)$ has genus g and k period-cycles. But the first is the obvious consequence of 2.4.2,(iii), whilst the second follows from 2.4.2,(iv) and our hypothesis (2).

Now we study the same problem for $\mathrm{sign}\tau$ = "-".

Theorem 3.1.3. *Let us suppose* N *is odd, and* $\mathrm{sign}\tau$="-". *Then, the pair of signatures* (σ,τ) *is an* N*-pair if and only if:*

(1) For each $i=1,\ldots,r$, m_i *divides* N, *and* $\mathrm{sign}\sigma$="-".

(2) $\mu(\sigma) = N\mu(\tau)$.

(3) There exist positive divisors $l_1,\ldots,l_{k'}$ *of* N *such that* $k=\sum_{j=1}^{k'} N/l_j$.

(4) If $g'=1$, *then* $N=l.c.m.(m_1,\ldots,m_r,l_1,\ldots,l_{k'})$.

Proof. The necessity of conditions (1), (2) and (3) follows in the same way that in the precedent theorem. Moreover, if Λ is an NEC group with $\sigma(\Lambda)=\tau$ and θ is an epimorphism from Λ onto Z_N whose kernel has signature σ, we write $\theta(x_i)=\zeta_i$, $\theta(e_j)=\eta_j$, $\theta(d_1)=\delta$. Then, by 2.2.3, $\#(\zeta_i)=m_i$. For $g'=1$ we have

(3.1.3.1) $$2\delta = -\left[\sum_{i=1}^{r} \zeta_i + \sum_{j=1}^{k} \eta_j\right], \quad \#(\zeta_i)=m_i, \quad \#(\eta_j)=l_j.$$

Let us define $M=l.c.m.(m_1,\ldots,m_r,l_1,\ldots,l_{k'})$. We must prove $M=N$. Since θ is surjective and the reflections in Λ belong to Γ,

$$N=|Z_N|=|<\zeta_1,\ldots,\zeta_r,\eta_1,\ldots,\eta_{k'},\delta>|=l.c.m.(M,\#(\delta)).$$

Hence, it is enough to see that $\#(\delta)$ divides M. But, since N is odd, $<\delta>=<2\delta>$. Thus it suffices to prove that $\#(2\delta)$ divides M, and this follows trivially from equality 3.1.3.1.

To prove the sufficiency we take an NEC group Λ with $\sigma(\Lambda)=\tau$. We must construct an epimorphism $\theta:\Lambda \longrightarrow Z_N$ such that $\sigma(\ker\theta)=\sigma$.

The definition of θ is different according with $g'=1$ or $g'>1$ and the parity of $S=N\left[\sum_{i=1}^{r} 1/m_i + \sum_{j=1}^{k'} 1/l_j\right]$. In any case we define:

$\theta(x_i)=N/m_i$, $\theta(e_j)=N/l_j$, $\theta(c_j)=0$, for $i=1,\ldots,r$, $j=1,\ldots,k'$.

If $g'=1$, $\theta(d_1)=\begin{cases} (N-S)/2 & \text{if S is odd.} \\ (2N-S)/2 & \text{if S is even.}\end{cases}$

If $g'>1$, we take $\theta(d_2)=1$, $\theta(d_i)=0$ for $i\geq 3$ and

$$\theta(d_1)=\begin{cases} (N-S-2)/2 & \text{if S is odd,} \\ (2N-S-2)/2 & \text{if S is even.}\end{cases}$$

It is easy to check that θ preserves relations. Also the surjectivity of θ is obvious if $g'>1$, since $\theta(d_2)=1$. For $g'=1$, we deduce from our hypothesis (4), that

$$|im\theta|\geq|<\theta(x_1),\ldots,\theta(x_r),\theta(e_1),\ldots,\theta(e_{k'})>|=l.c.m.(m_1,\ldots,m_r,l_1,\ldots,l_{k'})=N,$$

and we are done.

Finally, a straightforward computation using 2.2.3, 2.3.1 and 2.4.2 as in the previous theorem, shows that $\sigma(\ker\theta)=\sigma$. We leave the details to the reader.

Comment. The remainder of the paragraph deals with the case of even N, which is more involved. For example, see 2.1.3, the sign of σ does not determine that of τ. But, as we shall see, this is not the main difficulty. In what follows, given two NEC groups Γ and Λ, such that $\sigma(\Gamma)=\sigma$, $\sigma(\Lambda)=\tau$ and Γ is a normal subgroup of Λ with $\Lambda/\Gamma\cong Z_N$, Λ has signature 3.1.0.2 and we denote by p' the unique integer, $0\leq p'\leq s'$ such that c_{i0}, $i=1,\ldots,p'$, are just the canonical reflections of Λ which belong to Γ. This convention will be freely used without any other reference.

To avoid unnecessary repetitions in the arguments below we make the following:

Remarks 3.1.4. Let $\theta : \Lambda \longrightarrow Z_N$ be an epimorphism with kernel Γ.

(1) By the definition of p′,

$$\theta(c_{i0})=0 \text{ for } i=1,\dots,p';\ \theta(c_{i0})=N/2 \text{ for } i=p'+1,\dots,s'.$$

Moreover, by 2.4.3,

$$\theta(c_{i0})=\theta(c_{i,2l})=N/2;\ \theta(c_{i1})=\theta(c_{i,2l-1})=0,\ i=s'+1,\dots,k',\ l=1,\dots,s_i/2.$$

Thus, from now on, in order to define such a θ, we shall not write the images under it of the canonical reflections in Λ, since they are completely determined by the integer p′.

(2) As a consequence of (1), if $p'<k'$, there exists some canonical reflection $c\in\Lambda$ such that $\theta(c)=N/2$.

(3) There exists a word w of Λ with respect to Γ such that $\theta(w)=1$. In fact, let $f\in\Lambda$ be an element such that $\theta(f)=1$. Let $f=g_1\dots g_q$, where $g_1,\dots,g_q$ are canonical generators of Λ. Then

$$1=\theta(g_1)+\dots+\theta(g_q)=\theta(g_{i_1})+\dots+\theta(g_{i_s})=\theta(g_{i_1}\dots g_{i_s})$$

where $g_{i_1},\dots,g_{i_s}$ are those generators which are not in Γ. Now $w=g_{i_1}\dots g_{i_s}$ is the word we have looked for.

(4) If $p'<k'$ and $\theta(w)=N/2$ for some orientable word w, then $\text{sign}\sigma(\Gamma)=$"-". This is an obvious consequence of (2). We choose a reflection $c\in\Lambda$ with $\theta(c)=N/2$ and so $f=wc\in\Gamma$ is a nonorientable word.

(5) Conversely, if $p'<k'$ and $\text{sign}\sigma(\Gamma)=$"-", $\text{sign}\sigma(\Lambda)=$"+", then there exists an orientable word w such that $\theta(w)=N/2$.

Clearly the hypotheses imply $\theta(f)=0$ for some nonorientable word $f\in\Lambda$. Take $c\in\Lambda$ as in (2). Then $w=fc$ is orientable and $\theta(w)=N/2$.

(6) If $p'<k'$ and N is a multiple of 4, then $\text{sign}\sigma(\Gamma)=$"-". From (2) and (3) we have a reflection c and a word w in Λ verifying $\theta(w)=1$ and $\theta(c)=N/2$. Then $f=w^{N/2}c\in\Gamma$ is a nonorientable word. Hence, by 2.1.3, $\text{sign}\sigma(\Gamma)=$"-".

We are ready to prove for even N and signatures with sign "+", results analogous to 3.1.2 and 3.1.3.

Theorem 3.1.5. *Let us suppose* N *is even and* $\text{sign}\sigma=\text{sign}\tau=$"+". *Then* (σ,τ) *is an* N*-pair if and only if:*

(1) For each $i=1,\dots,r$, m_i *divides* N.

(2) $\mu(\sigma)=N\mu(\tau)$.

(3) There exist $0\le p'\le s'$ *and some positive divisors* $l_1,\dots,l_{p'}$ *of* N *such that:*

$$(3.1)\quad k=\sum_{j=1}^{p'} N/l_j+N/2\sum_{i=s'+1}^{k'} r_i/2.$$

(3.2) If $N\in 4\mathbb{Z}$, *then* $p'=k'$, $\{m_1,\ldots,m_r,l_1,\ldots,l_{k'}\}$ *has the elimination property, and for* $g'=0$, $N=l.c.m.(m_1,\ldots,m_r,l_1,\ldots,l_{k'})$.
(3.3) If $N\notin 4\mathbb{Z}$ *and* $p'=k'$, *condition* (3.2) *holds true.*
(3.4) If $N\notin 4\mathbb{Z}$ *and* $p'<k'$, *then each* m_i *and* l_j *divides* N/2. *Moreover, if* $g'=0$ *and* $k'=p'+1$, *then* $l.c.m.(m_1,\ldots,m_r,l_1,\ldots,l_{p'})=N/2$.

Proof. We begin with the "only if" part. Condition (1) follows from 2.4.4 and (2) is the Hurwitz Riemann formula. To prove (3) we choose $p'\leq s'$ such that $c_{10},\ldots,c_{p'0}$ are just those elements among $\{c_{10},\ldots,c_{s'0}\}$ that belong to $\ker\theta=\Gamma$. We also put $l_j=\#(\theta(e_j))$, $j=1,\ldots,p'$. Then each l_j divides $N=[\Lambda:\Gamma]$, and (3.1) is an immediate consequence of 2.4.4.

Let us show (3.2). The equality $p'=k'$ follows from our previous remark 3.1.4.(6), because signσ = "+". The relation

$$a_1b_1a_1^{-1}b_1^{-1}\ldots a_{g'}b_{g'}a_{g'}^{-1}b_{g'}^{-1}x_1\ldots x_re_1\ldots e_{k'}=1$$

implies

$$\sum_{i=1}^{r}\zeta_i+\sum_{j=1}^{k'}\eta_j=0 \quad \text{for } \zeta_i=\theta(x_i),\ \eta_j=\theta(e_j).$$

Since σ has no proper periods we deduce from 2.2.4 that $\#(\zeta_i)=\#(\theta(x_i))=\#(x_i)=m_i$. Thus, by 3.1.1, $\{m_1,\ldots,m_r,l_1,\ldots,l_{k'}\}$ has the elimination property. Let us suppose $g'=0$. Then, since $p'=k'$, we have

$$\mathbb{Z}_N=\langle\zeta_1,\ldots,\zeta_r,\eta_1,\ldots,\eta_{k'}\rangle$$

and so $N=l.c.m.(m_1,\ldots,m_r,l_1,\ldots,l_{k'})$. This proves (3.2).

Notice that we used the hypothesis $N\in 4\mathbb{Z}$ uniquely to deduce $p'=k'$. Hence (3.3) is also proved.

Finally we check (3.4). Observe that each m_i divides N/2. In fact, assume to the contrary that $m_i=2\alpha$ for some divisor α of N/2. Then $\#(\alpha\zeta_i)=2$, *i.e.*, $\alpha\zeta_i=N/2$. By (2) in 3.1.4, there exists a reflection $c\in\Lambda$ with $\theta(c)=N/2$. So $w=x_i^{\alpha}c\in\Gamma$ is a nonorientable word. Thus signσ=sign$\sigma(\Gamma)$="-" by 2.1.3, a contradiction. The same argument shows that each l_j divides N/2.

When $g'=0$ and $k'=p'+1$, the equality $x_1\ldots x_re_1\ldots e_{p'}e_{p'+1}=1$ implies

$$\mathbb{Z}_N=\langle\zeta_1,\ldots,\zeta_r,\eta_1,\ldots,\eta_{p'},N/2\rangle,$$

where the last N/2 corresponds to the canonical reflections in $\Lambda\backslash\Gamma$. So $l.c.m.(m_1,\ldots,m_r,l_1,\ldots,l_{p'},2)=N$. Moreover, all m_i and l_j are odd numbers, because they divide N/2. Hence $l.c.m.(m_1,\ldots,m_r,l_1,\ldots,l_{p'})=N/2$.

We divide the proof of the "if part" into four cases.

Case 1. $p'=k'$. Then condition (2) of 3.1.1 is fulfilled by

$$\{m_1,\ldots,m_r,l_1,\ldots,l_{k'}\}.$$

In fact this set has the elimination property, by 3.2 and 3.3. Moreover, from

(2) and (3.1), $2g-2=N\left[2g'-2+k'+\sum_{i=1}^{r}(1-1/m_i)-\sum_{j=1}^{k'}1/l_j\right]$ and so $\sum_{i=1}^{r}N/m_i+\sum_{j=1}^{k'}N/l_j$ is an even integer. Thus, the number of m_i or l_j being multiple of the maximum power of 2 dividing N is even.

Consequently, there exist some elements $\zeta_1,\ldots,\zeta_r,\eta_1,\ldots,\eta_{k'}\in Z_N$ with

$$\zeta_1+\ldots+\zeta_r+\eta_1+\ldots+\eta_{k'}=0;\ \#(\zeta_i)=m_i,\ \#(\eta_j)=l_j,\ \text{for } 1\le i\le r,\ 1\le j\le k'.$$

Even more, for $g'=0$ the elements $\{\zeta_1,\ldots,\zeta_r,\eta_1,\ldots,\eta_{k'}\}$ generate the whole group Z_N by 3.2 and 3.3.

So, if Λ is an NEC group with signature τ we have the following epimorphism $\theta:\Lambda\longrightarrow Z_N$, induced by

$$\theta(x_i)=\zeta_i;\ \theta(e_j)=\eta_j,\ \theta(c_{j0})=0,\ j=1,\ldots,k',\ i=1,\ldots,r;$$

$$\theta(a_n)=\theta(b_n)=\begin{cases}1 \text{ for } n=1,\\ 0 \text{ for } n=2,\ldots,g'.\end{cases}$$

We shall show that σ is the signature of $\Gamma=\ker\theta$. In fact $\sigma(\Gamma)$ has no proper periods by 2.2.4, its period-cycles are empty by 2.3.2, and from our hypothesis (3.1) and 2.4.4.(2), the number of them is k. The genus of $\sigma(\Gamma)$ is g in view of our condition (2) and 2.4.4.(3). It is also obvious that $\text{sign}\sigma(\Gamma)=\text{"}+\text{"}$ since the only proper generators of Λ reversing the orientation are the reflections, and they belong to Γ.

The second case $p'<k'$ (and so, by 3.2, $N\notin 4Z$) is divided into several steps; we write now $S=\sum_{i=1}^{r}N/m_i+\sum_{j=1}^{p'}N/l_j$. In all cases we shall construct an epimorphism $\theta:\Lambda\longrightarrow Z_N$ by mapping

$$x_i\longmapsto N/m_i,\ e_j\longmapsto N/l_j\ \text{for } i=1,\ldots,r;\ j=1,\ldots,p'.$$

For the other generators we distinguish:

Case 2. $p'<k'$, $g'\neq 0$.

The assignment

$$\theta(a_1)=2,\ \theta(b_1)=0,\ \theta(a_n)=\theta(b_n)=0,\ n=2,\ldots,g',$$

$$\theta(e_{p'+1})=N\text{-}S,\ \theta(e_j)=0,\ j=p'+2,\ldots,k',$$

and extended to the reflections of Λ as indicated in our remark 3.1.4.(1), induces a homomorphism $\theta:\Lambda\longrightarrow Z_N$ since the relevant relation

$$a_1b_1a_1^{-1}b_1^{-1}\ldots a_{g'}b_{g'}a_{g'}^{-1}b_{g'}^{-1}x_1\ldots x_re_1\ldots e_{k'}=1$$

becomes $0+N/m_1+\ldots+N/m_r+N/l_1+\ldots+N/l_{p'}+(N\text{-}S)+0=0\in Z_N$.

The surjectivity of θ is clear since

$$|\text{im}\theta|\ge|<\theta(a_1),\theta(c_{p'+1,0})>|=|<2,N/2>|=N.$$

We claim that the sign of the signature of $\Gamma=\ker\theta$ is "+". Using 3.1.4.(5) it suffices to see that $\theta(w)\neq N/2$ for each orientable word w.

But $\theta(w)$ is a linear combination, with integer coefficients, of $\theta(x_i)=N/m_i$, $\theta(e_j)=N/l_j$, $j=1,\ldots,p'$, and $\theta(e_{p'+1})=N\text{-}S$. Since all these numbers are even by our (3.4), whilst N/2 is odd, we deduce that $\theta(w)\neq N/2$.

Once this proved, a routinary computation, as in the first case, shows that σ is the signature of Γ, *i.e.*, (σ,τ) is an N-pair.

It only remains to define θ when $g'=0$, $p'<k'$.

Case 3. $g'=0$, $p'+1<k'$.

The same arguments as before prove that the assignment

$$\theta(e_{p'+1})=2,\ \theta(e_{p'+2})=N\text{-}(S+2),\ \theta(e_t)=0,\ p'+3\leq t\leq k'$$

induces an epimorphism $\theta:\Lambda \longrightarrow Z_N$ whose kernel has signature σ. The orientability of σ is clear since, as in the previous case, θ maps each orientable element to an even number. The surjectivity is obvious, because

$$\mathrm{im}\theta\supseteq<\theta(e_{p'+1}),\ \theta(c_{p'+1,0})>=<2,N/2>=Z_N.$$

Case 4. $g'=0$, $p'+1=k'$.

Here we define $\theta(e_{k'})=N\text{-}S$. It is clear that it induces a homomorphism $\theta:\Lambda \longrightarrow Z_N$. The orientability of $\ker\theta$ is again a trivial consequence of 3.1.4.(5) and, since the subgroup H of $\mathrm{im}\theta$ generated by $\theta(x_1),\ldots,\theta(x_r)$, $\theta(e_1),\ldots,\theta(e_{p'})$ has, by 3.4, order N/2, we obtain

$$|\mathrm{im}\theta|\geq|H+\theta(c_{k',0})|=l.c.m.(N/2,2)=N$$

i.e., θ is surjective. The standard arguments using 2.2.4, 2.3.2, 2.4.4 and our hypotheses (2) and (3.1) show that $\sigma(\ker\theta)=\sigma$. This completes the proof.

Comment. Our next results concern the same problem for the other possibilities for the signs of σ and τ. We have seen in the proof of the previous theorems that once a homomorphism $\theta:\Lambda \longrightarrow Z_N$ is given, it is straightforward to show the equality $\sigma(\ker\theta)=\sigma$ provided one can prove its surjectivity and that $\mathrm{sign}\sigma=\mathrm{sign}\sigma(\ker\theta)$. The remaining verifications can be easily done using 2.2.4, 2.3.2 and 2.4.4. Hence, from now on, we shall only check these two facts.

Theorem 3.1.6. *Let* N *be even,* $\mathrm{sign}\tau$ = "+" *and* $\mathrm{sign}\sigma$ = "-" *Then* (σ,τ) *is an* N-*pair if and only if:*

(1) For $i=1,\ldots,r$, m_i *divides* N.

(2) $\mu(\sigma)=N\mu(\tau)$.

(3) There exist $0\leq p'\leq s'$ *and some positive divisors* $l_1,\ldots,l_{p'}$ *of* N *such that*

(3.1) $k=\sum_{j=1}^{p'} N/l_j+N/2\sum_{i=s'+1}^{k'} r_i'/2.$

(3.2) $p'<k'$.

(3.3) If $g'=0$ *and* $k'=p'+1$, *then* $N = l.c.m.(m_1,\ldots,m_r,l_1,\ldots,l_{p'})$.

Proof. Let $\theta:\Lambda \longrightarrow Z_N$ be an epimorphism with $\sigma(\Lambda)=\tau$, such that $\Gamma=\ker\theta$ has signature σ. As before, we choose p' such that $c_{i0} \in \Gamma$ for $i \le p'$ and $c_{i0} \notin \Gamma$ for $p'+1 \le i \le s'$; we take $l_j = \#(\theta(e_j))$, $j=1,\dots,p'$. Now (1), (2), (3) and (3.1) are consequences of 2.4.4.

To prove (3.2) assume, by the way of contradiction, that $p'=k'$. Then Γ contains all canonical reflections of Λ, and so Γ does not contain a nonorientable word. Thus, by 2.1.3, $\text{sign}\sigma = \text{sign}\tau$, which is absurd.

Now we shall prove (3.3). Since Γ is a surface group we deduce from 2.2.4 and 2.3.2 that, for $1 \le i \le r$ and $1 \le j \le p'$, $\zeta_i = \theta(x_i)$ has order m_i and $\eta_j = \theta(e_j)$ has order l_j. Of course $c = c_{k'0}$ verifies $\theta(c) = N/2$.

All reduces to find an orientable word $w \in \Lambda$ with $\theta(w)=1$. In such a case, since $g'=0$ and $p'+1=k'$, $w \in \langle x_1,\dots,x_r,e_1,\dots,e_{p'+1}\rangle$. Moreover, from the relation $x_1 \dots x_r e_1 \dots e_{p'+1} = 1$ it follows that $e_{p'+1}$ is redundant, and so, $1 = \theta(w) \in \langle \zeta_1,\dots,\zeta_r,\eta_1,\dots,\eta_{p'}\rangle$, that is, $N = l.c.m.(m_1,\dots,m_r,l_1,\dots,l_{p'})$.

To find w we proceed as follows: from the surjectivity of θ, there exists a word $w_1 \in \Lambda$ such that $\theta(w_1)=1$. If w_1 is orientable, put $w=w_1$. Otherwise, we can choose a nonorientable word $w_2 \in \Gamma$, *i.e.*, $\theta(w_2)=0$, using 2.1.3, because $\text{sign}\sigma = $ "-". Now we take $w = w_1 w_2$.

Consequently we have proved that conditions in the theorem are necessary. To prove the converse we must construct θ. We first define

$$S = \sum_{i=1}^{r} N/m_i + \sum_{j=1}^{p'} N/l_j \in \mathbb{Z}_N.$$

Although we shall distinguish three cases, we take always $\theta : \Lambda \longrightarrow \mathbb{Z}_N$, with $\theta(x_i) = N/m_i$, $\theta(e_j) = N/l_j$ for $1 \le i \le r$, $1 \le j \le p'$ and θ transforms the reflections in Λ as indicated in 3.1.4.(1). Now,

Case 1. $g' \ne 0$. In this case we put

$$\theta(a_1)=1;\ \theta(b_1)=0;\ \theta(a_n)=\theta(b_n)=0,\ n=2,\dots,g'$$
$$\theta(e_{p'+1})=N\text{-}S;\ \theta(e_j)=0,\ j=p'+2,\dots,k'.$$

Case 2. $g'=0$, $p'+1<k'$. Then we choose

$$\theta(e_{p'+1})=1;\ \theta(e_{p'+2})=-(1+S);\ \theta(e_j)=0,\ j=p'+3,\dots,k'.$$

Case 3. $g'=0$, $p'+1=k'$. Here we define $\theta(e_{k'})=N\text{-}S$. It is obvious that in any case θ preserves relations. It is clear the surjectivity of θ in the first two cases since, respectively, $\theta(a_1)=1$ or $\theta(e_{p'+1})=1$.

In the third one, $\text{im}\theta = \langle \theta(x_1),\dots,\theta(x_r),\theta(e_1),\dots,\theta(e_{p'})\rangle$ and so, from (3.3)

$$|\text{im}\theta| = l.c.m.(m_1,\dots,m_r,l_1,\dots,l_{p'}) = N.$$

It only remains to check that $\ker\theta$ is nonorientable. It is enough to find a nonorientable word in $\ker\theta$ and use 2.1.3.

Since $p'<k'$ there exists some reflection $c \in \Lambda$ with $\theta(c)=N/2$. Hence we can take $w_1 = a_1^{N/2}c$ in case 1, and $w_2 = e_{p'+1}^{N/2}c$ in the second one. These are the words

we are looking for.

Finally, in the third case, Z_N is generated by $\theta(x_1),\dots,\theta(x_r)$, $\theta(e_1),\dots,\theta(e_{p'})$ and so there exists an orientable word $w\in\Lambda$ with $\theta(w)=1$. But then $w_3=w^{N/2}c$ is a nonorientable word in $\ker\theta$.

Notations and remarks 3.1.7. We want to characterize now the N-pairs (σ,τ) for even N and signτ = "-". In order to simplify notations we write:

(1) Given two families of integers $\{N/m_1,\dots,N/m_r\}$, $\{N/l_1,\dots,N/l_{p'}\}$ and given $\alpha=(\alpha_1,\dots,\alpha_r)$, $\beta=(\beta_1,\dots,\beta_{p'})$, for which

(3.1.7.1) $$g.c.d.(\alpha_i,m_i)=g.c.d.(\beta_j,l_j)=1$$

we put

$$S(\alpha,\beta)=\sum_{i=1}^{r}\alpha_iN/m_i+\sum_{j=1}^{p'}\beta_jN/l_j$$

without an explicit reference to the integers N/m_i, N/l_j in the symbol $S(\alpha,\beta)$. We shall say that the pair (α,β) is *order-preserving* with respect to $\{N/m_i,N/l_j|\ 1\le i\le r,\ 1\le j\le p'\}$ if it satisfies (3.1.7.1).

The reason for this terminology is clear: the classes modN of N/m_i and α_iN/m_i have both order m_i, and those of N/l_j and β_jN/l_j have order l_j.

(2) Suppose that sign$\sigma(\Lambda)$="-", $p'=k'$ and $\theta:\Lambda\longrightarrow Z_N$ is an epimorphism with kernel Γ. Let us write

$$\{x_1,\dots,x_r,e_1,\dots,e_{k'}\},\ \{d_1,\dots,d_{g'}\}$$

the sets of orientable (resp., glide reflections) canonical generators of Λ. Assume that $\theta(x_i)=\zeta_i$, $\theta(e_j)=\eta_j$ are even numbers in Z_N, $i=1,\dots,r$, $j=1,\dots,k'$, *i.e.*, they admit even representatives. Then:

(3.1.7.2) sign$\sigma(\Gamma)$="+" if and only if each $\theta(d_l)$ is odd, $l=1,\dots,g'$.

In fact, assume that $\theta(d_l)=u+NZ$ for even u, for some l. If u=0, then sign$\sigma(\Gamma)$="-" by 2.1.3. So $u\neq0$. Let us take a word $w\in\Lambda$ with $\theta(w)=1$. Since u is even, $w'=w^ud_l^{-1}$ is a nonorientable word in Γ. Again by 2.1.3 sign$\sigma(\Gamma)$="-", and so we have proved the "only if" part of the assertion.

Conversely, let us suppose that sign$\sigma(\Gamma)$="-". Then either there exists a nonorientable word $w\in\Gamma$, *i.e.*, $\theta(w)=0$, or some glide reflection w of Λ belongs to Γ. Since Z_N is abelian and $p'=k'$, we can assume that $w=w'd_1^{p_1}\dots d_{g'}^{p_{g'}}$ for some $w'\in\langle x_1,\dots,x_r,e_1,\dots,e_{k'}\rangle$, and integers $p_1,\dots,p_{g'}$, such that $p=p_1+\dots+p_{g'}$ is an odd number. From our assumption, $\theta(w')=\mu+NZ$ with μ even, and $0=\theta(w)=\mu+NZ+p_1\theta(d_1)+\dots+p_{g'}\theta(d_{g'})$. If we write $\theta(d_l)=(2\delta_l+1)+NZ$, for $l=1,\dots,g'$, we deduce that $\sum_{l=1}^{g'}p_l(2\delta_l+1)=-\mu(\mathrm{mod}N)$ is an even integer. Thus also p is even, a contradiction.

Theorem 3.1.8. *Let* N *be even and* signσ=signτ="-". *Then,* (σ,τ) *is an* N*-pair if and only if:*

(1) For each $i=1,\dots,r$, m_i *divides* N.

(2) $\mu(\sigma) = N\mu(\tau)$.

(3) There exist $0\le p'\le s'$ *and positive divisors* $l_1,\dots,l_{p'}$ *of* N *verifying:*

(3.1) $k = \sum_{j=1}^{p'} N/l_j + N/2 \sum_{i=s'+1}^{k'} r_i'/2.$

(3.2) If $p'=k'$ *there exists an order-preserving pair* (α,β) *with respect to* $\{N/m_1,\dots,N/m_r,N/l_1,\dots,N/l_{k'}\}$ *such that* $S(\alpha,\beta)$ *is even.*

(3.3) If $p'=k'$ *and* $g'=1$, *then* $l.c.m.(m_1,\dots,m_r,l_1,\dots,l_{k'})=N$.

(3.4) Assume $p'=k'$, $g'=2$, $N\in 4\mathbb{Z}$ *and every even* $S(\alpha,\beta)$ *is a multiple of* 4. *Then* N/m_i *or* N/l_j *are odd for some* i *or* j.

Proof. Let $p',l_1,\dots,l_{p'}$ be the numbers defined as in the last two theorems. Then conditions (1),(2),(3) and (3.1) are obvious consequences of 2.4.4.

Since $\Gamma=\ker\theta$ is a surface group, $\zeta_i=\theta(x_i)$ and $\eta_j=\theta(e_j)$, have respectively orders m_i and l_j. Let us write $H=<\zeta_1,\dots,\zeta_r,\eta_1,\dots,\eta_{p'}>\subseteq Z_N$.

We are going to prove (3.2). Of course $\zeta_i=\alpha_i N/m_i$ and $\eta_j=\beta_j N/m_j$ for some order-preserving pair (α,β). Since $1=d_1^2\dots d_{g'}^2x_1\dots x_re_1\dots e_{k'}$, we get $2[\theta(d_1)+\dots+\theta(d_{g'})]+S(\alpha,\beta)=0$ modN, and since N is even, $S(\alpha,\beta)$ is also even.

To see (3.3) note that $\theta(w)=0$ for either some nonorientable word or some glide reflection $w\in\Lambda$, and since Z_N is abelian, we can suppose $w=w_1d_1^{2a+1}$ for some integer a and some $w_1\in\theta^{-1}(H)$. Hence $0=\theta(w_1)+(2a+1)\delta$, where $\delta=\theta(d_1)$. On the other hand, from the relation $1=d_1^2x_1\dots x_re_1\dots e_{k'}$, we obtain $2\delta\in H$. Consequently, $\delta=-\theta(w_1)-2a\delta\in H$ and $Z_N=<H,\delta>=H$. In particular, $N=|H|=$ $=l.c.m.(m_1,\dots,m_r,l_1,\dots,l_{k'})$.

Finally, for 3.4, write $\delta_1=\theta(d_1)$, $\delta_2=\theta(d_2)$ and $\zeta_i=\alpha_iN/m_i$, $\eta_j=\beta_jN/l_j$, for the pair (α,β) we found in 3.2. By assumption, 4 divides both N and $S(\alpha,\beta)$. Suppose by the way of contradiction that N/m_i and N/l_j are even numbers. Then the same is true for ζ_i and η_j and, since $Z_N=<\delta_1,\delta_2,\zeta_1,\dots,\zeta_r,\eta_1,\dots,\eta_{k'}>$, either δ_1 or δ_2 must be odd. On the other hand, $2(\delta_1+\delta_2)+S(\alpha,\beta)=0$ modN and so $\delta_1+\delta_2$ is even. Hence both δ_1 and δ_2 are odd, and this contradicts 3.1.7.2. So, we have proved the necessity of our conditions.

To prove the sufficiency we must construct an epimorphism $\theta:\Lambda\longrightarrow Z_N$ whose kernel Γ has signature σ and $\sigma(\Lambda)=\tau$. The construction of θ shall depend on the values of g', $k'-p'$ and the parity of N/2. In each case we define $\theta(x_i)=\alpha_iN/m_i$ and $\theta(e_j)=\beta_jN/l_j$ for a suitable order-preserving pair (α,β). We begin with

Case 1. $p'=k'$, $g'\ge 3$.

We can choose (α,β) such that $S=S(\alpha,\beta)$ is even, and we define:

$$\theta(d_1)=0,\ \theta(d_2)=1,\ \theta(d_3)=-(S+2)/2,\ \theta(d_i)=0 \text{ if } 4\leq i\leq g'.$$

Since $\theta(d_2)=1$ and $\theta(d_1)=0$, θ is surjective, with nonorientable kernel.

Case 2. $p'=k'$, $g'=2$, $N\notin 4\mathbb{Z}$.

Take (α,β) such that $S(\alpha,\beta)=S$ is even. Then we define

$$S_1=\begin{cases} S-2 & \text{if } S\notin 4\mathbb{Z},\\ S+N-2 & \text{if } S\in 4\mathbb{Z},\end{cases}$$

and $\theta(d_1)=-1$, $\theta(d_2)=-S_1/2$. Clearly θ is an epimorphism because $1=\theta(d_1^{-1})$, and since $h=-S_1/2$ is even, $w=d_1^h d_2$ is a nonorientable word in Γ, *i.e.*, $\text{sign}\sigma(\Gamma)=$ "-".

Case 3. $p'=k'$, $g'=2$, $N\in 4\mathbb{Z}$.

If $S=S(\alpha,\beta)\in 2\mathbb{Z}\setminus 4\mathbb{Z}$ for some order-preserving pair (α,β) we define θ as above, with $S_1=S-2$. In the other case, by 3.2 and 3.4, there exists (α,β) such that $S=S(\alpha,\beta)$ is a multiple of 4 and, *e.g.*, N/m_1 is odd. Then m_1 is even and so α_1 is odd. In particular $\zeta_1=\alpha_1 N/m_1$ is odd and if we define $\theta(d_1)=-1$, $\theta(d_2)=-(S-2)/2$, then $w=d_1^{\zeta_1}x_1\in\Gamma$ is a nonorientable word, *i.e.*, $\text{sign}\sigma(\Gamma)=$ "-".

Case 4. $p'=k'$, $g'=1$.

As always, we can take by 3.2 an order-preserving pair (α,β) such that $S=S(\alpha,\beta)$ is even. Let us define $\theta(d_1)=-S/2$. Since, by 3.3, *l.c.m.*$(m_1,\ldots,m_r,l_1,\ldots,l_{k'})=N$, and $\text{im}\theta$ contains each $\alpha_i N/m_i$ and each $\beta_j N/l_j$, θ is surjective. In fact $\theta(w)=1$ for some orientable word $w\in\Lambda$. Consequently, $w_1=d_1w^{S/2}$ is a nonorientable word in $\ker\theta$.

Case 5. $p'<k'$.

Although we must distinguish several subcases we always define

$$\theta(x_i)=N/m_i,\ \theta(e_j)=N/l_j,\ 1\leq i\leq r,\ 1\leq j\leq p'$$

and we put

$$S=\sum_{i=1}^{r} N/m_i+\sum_{j=1}^{p'} N/l_j.$$

Subcase 5.1. $p'+1<k'$.

Then θ is given by $\theta(e_{p'+1})=1$, $\theta(e_{p'+2})=-(1+S)$, $\theta(d_i)=\theta(e_j)=0$ for $1\leq i\leq g'$, $p'+3\leq j\leq k'$. It is surjective, because $\theta(e_{p'+1})=1$, and $\sigma(\Gamma)$ has sign "-" since $d_1\in\Gamma$.

Subcase 5.2. $p'+1=k'$, $g'>1$.

Now we define $\theta(d_1)=0$, $\theta(d_2)=1$, $\theta(d_l)=0$, $l=3,\ldots,g'$, $\theta(e_{k'})=-(S+2)$. As before it is surjective, since $\theta(d_2)=1$. Moreover, $d_1\in\Gamma$ implies that $\sigma(\Gamma)$ has sign "-".

Subcase 5.3. $p'+1=k'$, $g'=1$.

In this situation we take $\theta(d_1)=N/2+1$, $\theta(e_{p'+1})=-(S+2)$. Here $k'>p'$ and so $\theta(c)=N/2$ for some reflection $c\in\Lambda$. Consequently $\theta(d_1c)=1$ and θ is surjective.

Even more, $w=d_1c$ is orientable. Hence $w'=w^{N/2}c$ is nonorientable, $\theta(w')=N/2+N/2=0$. Thus, by 2.1.3, the sign in the signature of $\ker\theta$ is "-".

So, in any possible case we have constructed an epimorphism $\theta:\Lambda \longrightarrow Z_N$ such that the sign of the signature of $\Gamma=\ker\theta$ is "-". It is now an easy task to check, using our assumptions (2), (3.1) and theorems 2.2.4, 2.3.2 and 2.4.4, that $\sigma(\Gamma) = \sigma$. In this way, the proof is finished.

Our last result in this section concerns the remaining case: signτ="-" and signσ="+".

Theorem 3.1.9. *Let* N *be even,* signτ="-" *and* signσ="+". *Then,* (σ,τ) *is an* N-*pair if and only if:*

(1) For each $i=1,...,r$, m_i *divides* N.

(2) $\mu(\sigma)=N\mu(\tau)$.

(3) There exist $0\le p'\le s'$ *and positive divisors* $l_1,...,l_{p'}$ *of* N *such that:*

(3.1) $k=\sum_{j=1}^{p'} N/l_j+N/2\sum_{i=s'+1}^{k'} r_i/2.$

(3.2) $M=l.c.m.(m_1,...,m_r,l_1,...,l_{p'})\neq N.$

(3.3) If $N=2^rA$, *where* A *is odd, then the set of those* m_i *and* l_j *which are multiple of* 2^r *has even cardinal.*

(3.4) There exists an order-preserving pair (α,β) *with respect to* $\{N/m_1,...,N/m_r,N/l_1,...,N/l_{p'}\}$ *such that all* α_iN/m_i *and* β_jN/l_j *are even numbers. Moreover, if* $p'=k'$ *and* $N\in 4Z$, *the number* $S(\alpha,\beta)$ *is a multiple of 4 if and only if* g' *is even.*

(3.5) If $p'<k'$, *then the numbers* m_i *and* l_j *are odd, and* $N\notin 4Z$.

(3.6) If $g'=1$ *and* $k'=p'$, *then* $M=N/2$.

Proof. We first suppose that (σ,τ) is an N-pair. Let us take p' such that

$$c_{10},...,c_{p'0}\in\Gamma;\ c_{p'+1,0},...,c_{s'0}\notin\Gamma,\ \Gamma=\ker\theta.$$

We also define l_j=order of $\theta(e_j)=\eta_j\in Z_N$, $j=1,...,p'$. Then each l_j divides N and conditions (1), (2), (3), (3.1) are immediate consequences of 2.4.4.

Since σ has no proper periods, we deduce from 2.2.4 that $\zeta_i=\theta(x_i)$ has order m_i. Thus, the subgroup $H=<\zeta_1,...,\zeta_r,\eta_1,...,\eta_{p'}>$ has order $M=l.c.m.(m_1,...,m_r,l_1,...,l_{p'})$. If M=N we would have $H=Z_N$. In particular, $\theta(d_1)=\theta(w)$ for some orientable word $w\in\Lambda$; but then $w_1=w^{-1}d_1\in\Gamma$ is a nonorientable word. This contradicts 2.1.3, because signσ="+". So (3.2) is proved.

Moreover, conditions (2) and (3.1) imply that

$$\sum_{i=1}^{r} N/m_i + \sum_{j=1}^{p'} N/l_j = N[g'-2+k'+r]+2(1-g)$$

is an even number. Thus the number of odd summands in the left hand side must be even. This proves (3.3).

Let us see (3.4). Since θ is surjective there exists a word $w \in \Lambda$ such that $\theta(w)=1$. Let us write $\delta_i=\theta(d_i)$. Then $\delta_i \neq 0$, because signσ="+". Thus $w^{-\delta_i}d_i \in \Gamma$ and it is a word. Using 2.1.3 once again, $w^{-\delta_i}$ must be nonorientable, and so w is nonorientable itself. Moreover, $w^{-\delta_i}$ being nonorientable, we deduce that δ_i is odd.

Also, $w^{-\zeta_i}x_i \in \Gamma$ and by 2.1.3, it must be orientable, since signσ="+". Thus ζ_i must be even. By the same reason, also η_j is even. But $\#(\zeta_i)=m_i$ and $\#(\eta_j)=l_j$. Thus $\zeta_i=\alpha_i N/m_i$, $\eta_j=\beta_j N/l_j$ for some integers α_i,β_j, *g.c.d.*$(m_i,\alpha_i)=$ $=$*g.c.d.*$(l_j,\beta_j)=1$. Consequently the pair (α,β) is order-preserving and

$\alpha_i N/m_i=\zeta_i$, $\beta_j N/l_j=\eta_j$ are even, $i=1,\dots,r$, $j=1,\dots,p'$.

Let us compute now the value of $S(\alpha,\beta)$ for $p'=k'$ and $N \in 4\mathbb{Z}$. From the relation $d_1^2\dots d_{g'}^2 x_1\dots x_r e_1\dots e_{p'}=1$ we deduce $2\sum_{i=1}^{g'}\delta_i+S(\alpha,\beta)=0$ modN.

Each δ_i is odd, and so $2\sum_{i=1}^{g'}\delta_i \in 4\mathbb{Z}$ if and only if $g' \in 2\mathbb{Z}$. Thus, since N is a multiple of 4, we are done. To prove (3.5) we shall see first that each m_i is odd. Let us suppose by the way of contradiction that some m_i is even. We know that $\zeta_i=\alpha_i N/m_i$, with *g.c.d.*$(\alpha_i,m_i)=1$. In particular α_i is odd, and so $\zeta_i=2^{r-h}u_i$ for some $1 \le h \le r$ and some divisor u_i of A (we use the notations in (3.3)). Consequently $N/2=2^{r-1}A \in <\zeta_i>$. That means $N/2=\theta(x_i^t)$ for some t. But $p'<k'$ implies there exists a reflection $c \in \Gamma$ with $\theta(c)=N/2$. Consequently $x_1^t c \in \Gamma$ is a nonorientable word. This contradicts 2.1.3, because sign$\sigma(\Gamma)$="+".

The same argument allows us to deduce that also $l_1,\dots,l_{p'}$ are odd integers.

Since θ is an epimorphism and $p'<k'$, there exist a word $w \in \Lambda$ and a reflection $c \in \Lambda$ such that $\theta(w)=1$ and $\theta(c)=N/2$. Then $w_1=w^{N/2}c \in \Gamma$. Since the sign of σ is "+", the word w_1 must be orientable by 2.1.3. Thus $w^{N/2}$ is nonorientable. In particular N/2 is odd.

Finally to prove (3.6) it is enough to check that $\delta=\theta(d_1) \notin H$ but $2\delta \in H$. In such a case, since $\{d_1,x_1,\dots,x_r,e_1,\dots,e_{p'}\}$ generates Λ we would have $\mathbb{Z}_N=H+<\delta>$ and so $M=N/2$.

If $\delta \in H$, there would exist an orientable word $w \in <x_1,\dots,x_r,e_1,\dots,e_{p'}>$ such that $\theta(d_1)=\theta(w)$, *i.e.*, $w^{-1}d_1$ would be a nonorientable word in Γ, which is absurd. On the other hand, the relation $d_1^2x_1...x_re_1...e_{p'}=1$ gives us $2\delta=-[\zeta_1+...+\zeta_r+\eta_1+...+\eta_{p'}]\in H$, as desired.

To prove the sufficiency we construct an epimorphism $\theta:\Lambda\longrightarrow \mathbf{Z}_N$ whose kernel Γ has signature σ. We separate the construction into two cases:

Case 1. $p'<k'$.

Then we know, by (3.5), that m_i, l_j, are odd and $N\notin 4\mathbf{Z}$. In particular, N/m_i and N/l_j are even numbers, and we define θ in the following way: let us denote $S=\sum_{i=1}^{r} N/m_i+\sum_{j=1}^{p'} N/l_j$ and take

$$\theta(x_i)=N/m_i,\ \theta(e_j)=N/l_j,\ i=1,...,r;\ j=1,...,p',\quad \theta(d_i)=1,\ i=1,...,g',$$
$$\theta(e_{p'+1})=-(S+2g'),\ \theta(e_l)=0,\ l=p'+2,...,k'.$$

The surjectivity of θ is evident. The main difficulty here is to prove that the signature of $\ker\theta$ has sign "+". But if this were not so, there would exist either a nonorientable word or a glide reflection $w\in\ker\theta$. Since $\mathbf{Z}_N$ is abelian we can suppose that w has the following form:

$$w = w_1d_1^{\varepsilon_1}...d_{g'}^{\varepsilon_{g'}}c^{\varepsilon}$$

where $w_1\in\langle x_1,...,x_r,e_1,...,e_{p'+1}\rangle$, $\varepsilon_1,...,\varepsilon_{g'},\varepsilon$ are integers, $\varepsilon=0$ or 1, $\varepsilon+\varepsilon_1+...+\varepsilon_{g'}=B$ is an odd integer, and $c\in\Lambda$ is a reflection with $\theta(c)=N/2$. Then

$$0=\theta(w_1)+(\varepsilon_1+...+\varepsilon_{g'})+\varepsilon N/2.$$

Clearly $D=\theta(w_1)$ is even, because each N/m_i, N/l_j, $S+2g'$ are even numbers. Since $0=D+(B-\varepsilon)+\varepsilon N/2$ (modN) and both D and N are even, $B+\varepsilon(N/2-1)$ is also even. This is false because B is odd and N/2-1 is even. Once this is proved, the standard argument shows that $\sigma(\Gamma)=\sigma$.

Case 2. $k'=p'$.

Here the epimorphism we are going to construct verifies the conditions of our remark 3.1.7.(2) and consequently the signature of the kernel has sign "+". The surjectivity will be also obvious since $\theta(d_1)$ will be equal to one (except for $g'=1$).

By (3.4) we have an order-preserving pair (α,β) such that α_iN/m_i and β_jN/l_j are even numbers. Then we take

$$\theta(x_i)=\alpha_iN/m_i,\ \theta(e_j)=\beta_jN/l_j,\ i=1,...,r,\ j=1,...,p'.$$

With the usual meaning, $S=S(\alpha,\beta)=\sum_{i=1}^{r}\alpha_iN/m_i+\sum_{j=1}^{p'}\beta_jN/l_j$. It only remains to define $\theta(d_1),...,\theta(d_{g'})$. If $g'>1$, we put $\theta(d_1)=...=\theta(d_{g'-1})=1$. Then, necessarily

(*) $$2(g'-1)+S+2\theta(d_{g'})=0\in\mathbf{Z}_N$$

and, in order to be in the conditions of 3.1.7.(2), $\theta(d_{g'})$ must be odd.

By the second part of (3.4), for $N\in 4\mathbf{Z}$, $S\in 4\mathbf{Z}$ if and only if g' is even.

Then we define $\theta(d_{g'})$ in the following way (also for $g'=1$):

$$\theta(d_{g'})=\begin{cases} -S/2-(g'-1) & \text{if } N\in 4\mathbb{Z} \text{ or } 2g'+S\in 4\mathbb{Z} \\ -S/2-(g'-1)+N/2 & \text{otherwise} \end{cases}$$

To finish the proof we must only check the equality $\mathrm{im}\theta=\mathbb{Z}_N$, when $g'=1$.

Let us write $H=\langle\theta(x_1),\dots,\theta(x_r),\theta(e_1),\dots,\theta(e_{p'})\rangle$. Notice that by (3.6) $|H|=N/2$. By its definition, $\delta=\theta(d_1)$ is odd. Thus $\delta\notin H$, because each $\theta(x_i)=\alpha_i N/m_i$ and $\theta(e_j)=\beta_j N/l_j$ are even numbers. Thus $H\subsetneqq\mathrm{im}\theta\subseteq\mathbb{Z}_N$, and $|H|=N/2$, *i.e.*, $\mathrm{im}\theta=\mathbb{Z}_N$. As always, it is very easy to check that σ is the signature of $\ker\theta$.

We finish this paragraph with an easy example which shows how the theorems we have proved can be used to determine the existence of automorphisms of Klein surfaces of given orders.

Example 3.1.10. Let us call a triple (g,k,N) p-*admissible* if there exists an orientable compact Klein surface with algebraic genus p, topological genus g, whose boundary has $k>0$ connected components, which admits an automorphism of order N.

We are going to prove that for odd N, the 4-admissible triples are

$$(g,k,N)=\begin{cases}(1,3,3)\\(0,5,3)\\(0,5,5)\\(2,1,5)\\(2,1,3)\end{cases}$$

Notice first that, in our previous notations, (g,k,N) is 4-admissible if and only if $2g+k=5$ and there exists an NEC signature τ such that (σ,τ) is an N-pair.

Consequently, for odd N, we deduce from 3.1.2 that (g,k,N) is 4-admissible if and only if:

(1) $2g+k=5$.

(2) There exist nonnegative integers g',k',r, and some divisors $m_1,\dots,m_r$, $l_1,\dots,l_{k'}$ of N, with $m_i\geq 2$ and $l_i\geq 1$, such that:

(2.1) $3=N\left[2g'-2+k'+\sum_{i=1}^{r}(1-1/m_i)\right]$,

(2.2) $k=\sum_{j=1}^{k'} N/l_j$,

(2.3) $\{m_1,\dots,m_r,l_1,\dots,l_{k'}\}$ has the elimination property,

(2.4) If $g'=0$, then $l.c.m.(m_1,\dots,m_r,l_1,\dots,l_{k'})=N$.

Notice that (2.2), together with $k\geq 1$, implies $k'\geq 1$, and from (2.1), we know $2g'+k'\leq 3/N+2<4$. We divide our calculations into two parts:

Case 1: $2g'+k'=3$.

Then, condition (2.1) becomes $3=N\left[1+\sum_{i=1}^{r}(1\text{-}1/m_i)\right]$ and since N is odd, we deduce N=3, r=0.

If k′=3, then g′=0 and we must look for divisors l_1,l_2,l_3 of 3 such that $l.c.m.(l_1,l_2,l_3)=3$, $\{l_1,l_2,l_3\}$ has the elimination property. Hence either $l_1=l_2=l_3=3$, or $l_1=l_2=3$ and $l_3=1$. By (2.2) and (1) we obtain either k=3 and g=1 or k=5 and g=0. Consequently we get (1,3,3) and (0,5,3) as 4-admissible triples.

If k′≠3, it is clearly k′=1=g′. Then we look for a divisor l_1 of 3 such that $\{l_1\}$ has the elimination property. That means, see Def.1, $l_1=1$ and so k=3, g=1, which does not provide a new admissible triple.

Case 2: $2g'+k'\leq 2$.

As we remarked before, $k'\geq 1$, and so we have: g′=0, k′=2, or g′=0, k′=1. Let us see first that we can reject the case g′=0, k′=2. Since $m_i\geq 2$ divides the odd number N, it is $m_i\geq 3$ for i=1,...,r, and so by (2.1),

$$3=N\sum_{i=1}^{r}(1\text{-}1/m_i)\geq 2rN/3.$$

Thus $rN\leq 4$, *i.e.*, N=3, r=1, a contradiction.

Hence, from now on we must only look for the triples (g,k,N) verifying conditions (1)-(2.4) with g′=0, k′=1.

Denote $L=\sum_{i=1}^{r}(1\text{-}1/m_i)$. From (2.1), L=3/N + 1. In particular, L>1 implies $r\geq 2$ and since each $m_i\geq 3$, we deduce $3/N+1=L\geq 1+1/3$. Consequently $N\leq 9$. In fact we are going to see that $N\leq 5$. For, notice that k′=1 and g′=0 implies $l.c.m.(m_1,\ldots,m_r)=l.c.m.(m_1,\ldots,m_r,l_1)=N$.

Hence, for N=7 or N=9, it is $m_1=N$, and so

$$3/N+1=L\geq(1\text{-}1/m_1)+(1\text{-}1/m_2)\geq(1\text{-}1/N)+2/3,$$

i.e., $N\leq 6$, absurd.

Finally, let us study N=3 and N=5. Since both are primes, $l_1=1$ or $l_1=N$, and so, by (2.2), k=N or k=1. Thus g=(5-N)/2 or g=2. For N=3, the triple (g,k,N)=(1,3,3) has already appeared. So it remains to deal with (2,1,3), (0,5,5) and (2,1,5).

Notice that since N is prime, $m_1=\ldots=m_r=N$ and so $3/N+1=L=\sum_{i=1}^{r}(1\text{-}1/m_i)=$ $=r(N\text{-}1)/N$, that is, $3+N=(N\text{-}1)r$. Thus r=3 for N=3 and r=2 for N=5. Consequently (2,1,3) is 4-admissible because g′=0, k′=1, r=3, $m_1=m_2=m_3=l_1=3$ verify conditions (2.1)-(2.4). In the same way, choosing g′=0, k′=1, r=2, $m_1=m_2=5$, and $l_1=1$ or $l_1=5$, we deduce that (0,5,5) and (2,1,5) are 4-admissible.

Comment. Of course, we have studied a very easy example. But the technique used shows that all examples (*i.e.*, p as big as desired, N odd or even), are simply handled with the help of a computer since the p-admissibility character of a triple (g,k,N) uniquely involves the existence of non-negative integers $g',k',r,m_1,\dots,m_r,l_1,\dots,l_{k'}$ (and also $p',s',r'_{s'+1},\dots,r'_{k'}$) verifying some arithmetical conditions, and we know, *a priori,* (*i.e.*, in terms of g, k and N), bounds for all of them.

3.2. THE MINIMUM GENUS AND MAXIMUM ORDER PROBLEMS

As was mentioned in the introduction, we solve here the minimum genus and maximum order problems. What makes our results deeper is that the study of this problem for orientable surfaces is handled separately for orientation preserving and orientation reversing automorphisms. By a surface in this section we mean a compact bordered Klein surface S of algebraic genus $p \geq 2$.

Let us represent S as a quotient H/Γ for some NEC group Γ. Each $f \in \mathrm{Aut}(S)$ generates a finite cyclic group of automorphisms of S isomorphic to Λ/Γ, where Λ is another NEC group containing Γ as a normal subgroup. Let us denote $C^-(\Lambda)$ the set of orientation reversing canonical generators of Λ. Then,

Proposition 3.2.1. *The following conditions are equivalent:*
(1) f *preserves orientation.*
(2) $C^-(\Lambda) \subseteq \Gamma$.
Proof. Let N be the order of f and let $f_1 = 1_H, \dots, f_N$ be the set of representatives of Λ/Γ. Then, if R′ is a fundamental region for Λ,

$$R = R' \cup f_2(R') \cup \dots \cup f_N(R')$$

is a fundamental region for Γ. Moreover, $R/\Gamma \approx H/\Gamma$ as topological spaces. Fix an orientation of R and consider the induced orientations of $f_i(R')$.
(1) ⇒ (2) Let us suppose that $g \in C^-(\Lambda) \setminus \Gamma$. Then we can assume that $f_2 = g$ and if $h = f^k$ is the image of Γf_2 under the isomorphism $\Lambda/\Gamma \cong \langle f \rangle$, the image of R′ corresponding to h is $f_2(R')$. Thus h, and so f, reverses orientation, a contradiction.
(2) ⇒ (1) Since $C^-(\Lambda) \subseteq \Gamma$ and Γ is a normal subgroup of Λ, the representatives $f_2,\dots,f_N$ can be taken as products of orientation preserving canonical generators of Λ. In particular, if Γf_i is the image of f under the isomorphism $\langle f \rangle \cong \Lambda/\Gamma$, f_i preserves the orientation. Consequently for $j \in \{1,\dots,N\}$, f_i is an orientation preserving map from the interior of $f_j(R')$ onto the interior of the region $f_i f_j(R')$ and of course if j runs over the set $\{1,\dots,N\}$, $f_i f_j$ runs over the set of all representatives of Λ/Γ. Moreover, since $C^-(\Lambda) \subseteq \Gamma$, the

identification of sides between $f_j(R')$ and $f_i f_j(R')$ is done by orientation preserving elements of Λ. Hence f preserves orientation.

Definition 1. Let $N \geq 2$ be an integer. Let us denote by $\mathscr{K}_+(N)$ (resp., $\mathscr{K}_-(N)$), the family of orientable (resp. nonorientable) compact Klein surfaces with nonempty boundary whose algebraic genus is bigger than 1, admitting an automorphism of order N. We put also

$$\mathscr{K}_+^+(N) = \{S \in \mathscr{K}_+(N) \mid \text{there exists an orientation preserving } f \in \mathrm{Aut}(S) \text{ of order } N\},$$
$$\mathscr{K}_+^-(N) = \{S \in \mathscr{K}_+(N) \mid \text{there exists an orientation reversing } f \in \mathrm{Aut}(S) \text{ of order } N\}.$$

Consider also

$$p_+(N) = \min\{p(S) \mid S \in \mathscr{K}_+(N)\}, \quad p_-(N) = \min\{p(S) \mid S \in \mathscr{K}_-(N)\},$$
$$p(N) = \min\{p_+(N), p_-(N)\},$$
$$p_+^+(N) = \min\{p(S) \mid S \in \mathscr{K}_+^+(N)\}, \quad p_+^-(N) = \min\{p(S) \mid S \in \mathscr{K}_+^-(N)\}.$$

Remarks and notations 3.2.2. (1) Our main goal is to compute $p_+(N)$, $p_-(N)$, $p(N)$, $p_+^+(N)$ and $p_+^-(N)$ as well as to find the topological types of the corresponding surfaces from $\mathscr{K}_+(N)$, $\mathscr{K}_-(N)$, $\mathscr{K}_+^+(N)$ or $\mathscr{K}_+^-(N)$ which achieve this minimum genus, that will be said to be N-*minimal*.

(2) Our strategy is rather obvious. A surface S can be represented as a quotient H/Γ for some surface NEC group Γ. If S has an automorphism of order N then $Z_N = \Lambda/\Gamma$ for some NEC group Λ with signature τ. The problem of finding a lower bound for the algebraic genus p of such a surface is equivalent, by the Hurwitz Riemann formula, to the problem of finding such a bound μ_0 for the area of NEC groups Λ standing up in the above presentation of Z_N. The last will be done using arithmetical considerations supported mainly by results of previous chapter and Lemma 3.2.4 below. Having such a bound found we shall look for NEC signatures τ_0 for which $\bar{\mu}(\tau_0) = \mu_0$ and using results of the previous section we shall be able to find all signatures σ_0 for which (σ_0, τ_0) is an N-pair what shall complete the proofs.

(3) Having a pair of signatures (σ, τ) we shall write $C^-(\tau) \subseteq \sigma$ if $C^-(\Lambda) \subseteq \Gamma$ for any NEC group Λ with signature τ containing an NEC group Γ with signature σ. We fix all notations above along all over the section.

We start with the following easy consequence of Lemma 3.2.1 and results of the previous section.

Corollary 3.2.3. *Let Γ be a surface NEC group such that $S = H/\Gamma$ is orientable. Let $f \in \mathrm{Aut}(S)$ be an element of order $N > 1$, and let Λ be an NEC group realizing* f, *i.e.*, $\langle f \rangle = \Lambda/\Gamma$. *As we know, for even* N,

$$\tau=\sigma(\Lambda)=(g';\pm;[m_1,\ldots,m_r];\{(\overset{s'}{-}),\ldots,(-),(2,\overset{r_i'}{\ldots},2)|s'+1\leq i\leq k'\})$$

and all r_i' *are even. Let* p' *be the integer for which* $c_{i0}\in\Gamma$ *for* $1\leq i\leq p'$ *and* $c_{i0}\notin\Gamma$ *for* $p'+1\leq i\leq s'$.

(1) If N *is odd, then* f *preserves orientation.*

(2) If N *is even and* $\text{sign}\tau=$"-", *then* f *reverses orientation.*

(3) If N *is even and* $\text{sign}\tau=$"+", *then* f *preserves orientation if and only if* $k'=p'$.

(4) If $N\in 4\mathbb{Z}$ *and* $\text{sign}\tau=$"+", *then* f *preserves orientation.*

Proof. (1) This is obvious, since $f^N=1_H$ preserves orientation.

(2) Since $\text{sign}\tau=$"-" and $\text{sign}\sigma=$"+" we have a glide reflection $d\in C^-(\Lambda)\setminus\Gamma$ and so f does not preserve orientation.

(3) Since $\text{sign}\tau=$"+", $C^-(\Lambda)$ equals the set of all canonical reflections of Λ. Thus if $p'=k'$ then $p'=s'=k'$, and so $C^-(\Lambda)\subseteq\Gamma$. Conversely, if $p'<k'$ by (2) in 3.1.4 Γ does not contain $C^-(\Lambda)$. Therefore (3) follows from 3.2.1.

(4) The equality $k'=p'$ holds true by (3.2) in 3.1.5 and so (4) follows from the previous assertion.

We shall need the following arithmetical

Lemma 3.2.4. *Let* $L>1$ *be an integer,* $L=p_1^{\alpha_1}\ldots p_t^{\alpha_t}$ *with* $p_1<\ldots<p_t$ *prime numbers. Let* R_L *be the set of pairs of positive integers* (x,y) *satisfying* *l.c.m.*(x,y)=L. *Consider the subset* R_L' *of pairs* $(x,y)\in R_L$ *such that* $x\neq 1\neq y$. *Then*

(1) $\delta(R_L)=\max\{1/x+1/y|(x,y)\in R_L\}=1+1/L$.

(2) $\delta(R_L')=\max\{1/x+1/y|(x,y)\in R_L'\}=\begin{cases}1/p_1+p_1/L & \text{if } \alpha_1=1,\ t>1,\\ 1/p_1+1/L & \text{if } \alpha_1>1 \text{ or } L=p_1,\end{cases}$

and this maximum is only attained at $(x,y)=(p_1,L/p_1)$ *in the first case and at* (p_1,L) *in the second one.*

Proof. Let us define $f:R_L\longrightarrow\mathbb{Q}:(x,y)\longmapsto 1/x+1/y$.

(1) If $(x,y)\in R_L$ and $d=$*g.c.d.*(x,y), the pair $(x'=x/d,y)\in R_L$ because *g.c.d.*(x',y) is equal to 1, and so *l.c.m.*$(x',y)=x'y=(xy)/d=$*l.c.m.*$(x,y)=L$. Moreover, $f(x',y)-f(x,y)=1/x'-1/x=(d-1)/x\geq 0$. Hence, in order to obtain $\delta(R_L)$ we can restrict our attention to pairs $(x,y)\in R_L$ with *g.c.d.*(x,y)=1, *i.e.* xy=L.

Obviously we can suppose $x\neq y$, since otherwise $1=$*g.c.d.*$(x,y)=x$, and $L=xy=1$, a contradiction. So, from now on, we assume $x<y$. This together with $xy=L$ implies $1\leq x<\sqrt{L}$. Hence $\delta(R_L)=\max\{1/x+x/L \mid 1\leq x<\sqrt{L},\ \textit{g.c.d.}(x,L/x)=1\}$. Since the derivative of $g:[1,\sqrt{L})\longrightarrow\mathbb{R}:x\longmapsto 1/x+x/L$ is negative everywhere, $g'(x)=-1/x^2+1/L<0$, we obtain $\delta(R_L)=g(1)=1+1/L$.

(2) The case $t=1$ is trivial: if $L=p_1^{\alpha_1}$ and $(x,y)\in R_L'$ with $x\leq y$, then necessarily x divides y and $L=l.c.m.(x,y)=y$. Thus

$$\delta(R_L')=1/L+\max\{1/x|x \text{ divides } L,\ x\neq 1\}=1/L+1/p_1.$$

Let us suppose from now on that $t\geq 2$. We construct the subsets

$$A=\{(x,y)\in R_L' \mid g.c.d.(x,y)=1\},\quad B=R_L'\backslash A.$$

We are going to compute the maxima of $f|A$ and $f|B$.

(a) To calculate the first it suffices to study $f|A'$ where $A'=\{(x,y)\in A|x<y\}$. But $(x,y)\in A'$ implies $xy=L$ and $f(x,y)=(y+x)/L$. Since $(y-x)^2=(y+x)^2-4L$ and $x<y$, f attains its maximum over A' at the point $(x_0,y_0)\in A'$ making biggest the difference y-x, *i.e.*, y_0 must be maximum and x_0 minimum. But $(x,y)\in A'$ implies that for $I=\{1,...,t\}$ and $J=\{i\in I|p_i \text{ divides } x\}$

$$x=\prod_{j\in J}p_j^{\alpha_j},\quad y=L/x.$$

Consequently, $x_0=\min\{p_j^{\alpha_j}|j\in I\}$, $y_0=L/x_0$, $\max f|A=1/x_0+x_0/L$.

(b) Let us take now $(x,y)\in B$ and $d=g.c.d.(x,y)$. Put now $x'=x/d$, $y'=y/d$. Then $l.c.m.(x',y')=x'y'=(xy/d)(1/d)=L/d$, *i.e.*, $(x',y')\in R_{L/d}$. Thus, from part (1),

$$1/x+1/y=1/d(1/x'+1/y')\leq 1/d\ \delta(R_{L/d})=1/d(1+d/L)=1/d+1/L.$$

In particular, since $d>1$ divides L, it is $d\geq p_1$, and so

$$f(x,y)\leq 1/p_1+1/L=f(p_1,L),\text{ and } (p_1,L)\in B.$$

In other words, $\max f|B=1/p_1+1/L$. To finish we need only compare both maxima.

Case 1. $\alpha_1=1$. Then $x_0=p_1$ and so

$$\max f|A = 1/x_0+x_0/L=1/p_1+p_1/L>1/p_1+1/L = \max f|B.$$

Thus, if $\alpha_1=1$ and $t>1$, we obtain

$$\delta(R_L')=\max f|A = 1/p_1+p_1/L.$$

Case 2. $\alpha_1>1$ or $L=p_1$. If $L=p_1$ it is $t=1$ and we have already proved that $\delta(R_L')=1/p_1+1/L$.

Suppose now $\alpha_1>1$ (and $t\geq 2$). Then $x_0\geq p_1+1$ and $L\geq p_1^2p_2^{\alpha_2}\geq p_1^2x_0$. Hence

$$\max f|B - \max f|A = 1/p_1+1/L-(1/x_0+x_0/L)=(1/L)[L/p_1-x_0+1-L/x_0].$$

But

$$L/p_1-x_0+1-L/x_0=(x_0-p_1)L/x_0p_1-x_0+1=(x_0-p_1-1)L/x_0p_1-x_0+1+p_1+L/x_0p_1-p_1=$$
$$=(x_0-p_1-1)[L/x_0p_1-1]+L/x_0p_1-p_1\geq L/x_0p_1-p_1\geq 0,$$

the inequalities because $x_0\geq p_1+1$ and $L\geq p_1^2x_0$. Whence, in this case we get $\delta(R_L')=\max f|B=1/p_1+1/L$. This completes the proof.

Our study splits naturally into several parts. Through all this section when we start to deal with an NEC group Λ it will be assumed to have a signature τ of the general form *i.e.*

$$\tau=(g;\pm;[m_1,...,m_r];\{(n_{i1},...,n_{is_i})|1\leq i\leq k\}).$$

Theorem 3.2.5. *Let* N *be a nonprime odd integer and* $N=p_1^{\alpha_1}\dots p_t^{\alpha_t}$ *for some prime numbers* $p_1<\dots<p_t$. *Then*

$$p_+^+(N)=p_+(N)=\begin{cases}(p_1-1)N/p_1 & \text{if } \alpha_1>1,\\ (p_1-1)(N/p_1-1) & \text{if } \alpha_1=1,\end{cases}$$

and the corresponding N*-minimal surface has* 1 *boundary component.*

Proof. By (1) in 3.2.3 $p_+^+(N)=p_+(N)$. Denote the right hand side of the formula by p and let us write $q=\begin{cases}N & \text{if } \alpha_1>1,\\ N/p_1 & \text{if } \alpha_1=1.\end{cases}$

Let $Z_N=\Lambda/\Gamma$, where Γ is an orientable bordered surface group and Λ is an NEC group with signature τ. We shall show that $\bar\mu(\tau)\geq\mu_0=(1-1/p_1-1/q)$ what in virtue of Hurwitz Riemann formula will show that $p_+(N)\geq p$. Since Γ has a period-cycle, $k\geq1$, and since N is odd and all its period-cycles are empty, the same holds true for τ by 2.3.1, and in addition signτ="+" by 2.1.2. Clearly $r>0$ since otherwise $\bar\mu(\tau)$ would be an integer and so greater than μ_0. If $g>0$ or $k>1$ then $\bar\mu(\tau)\geq 1-1/p_1>\mu_0$. So we can assume that $g=0$ and $k=1$. Now if $r\geq3$ then $\bar\mu(\tau)\geq1>\mu_0$ again, whilst if $r=1$ then $\bar\mu(\tau)<0$ and so τ is not an NEC signature. Consequently we can assume that $\tau=(0;+;[m_1,m_2];\{(-)\})$. But then by 3.1.2 *l.c.m.*$(m_1,m_2)=N$ and so by 3.2.4 $1/m_1+1/m_2\leq1/p_1+1/q$. Thus $\bar\mu(\tau)\geq\mu_0$ and if $\bar\mu(\tau)=\mu_0$ then

$$\tau=\tau_0=(0;+;[p_1,q];\{(-)\}).$$

Conversely it is a routine to check, using 3.1.2, that the signature

$$\sigma_0=(p/2;+;[-];\{(-)\})$$

is the only one with sign "+" for which (σ_0,τ_0) is an N-pair. Therefore $p_+(N)\leq p$ and also the last assertion is proved.

Comment. If N is odd then, by (1) in 3.2.3, $\mathcal{K}_+^-(N)$ is empty. In that case we write $p_+^-(N)=\infty$.

We shall compute now $p_+^+(N)$ and $p_+^-(N)$ for even N distinguishing between two possibilities: N is a multiple of 4 or not.

Theorem 3.2.6. *Let* $N\neq2$ *be even and* $N\notin4\mathbb{Z}$. *Then* $p_+^+(N)=p_+^-(N)=N/2-1$. *Moreover, any* N*-minimal surface* S *from* $\mathcal{K}_+^+(N)$ *has* 1 *boundary component whilst any such surface from* $\mathcal{K}_+^-(N)$ *has* N/2 *boundary components.*

Proof. Let $X=H/\Gamma$ be an orientable surface of algebraic genus p and let f be an automorphism of X of order N. Then $\langle f\rangle\cong Z_N=\Lambda/\Gamma$, for some NEC group Λ with signature τ. We shall show that $\bar\mu(\tau)\geq\mu_0=1/2-2/N$. In fact since Γ has a period-cycle, $k\geq1$. Assume first that all period-cycles of τ are empty. Then we can assume that $r>0$ since otherwise $\bar{\mu}(\tau)$ would be an integer. If $g>0$ or $k>1$ then $\bar{\mu}(\tau)\geq 1/2>\mu_0$. Therefore we can go further assuming that $g=0$ and $k=1$. Now if $r>2$ then $\bar{\mu}(\tau)\geq 1/2$ again, whilst if $r=1$ then $\bar{\mu}(\tau)<0$. Therefore $r=2$ and so $\tau=(0;+;[m_1,m_2];\{(-)\})$. But then by 3.1.5 *l.c.m.*$(m_1,m_2)=N$ and so by 3.2.4 we obtain $1/m_1+1/m_2\leq 1/2+2/N$. Thus $\bar{\mu}(\tau)\geq\mu_0$ and if $\bar{\mu}(\tau)=\mu_0$ then

$$\tau=\tau_1=(0;+;[2,N/2];\{(-)\}).$$

Now assume that a period-cycle of τ is nonempty. Then this period-cycle has an even length s. In particular if $g>0$ or $k>1$ then $\bar{\mu}(\tau)>1/2$. Therefore we can assume that $g=0$, $k=1$. But now if $s\geq 6$ then $\bar{\mu}(\tau)\geq 1/2$ whilst if $s=4$ then $\bar{\mu}(\tau)=0$ unless $r>0$; but in the last case $\bar{\mu}(\tau)\geq(1-1/m_1)\geq 1/2$. Thus we can assume that $s=2$. But then, since $\bar{\mu}(\tau)>0$, we have $r>0$ and if $r>1$ then $\bar{\mu}(\tau)\geq 1/2$. So let $\tau=(0;+;[m];\{(2,2)\})$. By 3.1.5 $m=N/2$ and in this case $\bar{\mu}(\tau)=\mu_0$. Let

$$\tau=\tau_2=(0;+;[N/2];\{(2,2)\})$$

be the corresponding signature.

Concluding we have shown that $\bar{\mu}(\tau)\geq\mu_0$ and the equality takes place just for $\tau=\tau_1$ and $\tau=\tau_2$ listed above. By the Hurwitz Riemann formula $p\geq N/2-1$ and so in particular this also holds for $p_+^+(N)$ and $p_+^-(N)$.

Now using 3.1.5 once more one can easily see that

$$\sigma_1=((N-2)/4;+;[-];\{(-)\})$$

is the unique signature with sign "+" for which (σ_1,τ_1) is an N-pair and in this case $C^-(\tau_1)\subseteq\sigma_1$. Thus $p_+^+(N)=N/2-1$ and a corresponding N-minimal surface has 1 boundary component.

Now consider the signature τ_2. Using 3.1.5 again we deduce that

$$\sigma_2=(0;+;[-];\{(-),\overset{N/2}{\ldots},(-)\})$$

is the unique signature for which (σ_2,τ_2) is an N-pair and in this case σ_2 cannot contain $C^-(\tau_2)$. Thus $p_+^-(N)=N/2-1$ and a corresponding N-minimal surface from $\mathcal{K}_+^-(N)$ has N/2 boundary components. This completes the proof.

Theorem 3.2.7. *Let* $N\in 4\mathbb{Z}$. *Then* $p_+^+(N)=N/2$ *and* $p_+^-(N)=N/2+1$. *Moreover any N-minimal surface S from* $\mathcal{K}_+^+(N)$ *has 1 boundary component and any such surface from* $\mathcal{K}_+^-(N)$ *has 2 if* $N\in 8\mathbb{Z}$ *and 4 boundary components otherwise.*

Proof. Let $X=H/\Gamma$ be an orientable surface of algebraic genus p and let f be an automorphism of X of order N. Then $<f>\cong Z_N=\Lambda/\Gamma$, for some NEC group Λ with signature τ. Assume first that f preserves orientation. Then by 3.2.1 $C^-(\Lambda)\subseteq\Gamma$. We shall show that $\bar{\mu}(\tau)\geq\mu_0=1/2-1/N$. In fact $k\geq 1$ and since Γ has only empty period-cycles the same holds true for τ by 2.3.2. Clearly $r>0$ since otherwise $\bar{\mu}(\tau)$ would be an integer, and so, if $g>0$ or $k>1$, then $\bar{\mu}(\tau)\geq 1/2>\mu_0$. Therefore we can assume that $g=0$ and $k=1$. Now if $r>2$ then again $\bar{\mu}(\tau)\geq 1/2$ whilst if $r=1$ then $\bar{\mu}(\tau)<0$. Hence $r=2$ and so we have to consider the case when $\tau=(0;+;[m_1,m_2];\{(-)\})$. But then by 3.1.5, *l.c.m.*$(m_1,m_2)=N$ and so by 3.2.4

$1/m_1+1/m_2 \leq 1/2+1/N$. Thus $\bar{\mu}(\tau)\geq\mu_0$ and if $\bar{\mu}(\tau)=\mu_0$ then

$$\tau=\tau_0=(0;+;[2,N];\{(-)\}).$$

In particular $p_+^+(N)\geq N/2$ by the Hurwitz Riemann formula.

Now using once more 3.1.5 it is easy to check that

$$\sigma_0=(N/4;+;[-];\{(-)\})$$

is the only NEC signature for which (σ_0,τ_0) is an N-pair. This leads us to conclude that $p_+^+(N)=N/2$ and any N-minimal surface from $\mathcal{K}_+^+(N)$ has 1 boundary component.

Now assume that f does not preserve orientation. This time we shall show that $\bar{\mu}(\tau)\geq 1/2$. By 3.2.1 Γ does not contain $C^-(\Lambda)$. By (4) in 3.2.3 signτ="-". If $g\geq 2$ or $k\geq 2$ then $\bar{\mu}(\tau)\geq 1$. So $g=1$ and $k=1$. If the period-cycle of τ is nonempty then $\bar{\mu}(\tau)\geq 1/2$ and if $\bar{\mu}(\tau)=1/2$ then $\tau=(1;-;[-];\{(2,2)\})$. But this case is impossible by (3.5) in 3.1.9. So assume that this period-cycle is empty. Then again $\bar{\mu}(\tau)\geq 1/2$ and this time equality takes place for

$$\tau=\tau_0=(1;-;[2];\{(-)\}).$$

We claim that the signature

$$\sigma_0=(g(N);+;[-];\{(-),\overset{k(N)}{\ldots},(-)\}),$$

where

$$k(N)=\begin{cases} 2 \text{ if } N\in 8\mathbb{Z}, \\ 4 \text{ otherwise}, \end{cases} \quad \text{and} \quad g(N)=\begin{cases} N/4 & \text{if } N\in 8\mathbb{Z}, \\ N/4-1 & \text{otherwise}, \end{cases}$$

is the only surface NEC signature with sign "+" for which (σ_0,τ_0) is an N-pair. In fact we are in case of 3.1.9. Let $l=N/k(N)$ be the integer given by (3.1) there. Then by (3.6) $l.c.m.(2,l)=N/2$. Now if $N\in 8\mathbb{Z}$ then $l=N/2$ and so $k(N)=2$. By the Hurwitz Riemann formula $g(N)=N/4$. So assume that $N\notin 8\mathbb{Z}$. Then $l=N/2$ or $l=N/4$. We claim that the former is impossible. In fact assume to a contrary that $l=N/2$. By (3.4) there exists an order preserving pair (α,β) with respect to $(N/2,N/l=2)$ and $S(\alpha,\beta)$ is not an multiple of 4. In this special case $\alpha=\{\alpha_0\}$, $\beta=\{\beta_0\}$, where α_0 and β_0 are odd integers. But then

$$S(\alpha,\beta)=(\alpha_0 N/2 + 2\beta_0)=2(\alpha_0 N/4 + \beta_0).$$

Since $N\notin 8\mathbb{Z}$, N/4 is odd and therefore $\alpha_0 N/4 + \beta_0$ is even as the sum of two odds. So $S(\alpha,\beta)$ is a multiple of 4, a contradiction. Concluding, if $N\notin 8\mathbb{Z}$ then $l=N/4$ and so $k(N)=4$. This implies $g(N)=N/4-1$ and the proof is complete.

We continue our program by computing $p_-(N)$ and the topological structure of the corresponding N-minimal surfaces from $\mathcal{K}_-(N)$ for nonprime odd N.

Theorem 3.2.8. *Let* N *be a nonprime odd integer and let* $N=p_1^{\alpha_1}\ldots p_t^{\alpha_t}$, *for prime numbers* $p_1<\ldots<p_t$. *Then* $p_-(N)=(p_1-1)N/p_1+1$, *and any* N*-minimal surface from* $\mathcal{K}_-(N)$ *has* 1 *if* $\alpha_1>1$ *and* 1 *or* p_1 *if* $\alpha_1=1$ *boundary components. In case* $\alpha_1=1$, *both possibilities actually occur.*

Proof. Let $X=H/\Gamma$ be a nonorientable surface of algebraic genus $p\geq 2$ and let f be an automorphism of X of order N. Then $<f> \cong Z_N=\Lambda/\Gamma$, for some NEC group Λ with signature τ. We shall show that $\bar{\mu}(\tau) \geq \mu_0 = 1 - 1/p_1$. Since N is odd and X is nonorientable, signτ="-" by 2.1.2. Moreover $k\geq 1$ and all period-cycles of τ are empty by 2.3.1. If $r=0$ then $\bar{\mu}(\tau)$ is an integer whilst if $g>1$ or $k>1$ or else $r\geq 2$ then $\bar{\mu}(\tau)\geq 1$. Therefore we can assume that $g=1$, $k=1$ and $r=1$ *i.e.* $\tau=(1;-;[m];\{(-)\})$. But then $\bar{\mu}(\tau)\geq\mu_0$ and if $\bar{\mu}(\tau)=\mu_0$ then

$$\tau=\tau_0=(1;-;[p_1];\{(-)\}).$$

In particular $p_-(N)\geq(p_1-1)N/p_1 + 1$ by the Hurwitz Riemann formula.

Now let $\sigma=(g(N);-;[-];\{(-),\overset{k(N)}{\ldots},(-)\})$ be a bordered surface signature for which (σ,τ_0) is an N-pair. We shall compute k(N). By 3.1.3 $k(N)=N/l$ for some l for which $l.c.m.(p_1,l)=N$. We see that for $\alpha_1>1$, $l=N$ and so $k(N)=1$ whilst for $\alpha_1=1$, either $l=N$ and $k(N)=1$ or $l=N/p_1$ and $k(N)=p_1$. Finally notice that both possibilities actually occur for suitable g(N) (determined by k(N)) since the conditions listed in 3.1.3 are also sufficient for (σ,τ_0) to be an N-pair. This completes the proof.

Theorem 3.2.9. *Let* $N\neq 2$ *be even. Then* $p_-(N)=N/2$ *and any N-minimal surface from* $\mathscr{K}_-(N)$ *has* N/2 *boundary components.*

Proof. Let Λ and Γ be as in the previous theorem. Here we shall show that $\bar{\mu}(\tau)\geq\mu_0=1/2-1/N$. For, assume first that signτ="-". If $k\geq 2$ or $g\geq 2$ then $\bar{\mu}(\tau)\geq 1$. So we can assume that $g=k=1$. If $r\geq 2$ or $s\geq 2$ then $\bar{\mu}(\tau)\geq 1/2$, and $\bar{\mu}(\tau)=0$ otherwise. Now let signτ="+". If $g\neq 0$ then, since $k\geq 1$, $\bar{\mu}(\tau)\geq 1$. So take $g=0$ and $k\leq 2$ since otherwise $\bar{\mu}(\tau)\geq 1$ again. But if $k=2$ then either a period-cycle is nonempty or $r>0$ and in these cases $\bar{\mu}(\tau)\geq 1/2$. Thus we can assume that $k=1$. Since Γ is nonorientable $s\neq 0$ by 2.1.3 (and consequently $s\geq 2$ since s is even). Now if $r\geq 2$ or $s\geq 4$ then $\bar{\mu}(\tau)\geq 1/2$. So $s=2$ and $r=1$ since for $r=0$ $\bar{\mu}(\tau)<0$. Therefore we can assume that $\tau=(0;-;[m];\{(2,2)\})$. But then $\bar{\mu}(\tau)=1/2-1/m\geq\mu_0$ and $\bar{\mu}(\tau)=\mu_0$ only for

$$\tau=\tau_0=(0;+;[N];\{(2,2)\}).$$

In particular $p_-(N)\geq N/2$ by the Hurwitz Riemann formula.

Now using theorem 3.1.6 we easily see that

$$\sigma_0=(1;-;[-];\{(-),\overset{N/2}{\ldots},(-)\})$$

is the unique signature of nonorientable surface NEC group for which (σ_0,τ_0) is an N-pair. This completes the proof.

Now let us fix two integers $k\geq 1$, $N\geq 2$. We define the following sets

$$\mathscr{K}_+(N,k)=\{S\in\mathscr{K}_+(N) \mid k(S)=k\}, \quad \mathscr{K}_-(N,k)=\{S\in\mathscr{K}_-(N) \mid k(S)=k\},$$

$$\mathscr{K}_+^+(N,k)=\{S\in\mathscr{K}_+^+(N)\mid k(S)=k\},\quad \mathscr{K}_+^-(N,k)=\{S\in\mathscr{K}_+^-(N)\mid k(S)=k\},$$

and the numbers

$$p_+(N,k)=\min\{p(S)\mid S\in\mathscr{K}_+(N,k)\},\quad p_-(N,k)=\min\{p(S)\mid S\in\mathscr{K}_-(N,k)\},$$

$$p_+^+(N,k)=\min\{p(S)\mid S\in\mathscr{K}_+^+(N,k)\},\quad p_+^-(N,k)=\min\{p(S)\mid S\in\mathscr{K}_+^-(N,k)\},$$

$$p(N,k)=\min\{p_+(N,k),p_-(N,k)\}.$$

Comment. We shall compute now these numbers for a prime N. For, we shall need to find the minimum number ρ of summands needed to express k as a sum of positive divisors (possibly equal to 1) of N. In case of prime N this decomposition is forced to be $k=QN+R=N+\overset{Q}{\ldots}+N+1+\overset{R}{\ldots}+1$, for some R in range $0\le R<N$. For arbitrary N there is not a such effective procedure. Of course, having the above numbers computed we shall have

$$p_+(N)=\min\{p_+(N,k)\mid k\ge 1\},\quad p_-(N)=\min\{p_-(N,k)\mid k\ge 1\},$$

$$p_+^+(N)=\min\{p_+^+(N,k)\mid k\ge 1\},\quad p_+^-(N)=\min\{p_+^-(N,k)\mid k\ge 1\},$$

and consequently we shall have solved the minimum algebraic genus problem for arbitrary N.

We shall need the following elementary

Lemma 3.2.10. *Let* $k\ge 1$, $N\ge 2$ *be integers, and* $k=QN+R$ *its euclidean division. If* $k=Q'N+R'$ *for non negative integers* Q' *and* R' *then* $Q+R\le Q'+R'$. *Moreover, if* $Q+R<Q'+R'$ *then* $(Q'+R')-(Q+R)\ge N-1$ *whilst if* $Q+R=Q'+R'$ *then* $Q=Q'$, $R=R'$.

Theorem 3.2.11. *Let* N *be an odd prime, and let* $k\ge 1$ *be an integer with decomposition* $k=QN+R$, *where* $0\le R<N$. *Then*

$$p_+^+(N,k)=p_+(N,k)=\begin{cases} N(Q+R-2)+1 & \text{if } Q+R\ge 3,\ R\ge 2,\\ N(Q+R-1) & \text{if } R=1,\ Q\ge 1 \text{ or } R=2,\ Q=0,\\ N(Q+R)-1 & \text{if } R=0,\ Q\ge 1 \text{ or } R=1,\ Q=0.\end{cases}$$

Proof. We start explaining in few words the strategy employed here and in the proofs of the next theorems. Let $S=H/\Gamma$ be a Klein surface of algebraic genus $p\ge 2$ and k $(k\ge 1)$ boundary components, having an automorphism of order N. Then there exists an NEC group Λ with signature τ containing the bordered surface NEC group Γ with signature σ as a normal subgroup with cyclic quotient Λ/Γ of order N. First we shall look for the lower bound $\mu_0(N,k)=\mu_0$ for $\bar{\mu}(\tau)$ that shall provide such a bound for p and consequently for $p_+^+(N,k)$ and $p_-(N,k)$. It happens that in all studied cases this lower bound is attained just for one signature $\tau=\tau_0$. Using results of the previous section we shall be able then to find a surface NEC signature σ_0 with k period-cycles for which (σ_0,τ_0) is an N-pair what will complete the proofs.

Notice first that since N is odd and in our case signσ="+" then also signτ="+" and all period-cycles of τ are empty. Moreover since N is prime, all proper periods of τ (if any) are equal to N, *i.e.*

$$\tau=(g;+;[N,\overset{r}{\dots},N];\{(-),\overset{k'}{\dots},(-)\}).$$

As a consequence of 3.1.2, $k=NQ'+(k'-Q')$ for some Q' in range $0\le Q'\le k'$ since N is prime.

Having made these preparations we start the proper proof. First notice that $p_+^+(N,k)=p_+(N,k)$ by 3.2.3. As the form of the statement suggests, the proof will be divided into certain parts.

Let $R\ge 2$ and $Q+R\ge 3$. We shall prove first that $\bar{\mu}(\tau)\ge\mu_0=Q+R-2$. By definition, $k=Q'N+(k'-Q')$. From 3.2.10 $k'\ge Q+R$ and so $\bar{\mu}(\tau)\ge k'-2\ge\mu_0$ as desired, and clearly the equality takes place just for

$$\tau=\tau_0=(0;+;[-];\{(-),\overset{Q+R}{\dots},(-)\}).$$

By the Hurwitz Riemann formula $p\ge N(Q+R-2)+1$ and in particular the same holds true for $p_+(N,k)$ and $p_+^+(N,k)$.

Conversely this bound is actually attained. For, it is enough to check that if $\sigma_0=((N-1)(R-2)/2;+;[-];\{(-),\overset{k}{\dots},(-)\})$, then the pair (σ_0,τ_0) satisfies the conditions of theorem 3.1.2 with $l_1=\dots=l_Q=1$, $l_{Q+1}=\dots=l_{Q+R}=N$. So it is an N-pair and $p_+^+(N,k)=p_+(N,k)=N(Q+R-2)+1$.

Now let $R=1$, $Q\ge 1$. We shall prove that $\bar{\mu}(\tau)\ge\mu_0=Q-1/N$. As in the previous case we deduce that $k'\ge Q+1$. If $k'>Q+1$ then $\bar{\mu}(\tau)\ge k'-2>\mu_0$ whilst if $k'=Q+1$ then by 3.2.10, $Q=Q'$ and $k'-Q'=R=1$. If $g>0$ then $\bar{\mu}(\tau)>Q+1>\mu_0$. So $g=0$ and we claim that $r>0$. In fact by (4) in 3.1.2, the set $\{N,\overset{r+1}{\dots},N,1,\overset{Q}{\dots},1\}$ has the elimination property. But then $\bar{\mu}(\tau)\ge\mu_0$ and the equality takes place just for

$$\tau=\tau_0=(0;+;[N];\{(-),\overset{Q+1}{\dots},(-)\}).$$

By the Hurwitz Riemann formula $p\ge NQ=N(Q+R-1)$ and in particular the same holds true for $p_+(N,k)$ and $p_+^+(N,k)$.

Conversely let $\sigma_0=(0;+;[-];\{(-),\overset{k}{\dots},(-)\})$. Then choosing $l_1=\dots=l_Q=1$, $l_{Q+1}=N$ in theorem 3.1.2, we conclude that (σ_0,τ_0) is an N-pair. Thus $p_+^+(N,k)=p_+(N,k)=N(Q+R-1)$.

Now let $R=2$, $Q=0$ (*i.e.* $k=2$). We shall show that $\bar{\mu}(\tau)\ge\mu_0=1-1/N$. For, notice first that $Q'=0$ since otherwise Γ would have at least $Q'N\ge N$ period-cycles. But then $k'=2$. Now if $g>0$ or $r\ge 2$ then $\bar{\mu}(\tau)\ge 1$ whilst if $g=0$ and $r=0$ then $\bar{\mu}(\tau)=0$. So $g=0$ and $r=1$ *i.e.*

$$\tau=\tau_0=(0;+;[N];\{(-),(-)\}).$$

We deduce from the Hurwitz Riemann formula that $p_+^+(N,k)$, $p_+(N,k)\ge N$. Moreover we easily deduce, using 3.1.2, that for $\sigma_0=((N-1)/2;+;[-];\{(-),(-)\})$, (σ_0,τ_0) is an N-pair and consequently $p_+^+(N,k)=p_+(N,k)=N$. Finally notice that in our case $N=N(Q+R-1)$ and thus the formula obtained in the previous case also covers this one.

Let R=0, Q≥1. We shall see that $\bar{\mu}(\tau)\geq\mu_0$=Q-2/N what as before shall imply that p_+(N,k)≥NQ-1. By lemma 3.2.10 k'≥Q and if k'>Q then k'-Q≥N-1 and consequently $\bar{\mu}(\tau)\geq$k'-2≥N+Q-3≥Q. Thus assume that k'=Q. But then by the second part of 3.2.10, Q'=Q and k'-Q'=R=0 *i.e.* k'=Q. If g>0 then $\bar{\mu}(\tau)\geq$Q and so we can assume that g=0. Now condition (4) in 3.1.2 says that the set $\{N,\overset{r}{\ldots},N,1,\overset{Q}{\ldots},1\}$ has the elimination property, whilst (3.2) in 3.1.2 gives *l.c.m.*$(N,\overset{r}{\ldots},N)$=N. Hence r≥2. In fact we can assume that r=2 since for r≥3, $\bar{\mu}(\tau)$>Q-2/N. Therefore

$$\tau=(0;+;[N,N];\{(-),\overset{Q}{\ldots},(-)\})$$

and $\bar{\mu}(\tau)=\mu_0$. Arguing as before the reader can easily complete the proof of this case.

So it remains to consider only the case R=1, Q=0, *i.e.* k=1. We shall show that $\bar{\mu}(\tau)\geq\mu_0$=1-2/N. In fact r>0 since otherwise $\bar{\mu}(\tau)$ would be an integer. If g>0 or k'≥2 then obviously $\bar{\mu}(\tau)\geq$1-1/N. So g=0 and k'=1. But then r≥2 since otherwise $\bar{\mu}(\tau)$<0. If r≥3 then $\bar{\mu}(\tau)\geq$1 and so we can assume that r=2. Then $\tau=\tau_0$=(0;+;[N,N];{(-)}) and $\bar{\mu}(\tau)$=1-2/N. By the Hurwitz Riemann formula p_+(N,k)≥N-1.

Conversely let σ_0=((N-1)/2;+;[-];{(-)}). Then choosing in 3.1.2 l_1=N we see that (σ_0,τ_0) is an N-pair and so p_+(N,k)=N-1. Finally notice that in our case N-1=N(Q+R)-1 and so the formula obtained in the previous case covers also this one.

The nonorientable case turns out to be simpler.

Theorem 3.2.12. *Let* N *be an odd prime and let* k≥1 *be an integer decomposed as* k=QN+R, *where* 0≤R<N. *Then*

$$p_-(N,k)=\begin{cases} N(Q+R-1)+1 & \text{if } R\geq 1,\ Q+R\geq 2,\\ N(Q+R) & \text{otherwise}.\end{cases}$$

Proof. Save the notations introduced in the proof of the previous theorem. Now since N is odd and signσ="-" also signτ="-".

First let R≥1 and Q+R≥2. By 3.2.10, k'=Q'+(k'-Q')≥Q+R and so $\bar{\mu}(\tau)\geq$k'-1 ≥ ≥Q+R-1. So by the Hurwitz Riemann formula p≥N(Q+R-1)+1 and in particular p_-(N,k)≥N(Q+R-1)+1.

Now let

$$\tau_0 = (1;-;[-];\{(-),\overset{Q+R}{\ldots},(-)\}),$$
$$\sigma_0 = (N(Q+R-1)-k+2;-;[-];\{(-),\overset{k}{\ldots},(-)\}).$$

Then choosing $l_1=\ldots=l_Q=1$, $l_{Q+1}=\ldots=l_{Q+R}$=N in 3.1.3 we obtain that (σ_0,τ_0) is an N-pair. Therefore p_-(N,k)=N(Q+R-1)+1.

Let R=0, Q≥1. We shall show that $\bar{\mu}(\tau) \geq Q-1/N$ what, as in the previous case, implies $p_-(N) \geq QN$. By 3.2.10, $k'=Q'+(k'-Q') \geq Q+R=Q$ and if $k'>Q$ then $k' \geq$ $\geq Q+N-1$ and so $\bar{\mu}(\tau) \geq k'-1 \geq Q+N-2>Q$. Therefore we can assume that $k'=Q$. But then using again 3.2.10 $k'-Q'=0$ *i.e.* $k'=Q'$. Moreover we can assume that $g=1$ since otherwise $\bar{\mu}(\tau) \geq Q+g-2 \geq Q$. We claim that $r=1$. In fact, if $r>1$, then $\bar{\mu}(\tau) \geq Q$. On the other hand, if $r=0$ we get by condition (4) in 3.1.3, *l.c.m.*$(l_1,\dots,l_Q)=N$ for some divisors $l_1,\dots,l_Q$ of N, and so from (3) in 3.1.3, $k=(N/l_1+\dots+N/l_Q)<QN$, a contradiction. Thus

$$\tau=\tau_0=(1;-;[N];\{(-),\overset{Q}{\dots},(-)\})$$

and $\bar{\mu}(\tau)=Q-1/N$. Now let

$$\sigma_0=(1;-;[-];\{(-),\overset{k}{\dots},(-)\}).$$

Then by 3.1.3 we deduce that (σ_0,τ_0) is an N-pair and so $p_-(N,k)=QN=N(Q+R)$.

Finally let Q=0, R=1, *i.e.* k=1. We shall show that $\bar{\mu}(\tau) \geq \mu_0=1-1/N$. Also now $r>0$ because otherwise $\bar{\mu}(\tau)$ is an integer. But then $\bar{\mu}(\tau) \geq \mu_0$ and if $\bar{\mu}(\tau)=\mu_0$ then

$$\tau=\tau_0=(1;-;[N];\{(-)\}).$$

By the Hurwitz Riemann formula $p_-(N) \geq N$ and to finish the proof it is enough to check that for $\sigma_0=(N;-;[-];\{(-)\})$, (σ_0,τ_0) verifies 3.1.3 and so it is an N-pair.

We found above $p_+(N)$, $p_-(N)$, $p(N)$, $p^+_+(N)$ and $p^-_+(N)$ for nonprime N. We obtain these values for odd prime N simply minimizing $p^+_+(N,k)$, $p^-_+(N,k)$ and $p_-(N,k)$ for k running over $\mathbb{N}$.

Corollary 3.2.13. *Let* N *be an odd prime. Then* $p^+_+(N)=p_+(N)=N-1$ *and* $p_-(N)=N$. *Moreover any* N*-minimal surface from* $\mathcal{K}^+_+(N)$, $\mathcal{K}_+(N)$ *or* $\mathcal{K}_-(N)$ *has* 1 *or* N *boundary components and both possibilities actually occur both in orientable and nonorientable cases.*

Proof. From 3.2.11 $p_+(N,1)=p^+_+(N,1)=p_+(N,N)=p^+_+(N,N)=N-1$ and from 3.2.12 $p_-(N,1)=p_-(N,N)=N$. The values k=1 and k=N correspond to Q=0, R=1 and Q=1, R=0, respectively *i.e.* for Q+R=1. We leave to the reader to inspect the respective formulas in 3.2.11 and 3.2.12 to see that for Q+R>1, $p_+(N,k)>N-1$ and $p_-(N,k)>N$, what will complete the proof.

It remains to find $p^+_+(2,k)$, $p^-_+(2,k)$ and $p_-(2,k)$.

Theorem 3.2.14. *Let* $k \geq 1$ *be an integer. Then*

$$p^+_+(2,k)=p^-_+(2,k)=p_+(2,k)=\begin{cases} k-1 & \textit{if } k \geq 3 \\ k+1 & \textit{otherwise} \end{cases}$$

In particular $p^{+}_{+}(2)=p^{-}_{+}(2)=p_{+}(2)=2$ *and any corresponding 2-minimal surface from* $\mathcal{K}^{+}_{+}(2)$ *and* $\mathcal{K}^{-}_{+}(2)$ *has* 1 *or* 3 *boundary components and both cases actually occur in orientation preserving and orientation reversing cases.*

Proof. The second statement is an obvious consequence of the first one. To see the first denote the right hand side of the formula above by p.

Observe that the algebraic genus p(S) of an orientable surface S having k=k(S) boundary components and an automorphism of order 2 is $\geq$p. In fact for $k\geq 3$, $p(S)=2g(S)+k(S)-1\geq k(S)-1=k-1=p$.

We have $p(S)\geq 2$. So if $k=k(S)=1$ then $p(S)\geq 2=k+1=p$, whilst if $k=k(S)=2$ then $p(S)\geq 3=k+1$ again. So $p^{+}_{+}(2,k)\geq p$, $p^{-}_{+}(2,k)\geq p$ and $p_{+}(2,k)\geq p$ and it remains only to prove that $p^{+}_{+}(2,k)\leq p$ and $p^{-}_{+}(2,k)\leq p$.

For consider the following surface NEC signatures:

$$\sigma_1 = (0;+;[-];\{(-),\overset{k}{\ldots},(-)\}) \text{ for } k\geq 3,$$
$$\sigma_2 = (1;+;[-];\{(-),\overset{k}{\ldots},(-)\}) \text{ for } k=1,2.$$

Let k=2Q+R, where $0\leq R\leq 1$ and consider the NEC signatures

$$\tau_1 = (0;+;[2,\overset{2-R}{\ldots},2];\{(-),\overset{Q+R}{\ldots},(-)\}) \text{ for } k\geq 3,$$
$$\tau_2 = (0;+;[2,\overset{k+2}{\ldots},2];\{(-)\}) \text{ for } k=1,2.$$

We claim that (σ_1,τ_1) and (σ_2,τ_2) are 2-pairs. In fact it is enough to check that for

$$p'=k'=Q+R,\ l_1=\ldots=l_Q=1,\ l_{Q+1}=\ldots=l_{Q+R}=2 \text{ for } k\geq 3,$$
$$p'=k'=1,\ l_1=1 \text{ for } k=2 \text{ and } p'=k'=1,\ l_1=2 \text{ for } k=1,$$

the conditions in 3.1.5 are fulfilled.

Finally using (3) in 3.2.3 we deduce that $p^{+}_{+}(2,k)\leq\begin{cases} k-1 & \text{for } k\geq 3\\ k+1 & \text{for } k=1,2\end{cases}$

i.e. $p^{+}_{+}(2,k)\leq p$.

Now we shall see that $p^{-}_{+}(2,k)\leq p$. Consider the NEC signatures

$$\tau_1=\begin{cases} (0;+;[-];\{(-),\overset{Q}{\ldots}(-),(2,2)\}) & \text{for } k\geq 3,\ R=1\\ (0;+;[-];\{(-),\overset{Q+1}{\ldots}(-)\}) & \text{for } k\geq 3,\ R=0\end{cases}$$

and

$$\tau_2=\begin{cases} (0;+;[-];\{(-),(2,2)\}) & \text{if } k=1\\ (0;+;[-];\{(2,2),(2,2)\}) & \text{if } k=2.\end{cases}$$

We claim that for σ_1 and σ_2 defined in the proof of the equality $p^{+}_{+}(2,k)=p$, (σ_1,τ_1) and (σ_2,τ_2) are 2-pairs. For, it is enough to check that conditions of 3.1.5 are fulfilled for

$$k'=Q+1,\ p'=Q,\ l_1=\ldots=l_Q=1, \text{ for } k\geq 3 \text{ and}$$
$$k'=2,\ p'=0 \qquad \text{for } k=1,2.$$

Finally p′ and k′ were so chosen that p′<k′ and so applying (3) in 3.2.3 we obtain

$$p^{-}_{+}(2,k)=\begin{cases} k-1 & \text{for } k\geq 3,\\ k+1 & \text{for } k=1,2.\end{cases}$$

The proof is complete.

Finally in a similar way as in the theorem above we prove the following

Theorem 3.2.15. *Let* $k \geq 1$ *be an integer. Then*

$$p_-(2,k) = \begin{cases} 2 & \text{if } k=1 \\ k & \text{if } k>1. \end{cases}$$

In particular $p_-(2)=2$, *and any 2-minimal surface from* $\mathscr{K}_-(2)$ *has 1 or 2 boundary components and both possibilities actually occur.*

Proof. The second part is obvious. Also if $S \in \mathscr{K}_-(2,k)$, then

$$p(S)=g(S)-1+k(S) \geq k(S)=k, \quad \text{and} \quad p(S) \geq 2.$$

This proves the trivial inequality

$$p_-(2,k) \geq \begin{cases} 2 & \text{if } k=1, \\ k & \text{if } k>1. \end{cases}$$

To see the converse let $k=2Q+R$ where $0 \leq R \leq 1$ and consider the NEC signatures

$$\sigma_1=(2;-;[-];\{(-)\})$$
$$\sigma_2=(1;-;[-];\{(-),\overset{k}{\ldots},(-)\}) \text{ if } k \geq 2.$$
$$\tau_1=(0;+;[2];\{(-),(-)\})$$
$$\tau_2=(0;+;[-];\{(-),\overset{Q}{\ldots},(-),(2,\overset{2(R+1)}{\ldots},2)\}) \text{ if } k \geq 2.$$

Both (σ_1,τ_1) and (σ_2,τ_2) are 2-pairs since they verify conditions in 3.1.6 with:

$$p'=1,\ l_1=2 \quad \text{if } k=1,$$
$$p'=Q,\ l_1=\ldots=l_{Q-1}=1,\ l_Q=2 \text{ if } k \geq 2.$$

This, in virtue of the Hurwitz Riemann formula, proves the result.

Given an integer N denote by $k_+^+(N)$, $k_+^-(N)$, $k_+(N)$ and $k_-(N)$ the number (not necessarily unique!) of connected components of the boundary of N-minimal surfaces from $\mathscr{K}_+^+(N)$, $\mathscr{K}_+^-(N)$, $\mathscr{K}_+(N)$ and $\mathscr{K}_-(N)$. For the reader's and our later convenience we collect results concerning "minimal genus problem" in the form of the following

Corollary 3.2.16. *Let* $N=p_1^{\alpha_1}\ldots p_t^{\alpha_t}$ *be an integer,* $N \geq 2$, *and* $p_1<\ldots<p_t$ *prime numbers. Then*

$$\textit{(i)} \quad p_+^+(N)=p_+(N)=\begin{cases} (p_1-1)((N/p_1)-1) & \alpha_1=1,\ t>1 \\ 2 & N=2 \\ (p_1-1)N/p_1 & \textit{otherwise} \end{cases}$$

(ii) $$p_+^-(N)=\begin{cases} - & N \textit{ is odd}, \\ N/2 - 1 & N \textit{ is even}, \alpha_1=1,\ t\neq 1, \\ N/2 + 1 & \textit{otherwise}, \end{cases}$$

(iii) $$p_-(N)=\begin{cases} (p_1-1)N/p_1+1 \quad N\neq 2, & N \textit{ is odd}, \\ (p_1-1)N/p_1 & N \textit{ is even } N\neq 2, \\ 2 & N=2 \end{cases}$$

(iv) $$p(N)=p_+(N)$$

(v) $$k_+(N)=k_+^+(N)=\begin{cases} 1 & N \textit{ is not prime} \\ 1 \textit{ or } N & N \textit{ is odd prime} \\ 1 \textit{ or } 3 & N=2 \end{cases}$$

(vi) $$k_+^-(N)=\begin{cases} - & N \textit{ is odd} \\ N/2 & N \textit{ is even}, \alpha_1=1,\ t>1 \\ 4 & N \textit{ is even}, \alpha_1=2 \\ 2 & N \textit{ is even}, \alpha_1\geq 3 \\ 1 \textit{ or } 3 & N=2 \end{cases}$$

(vii) $$k_-(N)=\begin{cases} 1 \textit{ or } p_1 & \alpha_1=1,\ p_1>2 \\ 1 & \alpha_1>1,\ p_1>2 \\ N/p_1 & p_1=2,\ \alpha_1+t>1 \\ 1 \textit{ or } 2 & N=2 \end{cases}$$

The results concerning $p_+^+(N,k)$, $p_+^-(N,k)$ and $p_-(N,k)$ *for prime* N can be summarized as follows:

Corollary 3.2.17. *Let* N *be prime and* $k\geq 1$. *Let us divide* $k=QN+R$, $0\leq R<N$. *Then:*

(i) $$p_+^+(N,k)=p_+(N,k)=\begin{cases} N(Q+R-2)+1 & Q+R\geq 3,\quad R\geq 2 \\ N(Q+R-1) & R=1,\ Q\geq 1 \quad \text{or} \quad R=2,\ Q=0 \\ N(Q+R)-1 & R=0,\ Q>1 \quad \text{or} \quad Q+R=1,\quad N\neq 2 \\ NQ+R+1 & Q+R=1,\quad N=2 \end{cases}$$

(ii) $$p_+^-(N,k)=\begin{cases} \infty & N \textit{ is odd} \\ k-1 & N=2,\ k\geq 3 \\ k+1 & N=2,\ k=1 \textit{ or } 2 \end{cases}$$

(iii) $$p_-(N,k)=\begin{cases} N(Q+R-1)+1 & R\geq 1,\quad Q+R\geq 2 \\ N(Q+R) & \textit{otherwise} \end{cases}$$

The so called "maximum order problem" is the following. Given an integer $p\geq 2$ find the maximum order N(p) that an automorphism of compact Klein surfaces of algebraic genus p can have. We finish this chapter with a detailed study of this problem. For each $p\geq 2$ we define

$$N_+^+(p)=\max\{N|\text{there exists } S\in\mathcal{K}_+^+(N) \text{ of algebraic genus } p\},$$

$$N_+^-(p)=\max\{N|\text{there exists } S\in\mathcal{K}_+^-(N) \text{ of algebraic genus } p\},$$

$$N_-(p)=\max\{N|\text{there exists } S\in\mathcal{K}_-(N) \text{ of algebraic genus } p\},$$

$$N_+(p)=\max\{N_+^+(p),N_+^-(p)\},\ N(p)=\max\{N_+(p),N_-(p)\}.$$

We compute these numbers:

Theorem 3.2.18. *Let* $p\geq 2$ *be an integer. Then*

(1) $N_-(p)=2p$.

(2) $N_+^+(p)=\begin{cases} 2(p+1) & \textit{if } p \textit{ is even} \\ 2p & \textit{if } p \textit{ is odd.}\end{cases}$

(3) $N_+^-(p)=\begin{cases} 2(p+1) & \textit{if } p \textit{ is even} \\ 2(p-1) & \textit{if } p \textit{ is odd.}\end{cases}$

(4) In particular,

$$N_+(p)=\begin{cases} 2(p+1) & \textit{if } p \textit{ is even} \\ 2p & \textit{if } p \textit{ is odd,}\end{cases} \quad \textit{and } N(p)=\begin{cases} 2(p+1) & \textit{if } p \textit{ is even} \\ 2p & \textit{if } p \textit{ is odd.}\end{cases}$$

Proof. Part (4) is the obvious consequence of the precedent ones.

(1) From 3.2.16, $p_-(2p)=p$, and so $N_-(p)\geq 2p$. On the other hand, this implies in particular $N_-(p)\neq 2$ and so, by 3.2.16, if p_1 is the smallest prime divisor of $N_-(p)$,

$$p\geq p_-(N_-(p))\geq (p_1-1)N_-(p)/p_1=(1-1/p_1)N_-(p)\geq N_-(p)/2.$$

This proves the converse inequality, $N_-(p)\leq 2p$.

We prove simultaneously (2) and (3). Firstly,

(a) $N_+^+(p)\leq 2(p+1)$ and $N_+^-(p)\leq 2(p+1)$.

In fact, let us denote by N, indistinctly, $N_+^+(p)$ or $N_+^-(p)$. Then, if p_1 is the smallest prime divisor of N, from 3.2.16 we deduce

$$p\geq p_+(N)\geq (p_1-1)(N/p_1-1),\ \textit{i.e.},\ N\leq p_1(p/(p_1-1)+1).$$

All reduces now to show that $p_1(p/(p_1-1)+1)\leq 2(p+1)$. Otherwise $p_1>2$ and $p(2-p_1)/(p_1-1)>2-p_1$. This implies $p<p_1-1$, and so $(p_1-1)((N/p_1)-1)\leq p<(p_1-1)$, *i.e.*, $N=p_1>2$ is prime. Using 3.2.16 once more, since N is an odd prime, $p_+(N)=p_1-1$, and we obtain

$$(p_1-1)>p\geq p_+(N)=p_1-1,\ \text{a contradiction.}$$

(b) Now we show $N_+^+(p)=N_+^-(p)=2(p+1)$ for even p.

In fact, since $p+1$ is odd, we apply 3.2.16 to deduce

$$p_+^+(2(p+1))=p=p_+^-(2(p+1)),\ \text{and so, } N_+^+(p)\geq 2(p+1),\ N_+^-(p)\geq 2(p+1).$$

This together with (a) gives us (b) and so (2) and (3) are proved for even p.

In what follows p is odd.

(c) Let us prove now that $N_+^+(p) \leq 2p$. Let us write $N=N_+^+(p)$. By (a) we know $N \leq 2(p+1)$ and if $N=2(p+1)$ we would have, using 3.2.16,

$$p \geq p_+^+(N)=p_+^+(2(p+1))=p+1, \text{ absurd.}$$

Consequently, all reduces to see that $N \neq 2p+1$.

Let p_1 be the smallest prime divisor of N. Then

$$p \geq p_+^+(N) \geq (p_1-1)((N/p_1)-1)$$

and so, if $N=2p+1$ we obtain

$$N/p_1 \leq (N-1)/2(p_1-1)+1, \ \textit{i.e.}, \ (N-p_1)/p_1 \leq (N-1)/2(p_1-1).$$

Clearing denominators and dividing by p_1^2 we get

$$2((N/p_1)-1)(1-1/p_1) \leq (N/p_1-1/p_1)$$

and so $N/p_1 \leq 2 + 2N/p_1^2 - 3/p_1$, *i.e.* $N/p_1(1-2/p_1) \leq 2 - 3/p_1$. Thus

$$N/p_1 \leq 2+(1/(p_1-2)) \leq 3.$$

Since N is odd, two possibilities can occur:

(i) $N=p_1$; (ii) $N/p_1=p_1=3$.

In the first case $p_+^+(N)=N-1$, and so

$$(N-1)/2=p \geq p_+^+(N)=N-1, \text{ a contradiction.}$$

In the second one $p=4$ is even, absurd. Thus (c) is proved.

(d) The proof of the converse inequality $N_+^+(p) \geq 2p$ is more involved than in the previous cases because $p_+^+(2p)$ is not p, but p-1.

Let us consider NEC signatures

$$\sigma=((p-1)/2;+;[-];\{(-),(-)\}) \text{ and } \tau=(0;+;[2,2p];\{(-)\}).$$

Then (σ,τ) is a 2p-pair since conditions in 3.1.5 are satisfied with $p'=k'=1$ and $l_1=p$. Hence there exist an NEC group Λ with signature τ and an epimorphism $\theta:\Lambda \longrightarrow \mathbb{Z}_{2p}$ whose kernel Γ has signature σ. Consequently, the orientable surface $S=H/\Gamma$ belongs to $\mathcal{K}_+(2p)$. Moreover, since $p'=k'$, we deduce from 3.2.3 that $S \in \mathcal{K}_+^+(2p)$. Also, $p(S)=2g(\sigma)+k(\sigma)-1=p$, and so $N_+^+(p) \geq 2p$.

It only remains to prove $N_+^-(p)=2(p-1)$ for odd p. By 3.2.16,

$$p_+^-(2(p-1))=p, \text{ and so } N_+^-(p) \geq 2(p-1).$$

Let us prove, to finish, the converse inequality. Let us denote $N_+^-(p)$ by N. Since $\mathcal{K}_+^-(N)$ is empty for odd N and by (a) $N_+^-(p) \leq 2(p+1)$, it is enough to prove:

$$\text{(i) } N \neq 2(p+1); \quad \text{(ii) } N \neq 2p.$$

The first is clear because p+1 is even and so $N=2(p+1)$ implies

$$p \geq p_+^-(N)=p_+^-(2(p+1))=p+2, \text{ a contradiction.}$$

Let us suppose $N=2p$. Then there exists a 2p-pair (σ,τ) with

$$\sigma=(g;+;[-];\{(-),\overset{p+1-2g}{\ldots\ldots},(-)\}), \ p+1-2g \neq 0.$$

We can write

$$\tau=(g';\pm;[m_1,\ldots,m_r];\{(-),\overset{s'}{\ldots},(-),(2,\overset{r'_i}{\ldots},2)|\ s'+1\le i\le k'\})$$

and since we are dealing with orientation preserving automorphisms we know by 3.2.3 that either signτ="+" and conditions in 3.1.5 hold true with $p'<k'$ or signτ="-" and conditions in 3.1.9 hold true. We claim:

(1) For every $i=1,\ldots,r$ there exists $b_i\in\mathbb{Z}$ such that $p=b_im_i$.

This is the immediate consequence of (3.4) in 3.1.5 if signτ="+". When signτ="-", then by (3.4) in 3.1.9 there exist $\alpha_1,\ldots,\alpha_r\in\mathbb{Z}$ such that $g.c.d.(\alpha_i,m_i)=1$ and α_i2p/m_i is even for $i=1,\ldots,r$. This implies that each m_i is odd and since it divides $2p$, it must divide p.

With the notations in 3.1.5 and 3.1.9, there exist integers p', $l_1,\ldots,l_{p'}$ verifying conditions listed there. We claim

(2) $s'=k'$, $p'\neq 0$, $r\neq 0$.

In fact, since $\bar\mu(\sigma)=2p\bar\mu(\tau)$ we obtain

$$\text{(3.2.18.1)}\qquad p\text{-}1=\bar\mu(\sigma)=2p[\alpha g'+k'\text{-}2+\sum_{j=1}^{r}(1\text{-}1/m_j)+1/2\sum_{i=s'+1}^{k'}r'_i/2]$$

where $\alpha=1$ if signτ="-" and $\alpha=2$ otherwise.

Substituting $x=\alpha g'+k'\text{-}2+r$, $l=\sum_{i=s'+1}^{k'}r'_i/2$, $m_j=p/b_j$ and $b=\sum_{j=1}^{r}b_j$, all of them are integers and $p\text{-}1=2p[x\text{-}b/p+l/2]$ or, equivalently, $2b\text{-}1=p(2x+l\text{-}1)$. Hence l is even, say $l=2u$ for some $u\in\mathbb{Z}$.

Condition (3.1) in 3.1.5 or 3.1.9 indistinctly, gives us

$$p+1\ge p+1\text{-}2g=\sum_{j=1}^{p'}2p/l_j+pl\ge 2pu.$$

Thus $u=0$, *i.e.*, $l=0$, and this proves $s'=k'$.

Moreover, since $p+1\text{-}2g\neq 0$, and $l=0$, we deduce $\sum_{j=1}^{p'}2p/l_j\neq 0$, and so $p'\neq 0$.

Finally, if $r=0$, we have $b=0$, and so $p\text{-}1=2px$, which is absurd. Now

(3) $\alpha g'+k'\le 1$.

Otherwise, since $r\neq 0$ we would have, from 3.2.18.1

$$p\text{-}1\ge 2p\sum_{j=1}^{r}(1\text{-}1/m_j)\ge 2p(1\text{-}1/m_1)\ge p,\ \text{which is absurd.}$$

Finally we get the contradiction we are looking for:

If signτ="-", then $g'\ge 1$ and $k'\ge p'>0$, *i.e.* $\alpha g'+k'=g'+k'\ge 2$.

If signτ="+", then $0<p'<k'$and $\alpha g'+k'\ge k'\ge 2$.

3.3. NOTES

Most results in section 1 are new, although in [14] and also in Harvey [61] it is solved the problem of the existence of automorphisms of given finite order in the cases of Klein surfaces without boundary and Riemann surfaces respectively. As far as we know, the first proofs of lemmas 3.1.1 and 3.2.4 are due to Harvey, [61].

The problem to find the minimum algebraic genus p(N) of those compact Riemann surfaces admitting an automorphism of order N was first solved by Wiman [127]. A modern proof is due to Harvey [61]. The existence of automorphisms of Riemann surfaces with big enough order was studied by Nakagawa [103]; concerning automorphisms of prime order we must quote the paper of Homma [68]. For Klein surfaces without boundary the question was solved by Hall [60] and also in [14], and the answer for Klein surfaces with nonempty connected boundary was given in [15]. The result for arbitrary compact Klein surfaces of algebraic genus bigger than or equal to 2 is established in [24]. In those papers it is also found, in the corresponding cases, the maximum order N(p) of automorphisms of surfaces with algebraic genus $p \geq 2$. May found in [96] the general answer for surfaces with nonempty boundary, without distinguishing the orientability of the involved automorphism. The first complete solution, Thm. 3.2.18, was given by J.A. Bujalance, [31], who also proved the "orientability criterion" 3.2.1. The minimum genus of unbordered nonorientable surfaces with given finite abelian group of automorphisms was obtained in [58].

Finally we remark that Cor. 3.2.17, which gives the numbers $p_{+}^{+}(N,k)$, $p_{+}^{-}(N,k)$ and $p_{-}(N,k)$ for prime N, is new. As far as we know, the same question remains still open if N is not prime.

CHAPTER - 4
Klein surfaces with groups of automorphisms in prescribed families

In 1.3.5 we showed that a group of automorphisms of a bordered Klein surface of algebraic genus $p \geq 2$ is finite. Here we show that such a group has at most $12(p-1)$ elements. Although this bound turns out to be attained for infinitely many values of p, this is no more true if we restrict attention to groups lying in certain specific classes. Results of this chapter concern the following problems:

Let $\mathscr{F}$ be a class of finite groups.

(1) Find the bound $N(p,\mathscr{F})$ for the order of a group of automorphisms G of a bordered Klein surface of algebraic genus $p \geq 2$ provided $G \in \mathscr{F}$.

(2) Having $N(p,\mathscr{F})$, describe those p for which this bound is attained.

(3) Describe the topological type of the corresponding Klein surfaces.

(4) Describe the algebraic structure of the corresponding groups of automorphisms.

We focus here attention on the following classes of abstract finite groups: all finite groups, soluble, supersoluble, nilpotent (p-groups) and abelian groups.

Throughout all the chapter, Klein surface and surface group will mean bordered compact Klein surface and bordered surface group, *i.e.* a surface NEC group with nonempty set of (empty) period-cycles, unless otherwise stated.

4.1. SURFACE-KERNEL FACTORS OF NEC GROUPS

Let X be a Klein surface of algebraic genus $p \geq 2$. Then by 1.2.3 $X = H/\Gamma$, for some surface group of algebraic genus p. Now if G is a group of automorphisms of X then, by 1.3.2, there exists an NEC group Λ such that $G = \Lambda/\Gamma$. Therefore from the Hurwitz Riemann formula it follows that the problem of finding the upper bound for the order of a group of automorphisms of a Klein surface is equivalent to the problem of finding the lower bound for the area of NEC groups Λ admitting surface groups as normal subgroups. We start with the following technical

Lemma 4.1.1. *Let Λ be an NEC group with area $<\pi/2$ admitting a surface group Γ as a normal subgroup. Then Λ has one of the following signatures*

$\sigma(\Lambda)$	$\mu(\Lambda)$
$(0;+;[-];\{(2,2,2,n)\})$	$\pi(n-2)/2n$ $(n\geq 3)$
$(0;+;[-];\{(2,2,3,3)\})$	$\pi/3$
$(0;+;[-];\{(2,2,3,4)\})$, $(0;+;[-];\{(2,2,4,3)\})$	$5\pi/12$
$(0;+;[-];\{(2,2,3,5)\})$, $(0;+;[-];\{(2,2,5,3)\})$	$7\pi/15$
$(0;+;[3];\{(2,2)\})$	$\pi/3$
$(0;+;[2,3];\{(-)\})$	$\pi/3$.

Proof. Let Λ have signature

$$(g;\pm;[m_1,...,m_r];\{(n_{i1},...,n_{is_i})|1\leq i\leq k\}).$$

If $k>2$ then $\mu(\Lambda)\geq 2\pi$. If $k=2$ and at least one period-cycle is nonempty then $\mu(\Lambda)\geq\pi/2$ whilst if both of them are empty then either $g\neq 0$ and so $\mu(\Lambda)\geq 2\pi$, or $r>0$ and then $\mu(\Lambda)\geq\pi$. Hence we can assume that $k=1$.

First suppose that the period-cycle is empty.
If $r=0$ then $\mu(\Lambda)$ is a multiple of 2π.
If $r=1$ then $g\neq 0$, since $\mu(\Lambda)>0$. But then $\mu(\Lambda)\geq\pi$.
If $r>2$ then $\mu(\Lambda)\geq\pi$.
So assume that $r=2$. If $g\neq 0$ then $\mu(\Lambda)\geq 2\pi$. Consequently $g=0$ and so $\mu(\Lambda)=2\pi(1\text{-}1/m_1\text{-}1/m_2)$. But then the only solution of the system of inequalities

$$0<1\text{-}1/m_1\text{-}1/m_2<1/4,\ m_1\leq m_2$$

is $m_1=2$, $m_2=3$. So using 0.2.6.4 we arrive to the last signature in the list above.

Now suppose that the period-cycle is nonempty. By 2.4.1 two consecutive periods are equal to 2. By 0.2.6.4 we can assume that $n_1=n_2=2$. If $g\neq 0$ then $\mu(\Lambda)\geq\pi$. So let $g=0$. If $r\geq 2$ then $\mu(\Lambda)\geq\pi$. If $r=1$ then $\mu(\Lambda)\geq\pi/2$ unless Λ has signature $(0;+;[3];\{(2,2)\})$. So assume that $r=0$. If $s\geq 5$ then $\mu(\Lambda)\geq\pi/2$. If $s\leq 3$ then $\mu(\Lambda)<0$. So Λ has signature $(0;+;[-];\{(2,2,m,n)\})$ and $\mu(\Lambda)=\pi(1\text{-}1/m\text{-}1/n)$. But the only solutions of the system of inequalities

$$0<1\text{-}1/m\text{-}1/n<1/2,\ n\leq m$$

are $m=2$, $n\geq 3$ or $m=3$, $n=3,4,5$. Interchanging the roles of m and n and using 0.2.6.4 we arrive to one of the remaining signatures.

A finite group G is said to be an *orientable (resp. nonorientable) surface kernel factor group of an NEC group* Λ if and only if G can be presented as a factor group Λ/Γ, where Γ is an orientable (resp. nonorientable) surface group.

Lemma 4.1.2. *A necessary and sufficient condition for a finite group* G *to be a* *surface kernel factor Λ/Γ, where Λ is an NEC group with signature*

$$(0;+;[-];\{(2,2,m,n)\})$$

is that G *can be generated by three elements* a,b,c *obeying nontrivially the relations*

(4.1.2.1) $$a^2=b^2=c^2=(ab)^m=(ac)^n=1.$$

Moreover if this is the case then Γ can be found such that the number k *of its period-cycles satisfies* $|G|=2kq$, *where* q *is the order of* bc, *and Γ can be assumed to be nonorientable if and only if* ab *and* ac *generate the whole group* G.

Proof. For the sake of technical convenience we shall prove the lemma for the case $m\leq n$. For, first let us see how a homomorphism θ from Λ onto G having a surface group Γ as the kernel must look like.

Λ is generated by elements c_0,c_1,c_2,c_3,c_4,e subject to the relations

(4.1.2.2) $$c_0^2=c_1^2=c_2^2=c_3^2=c_4^2=(c_0c_1)^2=(c_1c_2)^2=(c_2c_3)^m=(c_3c_4)^n=1,$$
$$e=1 \text{ and } c_4=ec_0e^{-1}$$

whilst nontrivial orientable elements in Γ have infinite order. Since Γ is a bordered surface group $\theta(c_i)=1$ for some i, by 2.3.6.1. Clearly $i\neq 0,3,4$ because in the other case $(c_3c_4)^2\in\Gamma$ is an orientable element of finite order (equal to n if n is odd and equal to n/2 if n is even). If $\theta(c_1)=1$ then $\theta(c_2)\neq 1$ since otherwise $c_1c_2\in\Gamma$ is an orientable element of order 2. Similarly if $\theta(c_2)=1$ then $\theta(c_1)\neq 1$. Notice that this case may occur if $m=2$ since otherwise $(c_2c_3)^2\in\Gamma$ is an orientable element of finite order. Moreover $\theta(c_2c_3)$ and $\theta(c_3c_4)$ have orders m and n respectively because, if this were not so, a power of c_2c_3 or c_3c_4 would be an orientable element of Γ of finite order. As a result G can be generated by

$$a=\begin{cases}\theta(c_3) \text{ if } \theta(c_1)=1\\ \theta(c_0) \text{ if } \theta(c_2)=1\end{cases},\quad b=\begin{cases}\theta(c_2) \text{ if } \theta(c_1)=1\\ \theta(c_1) \text{ if } \theta(c_2)=1\end{cases},\quad c=\begin{cases}\theta(c_0) \text{ if } \theta(c_1)=1\\ \theta(c_3) \text{ if } \theta(c_2)=1\end{cases}$$

and the relations (4.1.2.1) hold.

Now assume that Γ has k (empty) period-cycles. Then by theorem 2.3.3, $k=|G|/2q$, where q is the order of bc.

Finally if Γ is nonorientable then by theorem 2.1.3 a nonorientable word $w=w(c_0,c_\upsilon,c_3)$ belongs to Γ, where $\upsilon=1$ or $\upsilon=2$ according to $\theta(c_2)=1$ or $\theta(c_1)=1$. But w is nonorientable if and only if its length is odd. Given $i=0,\upsilon,3$ let $v_i=c_iw$. Then each v_i has even length and using obvious relations

$$c_0c_3=(c_3c_0)^{-1},\ c_\upsilon c_3=(c_3c_\upsilon)^{-1},\ c_0c_\upsilon=(c_3c_0)^{-1}(c_3c_\upsilon)$$

we conclude that $v_i=v_i(c_3c_0,c_3c_\upsilon)$ for $i=0,\upsilon,3$. So for every $i=0,\upsilon,3$, $\theta(c_i)=\theta(c_iw)=\theta(v_i)=v_i(\theta(c_3c_0),\theta(c_3c_\upsilon))$. Thus G is generated by ab and ac.

Conversely let G be a finite group generated by three elements a,b,c obeying the relations (4.1.2.1) and let q be the order of bc. Then the epimorphism θ from Λ onto G induced by the assignment

$$\theta(c_0)=\theta(c_4)=c,\ \theta(c_1)=1,\ \theta(c_2)=b,\ \theta(c_3)=a$$

has by theorem 2.2.3 a surface group Γ with $k=|G|/2q$ empty period-cycles as kernel. Moreover if ab and bc generate G then in particular $a=v(ab,ac)$ and so $w=c_3v(c_3c_2,c_3c_0)$ is a nonorientable word in Γ. Therefore Γ is nonorientable by theorem 2.1.3.

4.2. GROUPS OF AUTOMORPHISMS OF KLEIN SURFACES WITH MAXIMAL SYMMETRY

Theorem 4.2.1. *Let* G *be a group of automorphisms of a Klein surface* X *of algebraic genus* $p\geq 2$. *Then* $|G|\leq 12(p-1)$.

Proof. By 1.2.3 there exists a surface group Γ of algebraic genus p such that $X=H/\Gamma$ and, by 1.3.2, an NEC group Λ for which $G\cong\Lambda/\Gamma$. By 4.1.1 $\mu(\Lambda)\geq\pi/6$ and the bound is attained only for an NEC group Λ with signature $(0;+;[-];\{(2,2,2,3)\})$. So by the Hurwitz Riemann formula $|G|=\mu(\Gamma)/\mu(\Lambda)\leq$ $\leq 2\pi(p-1)/(\pi/6)=12(p-1)$.

A surface X of algebraic genus $p\geq 2$ admitting a group of automorphisms of order 12(p-1) (the maximum possible) is said to have *maximal symmetry.* From the proof of the previous theorem it follows that a finite group G of order 12(p-1) acts as a group of automorphisms on such a surface if and only if G is a surface kernel factor group of an NEC group Λ with signature

(4.2.1.1) $$(0;+;[-];\{(2,2,2,3)\}).$$

A finite group G will be said to be an M^**-group* if and only if it can be generated by three elements a,b,c obeying nontrivially the following relations

(4.2.1.2) $$a^2=b^2=c^2=(ab)^2=(ac)^3=1.$$

The order q of bc will be called *an index of* G and G will be said to be an M^**-group with index* q. As we shall see an M^*-group can have more than one index.

With the remark and definitions above the following theorem is an immediate consequence of lemma 4.1.2.

Theorem 4.2.2. *A finite group* G *is an* M^**-group with an index* q *if and only if* G *is the group of automorphisms of a Klein surface with maximal symmetry and* k *boundary components, where* $|G|=2kq$.

Remark 4.2.3. *Let* G *be an* M^**-group with index* q *and let* p *be the order of* abc. *Then* G *is also an* M^**-group with index* p.

Proof. Let $a'=a$, $b'=ab$, $c'=c$. Then a',b',c' show G as an M^*-group with index p.

Example 4.2.4.1. We will see that the group $G=D_6\cong D_3\times Z_2$ is the only M^*-group of order 12. In fact let a,b and c be the generators of G satisfying (4.2.1.2). Then $H=<a,c>$ is a dihedral group of order 6, normal in G. Now $bab^{-1}=a$ and $bcb^{-1}\in H$. The last is an element of order 2 and so it is equal to either a,c or aca. But $bcb^{-1}\neq a$ since in the other case $c=b^{-1}ab=a$, a contradiction. Thus either $bcb^{-1}=c$ and then $G\cong D_3\times Z_2$ with $(bc)^2=1$, or $bcb^{-1}=aca$ and then clearly bc has order 6 and so $G\cong D_6$. It is easy to check that both of mentioned presentations of the group D_6 as an M^*-group with indices 2 and 6 actually exist. In fact let $D_6=<x,y|x^2,y^2,(xy)^6>$ and choose a=xyx, b=yxy, c=y. Then a, b and c are elements of order 2 generating D_6 and ab, ac, bc and abc have orders 2,3,6 and 2 respectively. Therefore the assertion follows from 4.2.3.

Now let us see what are the possible topological types of Klein surfaces with maximal symmetry on which $G=D_6$ acts as the group of automorphisms. For, first notice that ab and ac cannot generate the whole group G. In fact if G has index 2 then $<ab,ac>$ is a dihedral group of order 6, whilst if G has index 6 then this group is a cyclic group of order 6. So such a surface X must be orientable by the second part of 4.1.2. By theorem 4.2.2 X has k=1 or k=3 boundary components. Finally 2=p=2g+k-1 implies that g=1 or g=0 according to k=1 or k=3 respectively. Therefore X is a sphere with three holes or a torus with one hole.

Example 4.2.4.2. Consider the group $G=S_4$. We will see that G is an M^*-group with indices 3 and 4. In fact the reader can easily check that needed presentations of G are the following

$$a=(1,2),\ b=(3,4),\ c=(1,4)$$

and

$$a=(1,2),\ b=(1,2)(3,4),\ c=(1,4).$$

So G acts on surfaces X and Y with maximal symmetry of algebraic genus p=3. By 4.2.2 X has k=4 boundary components. Now $3=p=\alpha g+k-1$ implies $\alpha g=0$ and so g=0 and thus X is a sphere with 4 holes. Similarly Y has k=3 boundary components and now $\alpha g=1$. So $\alpha=g=1$ and therefore Y is a real projective plane with 3 boundary components.

It can be shown that $G=S_4$ is the only M^*-group of order 24 and there are no topological types of Klein surfaces of algebraic genus 3 with maximal symmetry other than calculated above, but we shall not use this fact in other parts.

Now given a group G and an integer $m\geq 2$ let $\{G_n^{(m)}\}_{n\geq 2}$ denote its *m-Frattini series i.e* a series defined by

(4.2.4.1) $$G_2^{(m)}=G^m[G,G],\ \text{and}\ G_{n+1}^{(m)}=(G_n^{(m)})_2^{(m)},$$

where for any subgroup H of G, $H_2^{(m)}$ is the subgroup of H generated by all its commutators and m-powers. This series will play in this chapter an important role.

Theorem 4.2.5. *Let X be a Klein surface with maximal symmetry of algebraic genus p having k boundary components. Then given an odd integer m there exists a Klein surface X′ with maximal symmetry of algebraic genus* $p'=(p-1)m^p+1$, *having* $k'=km^{p-1}$ *boundary components. The surface X′ is orientable or not according with the orientability of the former surface X.*

Proof. Let $X=H/\Gamma$, where Γ is a surface group of algebraic genus p with k empty period-cycles and let G be the group of its automorphisms. Then $G=\Lambda/\Gamma$ where Λ is an NEC group with signature (4.2.1.1).

Consider the group $\Gamma'=\Gamma_2^{(m)}$. Then Γ' is a characteristic subgroup of Γ and so it is a normal subgroup of the group Λ. Since m is odd, all canonical reflections of Γ belong to Γ' and so $\Gamma/\Gamma' \cong Z_m \oplus \overset{p}{\ldots} \oplus Z_m$. Now each e_i induces in Γ/Γ' an element of order m. Thus each empty period-cycle in Γ induces, by theorem 2.3.1, m^{p-1} empty period-cycles in Γ'. As a result Γ' has km^{p-1} period-cycles all of them empty. By 2.2.3. Γ' has no proper periods and so it is a surface group. Moreover $|\Lambda/\Gamma'|=12(p-1)m^p$ and so $X'=H/\Gamma'$ is a Klein surface with maximal symmetry of algebraic genus $p'=(p-1)m^p+1$ having $k'=km^{p-1}$ boundary components. Finally by theorem 2.1.1, Γ and Γ' have the same sign. This completes the proof.

Combining the last theorem with the previous examples we obtain

Corollary 4.2.6. *The bound* 12(p-1) *for the order of a group of automorphisms of a Klein surface of algebraic genus* p *is attained for infinitely many values of* p *both in orientable and nonorientable cases.*

Remark 4.2.7. Choosing m=3 in theorem 4.2.5 and starting with a surface with maximal symmetry of algebraic genus 2, *e.g.* a sphere with 3 holes (see Ex. 4.2.4.1), we see that there are infinitely many M^*-groups of order 4×3^n that act on orientable Klein surfaces with maximal symmetry, whilst starting with a real projective plane with 3 boundary components that has maximal symmetry (see Ex. 4.2.4.2) we obtain infinitely many such groups of orders 8×3^n. A group of order $2^m\times 3^n$ is soluble by the theorem of Burnside. Therefore the bound 12(p-1) is attained for soluble groups for infinitely many values of p both in orientable and nonorientable cases.

The topological type (determined by the orientability, the topological genus and the number of boundary components) of a Klein surface X will be called *the species of* X.

We showed above that there is no shortage of M^*-groups. Moreover an M^*-group can act on different species with maximal symmetry as well as there are different species with maximal symmetry and with the same number of boundary components. The classification of all M^*-groups and of all species of Klein surfaces with maximal symmetry seems to be an enormous problem. Nevertheless certain pieces of it are more approachable. In the remainder of this section we will deal with the following two problems

Is there a bound (independent of p*) for the number of species with maximal symmetry within a single algebraic genus* p*?*

Given a positive integer k*, is there a bound* $\mu(k)$ *for the number of species with maximal symmetry having* k *boundary components?*

Let Ω be an abstract group given by the presentation

(4.2.7.1) $$<u,v,w|u^2,v^2,w^2,(uv)^2,(uw)^3>$$

and let Ω_0 be its subgroup generated by uv, uw. Ω_0 and Ω are known to be isomorphic to the modular and extended modular groups respectively. So in particular Ω_0 is a free product of two cyclic groups of orders 2 and 3. It is worth noting however that the last fact about Ω_0 can be easily deduced directly from the presentation (4.2.7.1) using the Reidemeister-Schreier algorithm for determining the presentation of a subgroup of a given group. The role of the group Ω in the study of M^*-groups follows from the following

Lemma 4.2.8. *A finite group* G *of order greater than* 6 *is an* M^**-group if and only if* G *is a homomorphic image of* Ω.

Proof. The necessity is obvious by the definition. So let $G\cong\Omega/K$ be a factor group of Ω of order greater than 6. If one of u,v,w,uv,uw belonged to K then Ω/K would be clearly a group of order ≤ 6. So the images a,b,c of u,v,w respectively show G as an M^*-group.

Corollary 4.2.9. *Let* H *be a normal subgroup of an* M^**-group* G *of index greater than* 6*. Then* G/H *is also an* M^**-group.*

We see that the study of M^*-groups lies in the study of factor groups of Ω. We will need later the following three technical lemmas describing certain subgroups of Ω.

Lemma 4.2.10. *Let* Ω' *be the commutator subgroup of the group* Ω. *Then* $\Omega/\Omega' = Z_2 \oplus Z_2$ *and* Ω' *is a group generated by two elements of order* 3. *In particular the order of* Ω'/Ω'' *divides* 9.

Proof. The first part of the lemma is obvious. Moreover it is easy to check that the subgroup H of Ω generated by $\alpha = uw$ and $\beta = vuwv$ is normal in Ω and $\Omega/H \cong Z_2 \oplus Z_2$. Therefore $H = \Omega'$.

We saw in 4.2.8 that a factor group of Ω of order greater than 6 is an M^*-group. In the next section we shall need the following result that describes normal subgroups of index 6.

Lemma 4.2.11. *Let* K *be a normal subgroup of* Ω *of index* 6. *Then* Ω/K *is the dihedral group* D_3 *of order* 6 *and* K *is a group generated by three elements of order* 2.

Proof. If one of u,w or uw belonged to K then $|\Omega/K| \leq 4$. Hence, the images $\bar{u}$ and $\bar{w}$ of u and w respectively in Ω/K generate a dihedral subgroup of order 6. So the first part follows.

Now let us see how the image $\bar{v}$ of v looks like. If $\bar{v} = \bar{w}$ then clearly $uw \in K$, a contradiction, whilst if $\bar{v} = \bar{u}\bar{w}\bar{u}$ then $(uw)^2 \in K$ and so $uw \in K$, a contradiction again. So $\bar{v} = \bar{u}$ or $\bar{v} = 1$ *i.e.* $uv \in K$ or $v \in K$ and thus K is the normal closure in Ω of uv or v, since each of the last is clearly a subgroup M for which $\Omega/M \cong D_3$. Let K_1 and K_2 be the subgroups of Ω generated by uv, wuvw, uwuvwu, and v, wvw, uwvwu, respectively. It is easy to check, using the relations (4.2.7.1), that both K_1 and K_2 are normal subgroups of Ω. Hence $K = K_1$ or $K = K_2$ and so we are done.

Lemma 4.2.12. *Let* N *be the normal closure of* $a = (uvw)^2$ *in* Ω. *Then* $\Omega/N \cong D_6$ *and* N *is the free group of rank* 2 *freely generated by* a *and* $b = (vwu)^2$.

Proof. First notice that Ω/N, the group with presentation

$$<u,v,w \mid u^2, v^2, w^2, (uv)^2, (uw)^3, (uvw)^2>$$

has order ≤ 12. In fact it is easy to check that the subgroup of Ω/N generated by v and uw is normal in Ω/N, it has order ≤ 6 and index ≤ 2. Thus $|\Omega/N| \leq 12$. On the other hand the assignment

$$u \longmapsto x(xy)^3, \quad v \longmapsto x, \quad w \longmapsto y$$

induces an epimorphism θ from Ω/N onto $D_6 = <x,y \mid x^2, y^2, (xy)^6>$. This proves the first part.

Obviously $a = (uvw)^2 \in \Omega_o$ and so N is a subgroup of Ω_o which is a free product of two cyclic groups A and B of orders 2 and 3 respectively. By the Kurosh subgroup theorem, N is a free product of a free group together with groups that are conjugate to A and B. But the last factors cannot appear since in the other case uv or uw would belong to N and Ω/N would have order ≤ 6. So N is a free group.

Now $b=uau\in N$ and so $H=<a,b>$ is a subgroup of N. We shall see that H is a normal subgroup of Ω. In fact it is easy to check that the following relations hold

$$\begin{aligned} &uau=b, && ubu=a,\\ (4.2.12.1)\qquad &vav=b^{-1}, && vbv=a^{-1},\\ &waw=a^{-1}, && wbw=a^{-1}b. \end{aligned}$$

Therefore H=N and it still remains to show that H is not cyclic. Assume to a contradiction that H is a cyclic group generated by c. Then $a=c^n$, $b=c^m$. Now the conjugation by u is an automorphism of H and so $ucu=c^{\pm 1}$, *i.e.* $a^{\pm 1}=c^{\pm n}=uc^nu=uau=b$. If $a=b$ then $wbw=a^{-1}b=1$, which implies $b=a=1$, absurd. If $a^{-1}=b$ then $waw=b$ and $wbw=b^2$. Thus $a=b^2=a^{-2}$. But then $a^3=1$ what is also false since N is free. This completes the proof.

Now let $A_n=<x,z|x^n,z^n,[x,z]>$ and $B_n=<y|y^n>$. We form the semidirect product K_n of A_n by B_n, where the action of B_n on A_n is given by

$$x^y=z^{-1}x, \quad z^y=z.$$

Then K_n has order n^3 and presentation

$$<x,y,z|x^n,y^n,z^n,z[y,x],[x,z],[y,z]>.$$

Clearly the relation $z^n=1$ is redundant because

$$1=yx^ny^{-1}=(yxy^{-1})^n=(z^{-1}x)^n=z^{-n}x^n=z^{-n}.$$

Since $z=[x,y]$ induces in K_n a central element it is easy to prove by induction that $(x^{-1}y)^k=x^{-k}y^kz^{k(k-1)/2}$. So in particular $(x^{-1}y)^n=z^{n(n-1)/2}$.

Now if n is odd $(x^{-1}y)^n=(z^n)^{(n-1)/2}=1$ whilst if n is even $(x^{-1}y)^n=z^{n/2}\neq 1$. Let

$$s_n=\begin{cases} n & \text{if n is odd,}\\ n/2 & \text{if n is even.}\end{cases}$$

Then clearly for every integer m dividing s_n

$$<x,y,z|x^n,y^n,z^m,z[y,x],[x,z],[y,z]>$$

is the presentation of a group K_{nm} of order n^2m. Now it is obvious that in K_{nm} $(x^{-1}y)^n=1$. The generator z can be eliminated and we obtain a new presentation of K_{nm}

$$<x,y|x^n,y^n,[x,y]^m,[x,[x,y]],[y,[x,y]]>.$$

Now by the previous lemma the following quotient of N

$$<a,b|a^n,b^n,[a,b]^m,[a,[a,b]],[b,[a,b]]>$$

is isomorphic to K_{nm} and, of course, $(a^{-1}b)^n=1$. Looking at the relations (4.2.12.1) we see that M, the normal closure in N of the set

$$\{a^n,b^n,[a,b]^m,[a,[a,b]],[b,[a,b]]\},$$

is a normal subgroup in Ω. Moreover $|\Omega/M|=|\Omega/N||N/M|=12n^2m$ and it is easy to check that $[a,b]=(vwu)(wv)^6(vwu)^{-1}$. Thus $\#(vw)=6\#([a,b])=6m$ and $\#(uvw)=2\#(a)=$

$=2n$. So using 4.2.3 we obtain

Theorem 4.2.13. *Given positive integers* n *and* m *with* m *dividing* s_n *there exists an* M^**-group* G_{nm} *of order* $12n^2m$ *with indices* 6m *and* 2n.

Now let $A_n = \langle x,z \mid x^{3n}, z^n, [x,z]\rangle$ and $B_n = \langle y \mid y^n\rangle$. Form the semidirect product L_n of A_n by B_n, where B_n acts on A_n according to the same formulas as in the previous case. Then L_n has order $3n^3$ and presentation

$$\langle x,y,z \mid x^{3n}, y^n, z^n, z[y,x], [x,z], [y,z]\rangle.$$

Now let m be an integer dividing s_n. Then the group L_{nm} with presentation

$$\langle x,y,z \mid x^{3n}, y^n, z^m, z[y,x], [x,z], [y,z]\rangle$$

has order $3n^2m$ and yx^{-1} has order 3n. The first claim is rather obvious. To see the second, notice that yx^{-1} induces in the quotient $L_{nm}/\langle z\rangle \cong \cong \langle x,y \mid x^{3n}, y^n, [x,y]\rangle$ an element of order 3n. On the other hand using induction one can prove that the relation

$$(yx^{-1})^k = y^k x^{-k} z^{-k(k-1)/2}$$

holds in L_{nm} for arbitrary k and so in particular $(yx^{-1})^{3n} = 1$ by the choice of m. Substituting $x=x'$, $y=y'x'$, $z=z'$ we obtain another presentation of L_{nm}:

$$\langle x',y',z' \mid x'^{3n}, (y'x')^n, z'^m, z'[y',x'], [x',z'], [y',z']\rangle,$$

in which x' and y' represent elements of order 3n. Moreover as before it can be shown by induction that $(x'^{-1}y')^{3n} = 1$. The generator $z' = [x',y']$ can be eliminated and we obtain the following presentation of the group L_{nm}:

$$\langle x',y' \mid x'^{3n}, (x'y')^n, [x',y']^m, [x',[x',y']], [y',[x',y']]\rangle.$$

Now, by the previous lemma, the quotient of N

$$\langle a,b \mid a^{3n}, (ab)^n, [a,b]^m, [a,[a,b]], [b,[a,b]]\rangle$$

is isomorphic to L_{nm}. Looking at the relations (4.2.12.1) we see that M, the normal closure in N of the set

$$\{a^{3n}, (ab)^n, [a,b]^m, [a,[a,b]], [b,[a,b]]\},$$

is a normal subgroup of Ω. Moreover $|\Omega/M| = |\Omega/N||N/M| = 36n^2m$. As earlier $\#(uw) = =6\#([a,b]) = 6m$ and $\#(uvw) = 2\#(a) = 6n$. So we obtain

Theorem 4.2.14. *Given positive integers* n *and* m *such that* m *divides* s_n *there exists an* M^**-group* H_{nm} *of order* $36n^2m$ *with indices* 6m *and* 6n.

The first application of the families of M^*-groups G_{nm} and H_{nm} just constructed is to solve in the negative the first of the problems stated above:

Corollary 4.2.15. *Let* r *be a positive integer. Then there exists an integer* p *such that there are at least* r *distinct species of Klein surfaces with maximal*

symmetry and algebraic genus p.

Proof. Let $t=6r$ and $p=2^t+1$. Given $k=1,...,r$ let $i(k)=2r+k$ and $j(k)=2(r-k)$. Then $j(k)<i(k)$ and $2i(k)+j(k)=t$. Now let $n(k)=2^{i(k)}$, $m(k)=2^{j(k)}$ and let $G_k=$ $=G_{n(k),m(k)}$ be the group constructed in theorem 4.2.13. Then $|G_k|=12n(k)^2m(k)=$ $=12\times 2^{2i(k)+j(k)}=12\times 2^t$ and so by theorem 4.2.2 each G_k acts as the group of automorphisms on a Klein surface with maximal symmetry of algebraic genus $p=2^t+1$ having $n(k)^2$ boundary components, because G_k has index $6m(k)$.

Lemma 4.2.16. *Let* G *be a finite group of order* ≥ 12 *generated by two elements* x *and* z *obeying the relations*

$$x^2=z^q=(xz)^3=1.$$

Then $|G|^2\geq q^3$.

Proof. For $q\leq 5$, $|G|^2\geq 12^2>5^3\geq q^3$. If $q=6$ then 6 divides the order of G. If $|G|\geq 18$ then the relation in question also holds true and so assume that $q=6$ and $|G|=12$. But then $H=\langle z\rangle$ is a subgroup of G of index 2 and so $xzx=z^\alpha$, for some α. But $z^{-1}xz^{-1}=xzx$. Hence $x=z^{\alpha+2}$ and so $G=\langle z\rangle$, a contradiction.

Thus we can assume that $q>6$. First we shall show that there exists an integer $n\leq l=|G|/q$, dividing q such that $xz^nx=z^n$. Let $H=\langle z\rangle$. Then $[G:H]=l$ and so the $l+1$ cosets xz^mxH, $0\leq m\leq l$, cannot be all distinct. Let n be the least positive integer such that $n\leq l$ and $xz^nx\in H$. Since $xz^qx\in H$, n divides q. Let $xz^nx=z^i$. We shall show that $z^i=z^n$. In fact

$$z^i=xz^nx=(xzx)^n=(z^{-1}xz^{-1})^n.$$

Conjugating by z^{-1} and z we obtain

$$z^i=(z^{-2}x)^n, \qquad z^i=(xz^{-2})^n.$$

Thus

$$xz^ix=x(z^{-2}x)^nx=(xz^{-2})^n=z^i=xz^nx$$

and therefore $z^i=z^n$.

Now let $N=\langle z^n\rangle$ and consider the group $L=G/N$. Then $|L|=|G|/|N|=|G|/(q/n)=$ $=nl\leq l^2$.

We will show that $|N|\leq l$. For, consider the transfer map τ from G onto N. Since N is a central subgroup, $\tau(g)=g^{nl}$ for each $g\in G$. Now notice that L is generated by the cosets $x'=xN$ and $x'z'=xzN$ that have orders 2 and 3 respectively. The first claim is rather obvious. In order to see the second one, assume first that $x'=1$, *i.e.* $x\in N$. Then $x=z^{q/2}$ and so $xz=z^{(q+2)/2}$. But since xz has order 3 this clearly implies that q divides $3(q+2)/2$ and so $q=2$, 3 or 6, a contradiction. Similarly if $x'z'=1$, *i.e.* $xz\in N$ then $xz=z^\alpha$. Thus $x=z^{\alpha-1}$ and since x is an element of order 2, $\alpha-1=q/2$ and so we are in the previous case. So 2 and 3 divide nl. Therefore $\tau(x)=\tau(xz)=1$ and thus $1=\tau(z)=z^{nl}$. Hence also q divides nl and so in particular $|N|=q/n\leq l$. Thus

$|G|=|L||N|\leq l^3$. Therefore, since $l=|G|/q$, $|G|^2\geq q^3$ as desired.

Corollary 4.2.17. *Let* G *be an* M^**-group of order* >12 *and with an index* q. *Then* $|G|^2\geq q^3$.

Proof. Consider G^+, the subgroup of G of index ≤2 generated by ab and bc. Then $|G|\geq|G^+|$ and by the lemma $|G^+|^2\geq q^3$. So the result follows.

Now we are ready to give the solution to the second problem

Corollary 4.2.18. *For each natural number* k *there are only finitely many species of Klein surfaces with maximal symmetry having exactly* k *boundary components.*

Proof. Let X be a Klein surface of genus p with maximal symmetry having k boundary components and let G be the group of automorphisms of X, with an index q. Then by theorem 4.2.2, $12(p-1)=|G|=2kq$. If $p\geq 3$ then, by the previous corollary, $|G|^2\geq q^3$ and so $q\leq 4k^2$. As a result $p=(kq/6)+1\leq 4k^3/6+1=2k^3/3+1$. Thus there are only a finite number of possibilities for the genus p and for each p there are at most two species with maximal symmetry and k boundary components (also for p=2, see Ex 4.2.4.1). This completes the proof.

4.3. SUPERSOLUBLE GROUPS OF AUTOMORPHISMS OF KLEIN SURFACES

The M^*-group D_6 is supersoluble and so the bound 12(p-1) is attained for this class of groups, for p=2. We will show in this section that given a non-negative integer n there is a Klein surface with maximal symmetry of algebraic genus $p=3^n+1$ that admits a supersoluble group of order 12(p-1) as the group of automorphisms and there are no other values of p for which this is so. Further, we give the classification of the species with maximal symmetry and supersoluble group as the group of automorphisms. In particular we show that all these species are orientable and this leads us to ask for the bound for the order of a supersoluble group of automorphisms of a nonorientable Klein surface. We show that a bound is 6(p-1) and it is attained if and only if $p\equiv 3\bmod(4)$ as well as we give the complete classification of the species of nonorientable Klein surfaces admitting supersoluble group of order 6(p-1) as a group of automorphisms. Finally the fact that this bound cannot be attained for surfaces of even algebraic genus leads us to looking for the bound for the order of a supersoluble group of automorphisms of a nonorientable Klein surface of even algebraic genus p. We prove that a bound is 4p. Moreover we characterize those values of p for which this bound is attained as well as we give a complete classification of the corresponding species.

We shall need the following, well known, results about supersoluble groups.

Lemma 4.3.1. *(a) The commutator subgroup G′ of a finite supersoluble group G is nilpotent.*

(b) If H is a nilpotent normal subgroup of a finite group G such that G/H′ is supersoluble, then G is also supersoluble.

(c) If G is a finite supersoluble group then G has a normal series:

$$1=G_0<G_1<\ldots<G_n=G$$

with cyclic factors of prime order of decreasing magnitude.

(d) An extension of a cyclic group by supersoluble is supersoluble.

Theorem 4.3.2. *Let G be an M^*-group. Then G is supersoluble if and only if $|G|=4\times 3^r$ for some integer r.*

Proof. Of course 12 divides |G|. By lemma 4.2.8 G is a homomorphic image of Ω. So G′, G/G′ and G′/G″ are homomorphic images of Ω', Ω/Ω' and Ω'/Ω'' respectively. Thus by 4.2.10 [G:G′] divides 4 and [G′:G″] divides 9.

First assume that G is supersoluble. By the previous lemma G′ is nilpotent. The group Ω' is generated by two elements of order 3, by 4.2.10. Thus G′ is a 3-group since it is a homomorphic image of Ω'. Therefore $|G|=4\times 3^r$.

Now assume that G is an M^*-group of order 4×3^r. We will show that G is supersoluble using induction on r.

We showed in 4.2.4.1 that the only M^*-group of order 12 is D_6 which is supersoluble. Thus the assertion is true for r=1.

Suppose then that r≥2. The group G is soluble by the theorem of Burnside. First we shall show that [G:G′]=4. Assume, by the way of contradiction, that [G:G′]=2. Then [G′:G″]=3 since in the other case G/G″ would be a group of order 18 and so it would be an M^*-group, by 4.2.9. A contradiction since 12 divides the order of an M^*-group as we mentioned at the beginning. So [G:G″]=6. But then, $G/G''=\Omega/K$, where, by 4.2.11 K is a group generated by elements of order 2. The group G″ is a homomorphic image of K and so $[G'':G^{(3)}]=2$. Therefore $\mathfrak{G}=G/G^{(3)}$ is an M^*-group of order 12 and so $\mathfrak{G}\cong D_6$. But then $\mathfrak{G}_{ab}=Z_2\oplus Z_2$, a contradiction since the last group is a homomorphic image of $G_{ab}=Z_2$. Therefore [G:G′]=4 and so G′ is a 3-group generated, by 4.2.10, by two elements of order 3. In particular G′ is nilpotent.

If r=2 then from the above |G′|=9 and so G′ is abelian *i.e.* G″=1. Therefore G=G/G″. So $G=\Omega/\Omega''$ since G/G″ is a homomorphic image of Ω/Ω'' and the order of Ω/Ω'' divides 36. But by the proof of 4.2.10 $\Omega' = <\alpha,\beta>$, where α=uw, β=vuwv and it is easy to check that, modulo Ω'', the following relations hold

$$u(\alpha\beta)u=(\alpha\beta)^{-1},\ v(\alpha\beta)v=\alpha\beta,\ w(\alpha\beta)w=(\alpha\beta)^{-1}.$$

So $G/G' \cong Z_2 \oplus Z_2$ and $G'=\Omega'/\Omega'' \cong Z_3 \oplus Z_3$ contains a subgroup $<\alpha\beta\Omega''>$ of order 3, invariant with respect to G. Therefore G is supersoluble by 4.3.1(d).

Finally assume that $r>2$. Then $G'' \neq 1$. By the induction hypothesis G/G'' is supersoluble. Thus, by lemma 4.3.1(b), G is also supersoluble.

The theorem above does not classify supersoluble M^*-groups, of course. This seems to be a rather difficult problem. Nevertheless it is possible to classify species of Klein surfaces with maximal symmetry and supersoluble group of automorphisms. The following corollary which is an easy consequence of the previous theorem and theorems 4.2.2, 4.2.13 and 4.2.14 is the first step in this direction.

Corollary 4.3.3. *Given positive integers* r *and* s *with* $r\geq 2$ *and* $r/2\leq s\leq r-1$ *there exists a Klein surface with maximal symmetry of algebraic genus* $p=3^{r-1}+1$ *with* $k=3^s$ *boundary components that admits a supersoluble group of order* 12(p-1) *as the group of automorphisms.*

We shall see that these sufficient conditions for the existence of a Klein surface in question are actually also necessary. For, we shall need some technical results concerning the lower central series of the Sylow 3-subgroup of a supersoluble M^*-group that we state in the next lemma. In what follows G is a supersoluble M^*-group generated by three elements a, b and c obeying nontrivially the relations:

(4.3.3.1) $$a^2=b^2=c^2=(ab)^2=(ac)^3=(bc)^q=1,$$

P will be the Sylow 3-subgroup of G and $\gamma_i=\gamma_i(P)$ is the lower central series of P, *i.e.* a series defined by

$$\gamma_1(P)=P,\ \gamma_{n+1}(P)=[P,\gamma_n(P)].$$

Lemma 4.3.4. *Let* G *be a supersoluble* M^**-group of order* 4×3^r $(r\geq 2)$. *Then*

(a) P *is the normal subgroup of* G *generated by two elements* $x=ab(cb)^3$, $z=(bc)^2$.

(b) *If* $xz^mx^{-1}=z^n$ *for some integers* m *and* n *then* $z^m=z^n$.

(c) $\gamma_1/\gamma_2\cong Z_3\oplus Z_3$ *and for* $i\geq 2$ *the exponent of the quotient* γ_i/γ_{i+1} *is at most* 3.

(d) *If for some integers* d *and* j, $z^d\in\gamma_j$, $z^d\notin\gamma_{j+1}$ *and* $[\gamma_j:\gamma_{j+1}]=3$ *then* $[\gamma_{j+1}:\gamma_{j+2}]\leq 3$.

(e) *The order* q *of* bc *equals* 2×3^t *for some* t, *and if* $t\geq 2$ *then the class of nilpotency* $c=cl(P)\geq t+1$.

Proof. (a) In the proof of theorem 4.3.2 we showed that $[G:G']=4$ and so $G'=P$. By lemma 4.2.10 G′ is generated by $\alpha=ac$ and $\beta=bacb$ and it is easy to check that $\alpha=xz^2$, $\beta=xz$. Moreover $z=\beta^{-1}\alpha$ and $x=\beta\alpha^{-1}\beta$. So (a) follows.

(b) The proof is similar to the proof of part of lemma 4.2.16. In fact using the relations (4.3.3.1) we check that

$$xz^m x^{-1}=ab(bc)^{2m}ab \quad \text{and} \quad (ab)(bc)(ab)=(cb)(ab)(cb).$$

So

$$(bc)^{2n}=z^n=xz^m x^{-1}=(ab)(bc)^{2m}(ab)=((ab)(bc)(ab))^{2m}=((cb)(ab)(cb))^{2m}.$$

Conjugating by cb and $(cb)^{-1}$ we obtain

$$(bc)^{2n}=((cb)^2(ab))^{2m}=((ab)(cb)^2)^{2m}.$$

Thus

$$(ab)(bc)^{2n}(ab)=(ab)((ab)(cb)^2)^{2m}(ab)=((cb)^2(ab))^{2m}=(bc)^{2n}=(ab)(bc)^{2m}(ab).$$

Hence $(bc)^{2n}=(bc)^{2m}$ and therefore $z^n=z^m$.

(c) P is a 3-group of order 3^r ($r\geq 2$) generated by two elements of order 3. So γ_1/γ_2 is a factor group of $Z_3\oplus Z_3$. On the other hand P has a normal subgroup N of index 9. Since P/N is abelian, $\gamma_2=P'\subseteq N$. As a consequence $|\gamma_1/\gamma_2|\geq 9$. So $\gamma_1/\gamma_2\cong Z_3\oplus Z_3$ and this completes the proof of the first part of the assertion.

Now using the obvious formula

$$[fg,u]=[g,u]^f[f,u]\equiv[g,u][f,u] \bmod \gamma_{i+2}$$

that holds for any $u\in\gamma_i$ and $f,g\in P$, and the induction on i, one can easily prove the second part.

(d) γ_j is generated by γ_{j+1} and z^d. So by the theorem of P.Hall γ_{j+1} is generated by γ_{j+2} and the commutators $[z^d,z]$, $[z^d,x]$. But $[z^d,z]=1$, of course, and so $\gamma_{j+1}/\gamma_{j+2}$ is cyclic. Thus (d) follows from (c).

(e) Clearly $z=(bc)^2\in P$. So $q=3^t$ or $q=2\times 3^t$. But the first case is impossible since otherwise $bc\in P$ and so $ab=ac(bc)^{-1}$ would be an element of order 2 in P. So the first part follows. Now consider the elements $A=(acb)x(acb)^{-1}$, $B=(cb)x(cb)^{-1}$ of P. It is easy to check using the relations (4.3.3.1) that $A=(cab)^2$, $B=(acb)^2$ and $[B^{-1},A]=(bc)^6=z^3$. Moreover $AB=[z^{-1},A^{-1}]\in\gamma_2$. So $z^3=[B^{-1},A]=[B^{-1},AB]\in\gamma_3$. Now using (c) and induction on i we show that $z^{3^{t-1}}\in\gamma_{t+1}$ and therefore $\gamma_{t+1}\neq 1$.

Theorem 4.3.5. *Let* G *be a supersoluble* M^**-group of order* 4×3^r, *where* $r\geq 2$, *with an index* q. *Then* $q=2\times 3^t$ *for some integer* t *such that* $1\leq t\leq r/2$.

Proof. From (a) and (e) of the previous lemma we have that $P=G'$ is a 3-group of order $3^r\geq 9$ generated by two elements $\alpha=ac$, $\beta=bacb$, both of order 3, and also $q=2\times 3^t$. Notice that $t\neq 0$ since in the other case, by (a) above, P would be a cyclic group of order 3. So it remains to prove that $t \leq r/2$.

First we will see that the assertion holds for $r \leq 4$. In fact it is obvious for $r=2$. If $r=3$ then by the part (c) of the previous lemma $\gamma_3=1$ and so $cl(P)=2$. Hence, by (e) of 4.3.4 $t=1$. Finally let $r=4$ and let P_0 be a normal subgroup of G of order 3. Applying the argument above to G/P_0 we get $t \leq 2 = r/2$.

Now assume to a contrary that G is an M^*-group of the smallest order 4×3^r with an index $q=2 \times 3^t$, where $t > r/2$. Let G_0 be a normal subgroup of G of order 3. G/G_0 has order $4 \times 3^{r-1}$ and an index 2×3^s, where $s=t-1$ or t. By induction hypothesis $s \leq (r-1)/2$. So $s=t-1$. But then $r/2 < t \leq (r+1)/2$ and thus r can be assumed to be odd and equal to $2t-1$.

Let $c=cl(P)$ be the class of nilpotency of P and let as before $\{\gamma_i\}$ be the lower central series of P. Clearly each γ_i is normal in G. By 4.3.4(e), $c \geq t+1$. By 4.3.4(c) γ_c has exponent 3. So $\tilde{G}=G/\gamma_c$ has index $\tilde{q}=2 \times 3^t$ or $2 \times 3^{t-1}$. But if the first were the case then by induction hypothesis again $t \leq (r-1)/2=(2t-2)/2$, a contradiction. So $\tilde{q}=2 \times 3^{t-1}$. Now if $|\gamma_c| \geq 3^2$ then $|\tilde{G}| \leq 4 \times 3^{2t-3}$ and so using again induction hypothesis $t-1 \leq (2t-3)/2$. Thus $|\gamma_c|=3$. If $[\gamma_{c-1}:\gamma_c]=3$ then $z^{3^{t-2}} \in \gamma_{c-1}$, since in the other case applying the inductive hypothesis for G/γ_{c-1} we get $t-1 \leq (2t-3)/2$, once more. But then γ_{c-1} would be a normal subgroup of G generated by $z^{3^{t-2}}$ and by 4.3.4(b), γ_{c-1} would be contained in the center of P and as a consequence γ_c would be trivial. If $[\gamma_{c-1}:\gamma_c] \geq 3^3$ then applying induction hypothesis to G/γ_{c-1} we obtain $t-2 \leq (2t-5)/2$. At this point we have the following information about the lower central series of P

$$P \overset{9}{\text{---}} \gamma_2 \overset{3}{\text{---}} \gamma_3 \text{---} \cdots \text{---} \gamma_{c-1} \overset{9}{\text{---}} \gamma_c \overset{3}{\text{---}} 1$$

Now, since $c \geq t+1$, not all remaining factors have order ≥ 9. Otherwise $3^{2t-1}=|P| \geq 9 \times 9^{c-2}=3^{2(c-1)} \geq 3^{2t}$, a contradiction.

Let m be the largest integer so that $[\gamma_m:\gamma_{m+1}]=3$ and $[\gamma_j:\gamma_{j+1}] \geq 9$ for $m < j \leq c-1$. Then by the induction hypothesis G/γ_m has index at most $d=2 \times 3^{t-1-(c-m)}$ and so z^d is an element of γ_m. But z^d has order 3^{c-m+1} and γ_{m+1} has exponent at most 3^{c-m}. Thus in particular $z^d \notin \gamma_{m+1}$, a contradiction since from the one side $[\gamma_{m+1}:\gamma_{m+2}] \geq 9$ whilst from the other one $[\gamma_{m+1}:\gamma_{m+2}] \leq 3$ by 4.3.4(d). This completes the proof.

Lemma 4.3.6. *A Klein surface with maximal symmetry admitting a supersoluble group as the group of automorphisms is orientable.*

Proof. Let G be a supersoluble M^*-group. Then G is generated by three elements a,b and c of order 2 obeying the relations

$$(ab)^2=(bc)^3=1.$$

Now, if G were acting on a nonorientable Klein surface then, by 4.1.1, it would be generated by ab and bc and so in particular $|G/G'|$ would divide 6, a contradiction since $[G:G']=4$ as we proved in 4.3.2.

Using 4.2.4.1, 4.3.3 and then combining the last two results with theorem 4.2.2, we obtain the following theorem classifying species of Klein surfaces with maximal symmetry and supersoluble group of automorphisms.

Theorem 4.3.7. *(1) There are only two species of Klein surfaces of algebraic genus* $p=2$ *with maximal symmetry and supersoluble group of automorphisms of order* $12(p-1)=12$*: a sphere with 3 holes and a torus with one hole.*
(2) The necessary and sufficient condition for the existence of a Klein surface X *of algebraic genus* $p>2$ *with maximal symmetry, supersoluble group of automorphisms of order* $12(p-1)$ *and with* k *boundary components is that*

(a) $p-1=3^r$ *for some* $r\geq 1$,

(b) $k=3^s$ *for some integer* s *such that* $(r+1)/2\leq s\leq r$.

In such a case X *is necessarily orientable.*

We have shown in 4.3.6 that a Klein surface with maximal symmetry and supersoluble M^*-group of automorphisms is orientable. In the rest of this section we deal with the maximal supersoluble groups of automorphisms of nonorientable Klein surfaces. We will need the following technical results:

Lemma 4.3.8. *A finite nilpotent group* G *cannot be generated by three elements* a,b *and* c, *of order 2 such that* ab *and* bc *generate the whole group* G *and* ab, bc *and* ac *have orders 2,* k *and* l *respectively, with* k *and* l *greater than 2.*

We precede the proof by the following easy

Lemma 4.3.9. *Let* $\langle x,y|x^2,y^2,(xy)^k\rangle$ *be the presentation of the dihedral group* G *of order* 2k *and let* $z\in G$ *be an element of order 2 such that* xz *and* yz *have order 2. Then* k *is even and* $z=(xy)^{k/2}$.

Proof. The only elements of order 2 in G are those represented as $(xy)^\alpha x$ $0\leq\alpha\leq k-1$ and $(xy)^{k/2}$, when k is even. So we will be done when we show that z cannot have the first form. If this were not so then, since $xz=(yx)^\alpha$ has order 2, k is even and $\alpha=k/2$. But also $yz=(yx)^{\alpha+1}$ is an element of order 2. Thus $\alpha+1=k/2$ and therefore $\alpha=\alpha+1$, a contradiction.

Proof of lemma 4.3.8.

Assume, by the way of contradiction, that such a group G exists. A finite nilpotent group is the product of its Sylow subgroups and G is generated by elements of order 2. Thus G is a 2-group, say of order 2^N, and $k=2^m$, $l=2^n$, where $m\geq 2$ and $n\geq 2$. Changing the role of the generators a and b, if necessary, we can assume that $k \leq l$. Obviously $N \geq 3$ since G is not abelian.

If $N=3$ then $|G|=8$. The group G cannot be cyclic and so ac has order 4 and $G=<a,c>$ is a dihedral group. But then $b=(ac)^{\alpha}a$ or $b=(ac)^2$. We shall show that the former is impossible. In fact since ab has order 2, $\alpha=2$. But then $bc=(ac)^3$, and $ab=(ac)(bc)^{-1}=(ac)^{-2}$. Thus G is a cyclic group of order 4, a contradiction. Hence $b=(ac)^2$, but then bc has order 2.

Now let $N=4$. If ac has order 8 then $G \cong D_8$ and we arrive to a contradiction as in case $N=3$. So we can assume that bc and ac have order 4. Then $H=<a,c>$ is a normal subgroup of G as a subgroup of index 2. Clearly $(bc)^2 \in H$. But then $(bc)^2=(ac)^2$ or $(bc)^2=(ac)^{\alpha}a$ for some $0 \leq \alpha \leq 3$. In the first case $K=<(bc)^2>$ is a normal subgroup of G. In fact $b(bc)^2b=(cb)^2=(bc)^2$, $c(bc)^2c=(cb)^2=(bc)^2$, $a(bc)^2a=a(ac)^2a=(ca)^2=(ac)^2=(bc)^2$. Moreover G/K is a factor group of $Z_2 \oplus Z_2$. Thus $|G| \leq 8$, a contradiction.

If $(bc)^2=a$ then $ac=bcb$ and so $(ac)^2=1$, a contradiction.

If $(bc)^2=aca$ then $cbc=baca=abca$ and so $(bc)^2=1$, a contradiction.

If $(bc)^2=(ac)^2a$ then $bcb=(ac)^3$ and so $(ac)^2=1$, a contradiction again.

Finally if $(bc)^2=(ac)^3a$ then $bcb=1$ and so $c=1$, a contradiction once more. So $N \geq 5$. In all considerations below we shall assume that N is minimal.

First assume then that $k=l=4$, and let H be a normal subgroup of G of order 2. Consider the quotient $\mathfrak{G}=G/H$. By the minimality of N, one of the elements $a,b,c,ab,(bc)^2$, or $(ac)^2$ belongs to H. But then this element would become in $\mathfrak{G}$ a relation that collapses it to a group of order ≤ 8, a contradiction.

Now let $k=4$ and $l \geq 8$. Consider, as in the first case, the normal subgroup H of G of order 2 and let $\mathfrak{G}$ be the quotient. By the minimality of N at least one of a,b,c,ab, or $(bc)^2$ belongs to H. But if a,c or ab belonged then it would produce in $\mathfrak{G}$ a relation collapsing it to a group of order ≤ 8. If b belonged to H then, since H is normal in G, $(bc)^2$ would be a relation in the former group against $k=4$. Finally if none of the elements considered before belonged to H then again by the minimality of N, $(bc)^2$ would belong to H. But then it would collapse $\mathfrak{G}$ to a dihedral group with the presentation

$$<\tilde{a},\tilde{b},\tilde{c} \mid \tilde{a}^2,\tilde{b}^2,\tilde{c}^2,(\tilde{a}\tilde{b})^2,(\tilde{b}\tilde{c})^2,(\tilde{a}\tilde{c})^s>,$$

where $s=l$ or $s=l/2$ ($\tilde{a}$, $\tilde{b}$, and $\tilde{c}$ are the images of a,b, and c respectively under the canonical projection). By Lemma 4.3.9. $\tilde{b}=(\tilde{a}\tilde{c})^{s/2}$, and so $\tilde{b}(\tilde{a}\tilde{c})^{s/2}=1$. Thus $b(ac)^{s/2}$ belongs to H. As a result $b(ac)^{s/2}=1$ or $b(ac)^{s/2}=(bc)^2$. In the first case $cb=c(ac)^{s/2}$ whilst in the second $bc=c(ca)^{s/2}$. In both cases bc has order 2, a contradiction.

Finally assume that $8 \leq k \leq l$, and let us choose k minimal. Let as before H be a normal subgroup of G of order 2 and let $\mathfrak{G}$ be the corresponding quotient. If c belonged to H then $aca^{-1}=c$, *i.e.* $l=2$, absurd. Similarly if a or b belongs to H. If ab belonged to H then the quotient would be a cyclic group generated by $\tilde{b}\tilde{c}$, the product of two elements of order 2, a contradiction once more. Now since $k,l\geq 8$, $\tilde{G}$ would be a group satisfying the hypothesis of the lemma, what contradicts the minimality of N. This completes the proof.

Now we are ready to prove the following

Theorem 4.3.10. *Let* G *be a supersoluble group of automorphisms of a nonorientable Klein surface* X *of algebraic genus* $p\geq 3$. *Then* $|G|\leq 6(p-1)$. *There exists a nonorientable surface of algebraic genus* 2, *a real projective plane with two holes, which admits a supersoluble group of automorphisms of order* 8.

Proof. Let $X=H/\Gamma$, where Γ is a nonorientable surface group of algebraic genus $p\geq 3$ and let $G=\Lambda/\Gamma$ be a supersoluble group of automorphisms of X. We will show that $\mu(\Lambda)\geq \pi/3$, what by the Hurwitz Riemann formula will complete the first part of the proof.

By lemma 4.1.1 we have to rule out only NEC groups Λ with signatures:

$$(0;+;[-];\{(2,2,2,k)\}),$$

where k=3,4 or 5.

If k=3, then $|G|=12(p-1)$ against 4.3.6. Let k=4. Then by 4.1.2 G is generated by three elements a,b and c of order 2 obeying the relations

(4.3.10.1) $$(ab)^2=(ac)^4=1$$

and in addition ab and ac generate the whole group G. By 4.3.1(c) G possesses a normal subgroup H of odd order such that G/H is a 2-group. Clearly none of a,b,c,ab,$(ac)^2$ belongs to H. Also $bc\notin H$ and by lemma 4.3.8 $(bc)^2\in H$. So G/H is a dihedral group of order 8 generated by the images of ab and bc. We shall show that G=G/H. In fact if this were not so then by 4.3.1(c) the group G can be assumed to have order 8q, where q is an odd prime. G contains a normal subgroup of index 8 with a dihedral quotient as we just showed. By 4.3.1(c) G contains also a normal subgroup of index 4. So $|G/G'|=4$ and thus G' is a group of order 2q. By 4.3.1(a) G' is nilpotent and so it is a direct product of two cyclic groups M and N of orders 2 and q respectively. Each of these groups is characteristic in G' and hence normal in the whole group G. Let $\tilde{G}=G/M$. If $a\in M$ then, since M is normal in G, cac=a and thus $(ca)^2=1$, absurd. If at least one of b,c belongs to M then $|\tilde{G}|\leq 8$. Finally if $ab\in M$, then $\tilde{G}$ would be generated by the image of ac and so $|\tilde{G}|\leq 4$. As a result $(ac)^2\in M$. But then $\tilde{G}$ would be a dihedral group of order 4q generated by three elements $\tilde{a},\tilde{b},\tilde{c}$ of order 2 such that $\tilde{a}\tilde{b}$, $\tilde{a}\tilde{c}$ and $\tilde{b}\tilde{c}$ have orders 2, 2 and 2q respectively, where $\tilde{a}$, $\tilde{b}$ and $\tilde{c}$ are the images in $\tilde{G}$ of a,b and c respectively. By 4.3.9 $\tilde{a}=(\tilde{b}\tilde{c})^q$ and so $a(bc)^q\in M$. Hence $a(bc)^q=1$ or $a(bc)^q=(ac)^2$. In the first case $ca=c(bc)^q$ whilst in the second $ac=c(cb)^q$. Thus ac would have order 2, a contradiction. Hence $G=G/H=D_4$.

But then the corresponding Klein surface has genus 2. Arguing as in Ex. 4.2.4.1 the reader can easily check that it is a real projective plane with two holes. This also completes the last part of the theorem.

Now consider the case k=5. By 4.1.2 G is generated by three elements a,b and c of order 2 that obey the relations

$$(ab)^2=(ac)^5=1,$$

and in addition ab and ac generate the whole group G. So in particular $|G/G'|$ would divide 10. On the other hand it is easy to see that 4 divides $|G/G'|$. In fact a and b generate a subgroup of G of order 4 and so in particular 4 divides the order of G. Hence by 4.3.1(c) G contains a normal subgroup of index 4. This implies that 4 divides $|G/G'|$ as desired. A contradiction that completes the proof.

Lemma 4.3.11. *Let* G *be a group with presentation*

$$<x,y,z|x^2,y^2,z^2,(xy)^2,(xz)^{2r},(yz)^q,\ y(xz)^2y(xz)^{2k}>,$$

where r *is an odd prime,* $1 \le k \le r-1$ *and* 4 *divides* q. *Assume that* xy *and* xz *generate the whole group* G. *Then* $(xz)^2=1$ *in* G *and* G *is the dihedral group* D_q.

Proof. Assume to get a contradiction that $a=(xz)^2 \neq 1$. Then $H=<a>$ is a subgroup of G of order r. Since $yay^{-1}=a^{-k}$, H is a normal subgroup of G. Clearly G/H is the dihedral group D_q with presentation

$$<\tilde{x},\tilde{y},\tilde{z}\,|\,\tilde{x}^2,\tilde{y}^2,\tilde{z}^2,(\tilde{x}\,\tilde{y})^2,(\tilde{x}\,\tilde{z})^2,(\tilde{y}\,\tilde{z})^q>.$$

By 4.3.9 $\tilde{x}=(\tilde{y}\,\tilde{z})^{q/2}$. So $x(yz)^{q/2}\in H$. Now, if $x(yz)^{q/2}=1$, then clearly $(xz)^2=1$. So let $x(yz)^{q/2}=(xz)^{2\alpha}$ for some α in range $1\le\alpha\le r-1$. Let $\phi:G\longrightarrow \mathrm{Aut}H$ be the homomorphism induced by the action of G on H given by the conjugation. We have $x=(xz)^{2\alpha}(zy)^{q/2}$. Since ϕ is a homomorphism and AutH is abelian we have $\phi(x)=\phi(x)^{2\alpha}\phi(y)^{q/2}\phi(z)^{2\alpha+q/2}$. Now $\phi(x)^2=\phi(y)^2=\phi(z)^2=1_H$. So since q/2 is even, we see that, for each α, $\phi(x)=1_H$. Hence $(zx)^2=x(xz)^2x=\phi(x)((xz)^2)=(xz)^2$, and so $(xz)^4=1$. But the last relation and $(xz)^{2r}=1$ imply that $(xz)^2=1$. A contradiction that proves the first part. The second part is now obvious.

Lemma 4.3.12. *Let* N *be an even positive integer and let* G_N *be a group with presentation*

$$<x,y,z|x^2,y^2,z^2,(xy)^2,(xz)^6,(yz)^N,y(xz)^2y(xz)^4,(yz)^{N/2}(xz)^2x>.$$

If $N\equiv 2\mathrm{mod}(4)$, *then* G_N *is a supersoluble group of order* 6N, *generated by* xy *and* xz. *In particular* $(xz)^2\neq 1$. *Moreover* $x,y,z\notin(G_N)'$.

Proof. We see from the defining relations of G_N that $H=<(xz)^2>$ is a normal subgroup of G_N and G_N/H is a dihedral group of order 2N. In particular G is supersoluble by 4.3.1(d) and $|G|\le 6N$. Thus we will be done if we construct a group of order 6N generated by three elements A,B and C obeying relations listed in the presentation of G_N, such that AB and AC generate G.

For, we start with $D_N = <x,y|x^2,y^2,(xy)^N>$ and form a semidirect product G of $Z_3 = <w|w^3>$ by D_N subject to the action

$$w^x = w,\ w^y = w^{-1}.$$

Then $|G| = 6N$ and it is easy to check that

$$A = (xy)^{N/2},\ B = x,\ C = wy$$

verify the relations in question and AB, AC generate G. Finally the images $\bar{y}$ and $\bar{z}$ in $D_N = G_N/H$ of y and z respectively, generate it, and $\bar{x} = (\bar{y}\bar{z})^{N/2}$. So $\bar{x},\bar{y},\bar{z} \notin (D_N)'$, the commutator subgroup of D_N and so $x,y,z \notin (G_N)'$.

Lemma 4.3.13. *Let* G *be a group of order* 4×3^r *generated by three elements* a,b *and* c *of order* 2 *such that the following relations hold nontrivially:*

(4.3.13.1) $$(ab)^2 = (ac)^6 = (bc)^{2\times 3^m} = 1.$$

Suppose that ab *and* bc *generate the whole group* G, *and* $a,b,c \notin G'$. *Then* G *is supersoluble.*

Proof. First we shall see that G′ is generated by two elements of orders 3 and 3^m. For let Λ be an NEC group with signature $(0;+;[-];\{(2,6,2\times 3^m)\})$ and let $\theta:\Lambda \longrightarrow G$ be the epimorphism defined by

$$\theta(c_0) = \theta(c_3) = b,\ \theta(c_1) = a,\ \theta(c_2) = c,\ \theta(e) = 1.$$

Then $\ker\theta$ is a surface group without period-cycles by 2.3.6.2. Let $\pi:G \longrightarrow G_{ab}$ be the canonical projection and $\tilde{\theta} = \pi\theta$. We shall see how $\Gamma = \ker\tilde{\theta}$ looks like. First $\Lambda/\Gamma = G/G' = Z_2 \oplus Z_2$. By 2.3.6.2 Γ has no period-cycles and by 2.2.4 it has two proper periods equal to 3 and 3^m respectively. So $\sigma(\Gamma) = (g;\pm;[3,3^m];\{-\})$. By the Hurwitz Riemann formula we obtain $\alpha = g = 1$ and therefore

$$\sigma(\Gamma) = (1;-;[3,3^m];\{-\}).$$

Clearly $\theta|\Gamma$ establishes an epimorphism $\Gamma \longrightarrow G'$ whose kernel is a surface group (equal to the kernel of θ not restricted). So by 2.2.4, $\tilde{x}_1$, $\tilde{x}_2$, the images of the canonical generators of Γ, are elements of orders 3 and 3^m respectively. Moreover since the order of G′ is odd, the relation $\tilde{d}_1^2\tilde{x}_1\tilde{x}_2 = 1$ implies that $\tilde{d}_1$ is redundant. Concluding, G′ is a 3-group of order 3^n, say, generated by two elements of orders 3 and 3^m, as desired.

We will prove the assertion by induction on n.

If $n=1$ then G is supersoluble, of course, whilst if $n=2$ then $|G'| = 9$ and so in particular G′ is abelian. If G′ is cyclic there is nothing to do. Otherwise $G' = Z_3 \oplus Z_3$. Although this group is characteristically simple, the reader can check that any subgroup of GL(2,3) being a homomorphic image of $Z_2 \oplus Z_2$ has in $Z_3 \oplus Z_3$ an invariant subgroup of order 3 and so G is supersoluble in virtue of the previous case ($n=1$) and 4.3.1(d).

So let $n>2$. If G' is nonabelian then $|G''|\neq 1$ and so G/G'' is a group of order $4\times 3^{n'}$ for some $n'<r$. Now, either $(ac)^2\in G''$ and G/G'' is a dihedral group which is supersoluble, or else $(ac)^2\notin G''$ and G/G'' is supersoluble by the inductive hypothesis. But then G is supersoluble by 4.3.1(b).

So we can assume that G' is abelian. If G' is cyclic there is nothing to do; G is clearly supersoluble by 4.3.1(d). So we can assume that $G'=\mathbb{Z}_3\oplus\mathbb{Z}_{3^{n-1}}=$ $=\langle a,b|a^3,b^{3^{n-1}},[a,b]\rangle$. But then $H=\langle b^{3^{n-2}}\rangle$ is clearly a characteristic subgroup of G', and so is normal in G. Now G/H is supersoluble by the inductive hypothesis. Therefore G is supersoluble by 4.3.1(d).

Theorem 4.3.14. *Let* G *be a supersoluble group of automorphisms of a nonorientable Klein surface* X *of algebraic genus* $p\geq 3$, *with* k *boundary components, of order* $6(p-1)$. *Then* $p\equiv 3\mathrm{mod}(4)$, G *is an extension of a 3-group by a dihedral group and* k *is a nontrivial power of* 3 *dividing* $6(p-1)$.

Proof. Let $X=H/\Gamma$ and $G=\Lambda/\Gamma$. By the Hurwitz Riemann formula $\mu(\Lambda)=\pi/3$ and by 4.1.1 Λ has one of the following signatures:

$$(0;+;[2,3];\{(-)\}),\ (0;+;[3];\{(2,2)\}),$$
$$(0;+;[-];\{(2,2,3,3)\}),\ (0;+;[-];\{(2,2,2,6)\}).$$

What we are going to show first is that Λ cannot have any of the first three signatures.

If Λ has the first signature, then c_0 is the only nonorientable canonical generator of Λ and, since Γ has nonempty set of period-cycles, $c_0\in\Gamma$. So by 2.1.3 Γ cannot be nonorientable.

Now assume, that Λ has the second signature. Then Λ is generated by c_0, c_1, c_2, e, x subject to the relations

$$x^3=1,\ c_0^2=c_1^2=c_2^2=(c_0c_1)^2=(c_1c_2)^2=1,\ xe=1,\ ec_0e^{-1}=c_2.$$

By Remark 2.3.6.2 a reflection c_i of Λ belongs to Γ.

If one of c_0, c_2 belongs to Γ then the other one also does. Assume first that this is the case. Then c_1 does not belong to Γ since otherwise Γ would have a nonempty period-cycle by 2.3.2. We see that the generator e is redundant and so G is generated by two elements $\tilde{c}_1$ and $\tilde{x}$, the images of c_1 and x respectively. The element $\tilde{c}_1$ has order 2, of course, and $\tilde{x}$ has order 3 by 2.2.4. Now since Γ is nonorientable, a nonorientable word $w=w(x,c_1)$ belongs to Γ. But clearly w is nonorientable if and only if the exponent sum of c_1 in w is odd. But if this is so then $\tilde{c}_1$ becomes redundant in G_{ab} and thus G_{ab} is a factor group of $\mathbb{Z}_3$. On the other hand $|G|$ is even and by 4.3.1(c) G contains a subgroup of index 2 and so in particular 2 divides the order of G_{ab}. A contradiction.

Similarly if c_1 belongs to Γ then none of c_0, c_2 belongs. But c_2 and e are clearly redundant, and so G is generated by the images of c_0 and x and we get a contradiction as before.

A group with the third signature can be ruled out in a similar manner as a group with signature (0;+;[-];{(2,2,2,5)}) in theorem 4.3.10.

So it remains to deal only with an NEC group Λ with signature (0;+;[-];{(2,2,2,6)}). By 4.1.2 G is generated by three elements a, b and c of order 2 obeying nontrivially the relations:

(4.3.14.1) $$(ab)^2=(ac)^6=1$$

and ab and ac generate the whole group G. Let q be the order of bc. Clearly $<a,b>$ is a subgroup of G of order 4 and in particular 4 divides $|G|$ and thus 2 divides p-1 *i.e.* p is odd.

Now we shall show that 4 does not divide p-1, *i.e.* 8 does not divide $|G|$. In fact let 2^α be the greatest divisor of $|G|$ of this form and assume to get a contradiction that $\alpha>2$. Let G be such a group of smallest possible order. By 4.3.1(c) there exists a normal subgroup H of G such that $|G/H|=2^\alpha\times 3^\beta$, where $(|H|,6)=1$. Clearly relations (4.3.14.1) still hold in G/H for the images of a, b, and c, and the images of ab and ac generate G/H. So by the minimality of G, H=1, *i.e.* $|G|=2^\alpha\times 3^\beta$. Now by 4.3.1(c) there exists a normal subgroup K of G of order 3. None of a,b,c,ab belongs to K. Therefore using again the minimality of G we deduce that $(ac)^2\in K$. But then $K=<(ac)^2>$, and so $b(ac)^2b$ is equal to $(ac)^2$ or to $(ac)^4$. So if we show that 4 divides q we shall get a contradiction with lemma 4.3.11 applied for r=3. In order to see that 4 divides q consider a normal subgroup L of G of odd order such that G/L is a 2-group. Let $q=2^\beta s$, where s is odd. Notice that a,b,c induce in G/L elements $\bar{a},\bar{b},\bar{c}$ of order 2, such that $\bar{a}\bar{b},\bar{a}\bar{c},\bar{b}\bar{c}$ have orders 2,2 and 2^β respectively. So $|G/L|=2^{\beta+1}$. Since $\alpha>2$, we get $\beta\geq 2$. So 4 does not divide p-1 and consequently $p\equiv 3\bmod(4)$.

Using 4.3.1(c) once more we obtain that G contains a normal subgroup of index 4. On the other hand G_{ab} is generated by elements of order 2 as well as by the images of ab and ac. Thus $G_{ab}=Z_2\oplus Z_2$.

Now G' is nilpotent by 4.3.1(a) and so it is a direct product of its Sylow subgroups each of which is a characteristic subgroup of G' and so a normal subgroup of G. Let P be the Sylow 3-subgroup of G, say of order 3^m. We see that $(ac)^2\in P$ and so $\tilde{G}=G/P$ is a dihedral group.

Finally let $q=3^n r$, where $(r,3)=1$. Then bc induces in $\tilde{G}$ an element of order r and $|\tilde{G}|=2r$. Hereby X has $k=|G|/2q=3^m\times 2r/3^n\times 2r=3^{m-n}$ boundary components. Notice that $k\neq 1$. Otherwise $|G|=2q$ and so $G=D_q$ is generated by b and c. Hence $a=(bc)^{q/2}$ or $a=(bc)^\gamma b$. But, if the first were the case, ac would have order 2, a contradiction. In the second one $ab=(bc)^\gamma$ and thus $\gamma=q/2$. So $ac=(bc)^{(q+2)/2}$, *i.e.* q divides 6 and so $|G|\leq 12$. On the other hand $|G|=6(p-1)$ and $p\geq 3$. So $|G|=12$ and thus q=6. But then $ac=(bc)^4$ and therefore

$(ac)^3=(bc)^{12}=1$, a contradiction again.

Theorem 4.3.15. *There exists a nonorientable Klein surface of algebraic genus* $p\geq 3$ *having* k *boundary components and a supersoluble group of automorphisms of order* 6(p-1) *if and only if*

(i) $p\equiv 3\mathrm{mod}(4)$.

(ii) k *divides* 6(p-1) *and* $k=3^n$ *for some* $n\geq 1$.

Proof. The "only if" part is the immediate consequence of the last theorem. Conversely we can write $p-1=2\times 3^m\times M$, where $m\geq 0$ and $(M,6)=1$. Since k divides 6(p-1), $N=2\times 3^{m-n+1}$ is an integer. All we need is to construct a supersoluble group G of order $4\times 3^{m+1}\times M$ generated by three elements A,B,C of order 2 such that AB, AC and BC have orders 2,6 and MN and AB,AC generate the whole group G. Once this will be done, G will be the nonorientable surface kernel factor Λ/Γ, where Λ is an NEC group with signature

$$(0;+;[-];\{(2,2,2,6)\})$$

and we finish in virtue of 4.1.2 and the Hurwitz Riemann formula.

First we consider the supersoluble group G_N of order $4\times 3^{m-n+2}$ constructed in 4.3.12. By results of chapter 2, G_N can be represented as Λ_1/Γ, where Λ_1 is an NEC group with signature $(0;+;[-];\{(2,6,N)\})$ and Γ is a nonorientable surface group without period-cycles.

We claim that there exists a subgroup Γ_1 of Γ such that:

(i) Γ_1 is a nonorientable surface group without period-cycles,

(ii) Γ_1 is normal in Λ,

(iii) $|\Gamma/\Gamma_1|=3^s$, where $s\geq n-1$.

In fact it is sufficient to take $\Gamma_1=\Gamma_l^{(3)}$, the lth 3-Frattini subgroup (see 4.2.4.1) for l large enough.

Let $\mathfrak{G}=\Lambda_1/\Gamma_1$ and $H=\Gamma/\Gamma_1$. Then $\mathfrak{G}/H\cong G_N$ and in particular $\mathfrak{G}$ has order $4\times 3^{m-n+2+s}\geq 4\times 3^{m+1}$. Now let A,B and C be the images of c_0,c_1 and c_2 in $\mathfrak{G}$. Then by results of Chapter 2 they are elements of order 2 such that AB,AC and BC have orders 2,6 and N, respectively, and AB,AC generate $\mathfrak{G}$. Clearly $A,B,C\notin\mathfrak{G}'$ since their images $\bar{A},\bar{B},\bar{C}$ in G_N do not belong to $(G_N)'$.

Therefore $\mathfrak{G}$ is supersoluble by lemma 4.3.13 and by 4.3.1(c) $\mathfrak{G}$ has an invariant series

$$\mathfrak{G}=\mathfrak{G}_0\overset{2}{—}\mathfrak{G}_1\overset{2}{—}\mathfrak{G}_2\overset{3}{—}\mathfrak{G}_3\overset{3}{—}\cdots\overset{3}{—}\mathfrak{G}_{m-n+2+s}=1.$$

Clearly $H\subseteq\mathfrak{G}_2$. Let $H_i=H\cap\mathfrak{G}_i$, $i=2,\dots,m-n+2+s$. Then it is easy to see that for some i_0, $G_1=\mathfrak{G}/H_{i_0}$ is a group of order $4\times 3^{m+1}$.

Let $\psi:\Lambda_1\longrightarrow\mathfrak{G}$ be the natural epimorphism. Denote the inverse image of H_{i_0} in Λ_1 by Γ_2. Then Γ_2 is a nonorientable surface group without period-cycles as a subgroup of Γ of odd index. Moreover $\Lambda_1/\Gamma_2=\psi(\Lambda_1)/\psi(\Gamma_2)=\tilde{G}/H_{i_0}=G_1$. As a consequence G_1 is a group of order $4\times 3^{m+1}$ generated by three elements a,b and c of order 2 such that the relations (4.3.14.1) hold nontrivially, ab, ac generate the whole group G_1 and bc has order N. The group G_1 is supersoluble as a factor group of $\tilde{G}$. Let us represent G_N and G_1 as the factors F/R_N and F/R respectively, where F is a free group on x, y and z and let $K=\langle w|w^M\rangle$. We claim that the formulas

$$w^x=w,\ w^y=w^{-1},\ w^z=w^{-1}$$

induce an action of G_1 on K. The group G_N is a homomorphic image of G_1 and so in particular $R\subseteq R_N$. But the reader can easily check that for each defining relator r of G_N $w^r=w$. Thus this holds also for any element of R_N and so in particular for any element of R as we wanted.

Now we form the semidirect product G of K by G_1. Then it is easy to check that

$$A=a,\ B=b,\ C=wc$$

are elements of order 2 such that AB, AC, BC have orders 2, 6 and NM and in addition AB, AC generate the whole group G. Also G is supersoluble by 4.3.1(d). Thus G is the group we have looked for.

As we saw in 4.3.14, a nonorientable Klein surface admitting a supersoluble group of the maximal possible order as a group of automorphisms has odd algebraic genus unless p=2. This fact leads us to study surfaces of even algebraic genus.

Theorem 4.3.16. *Let X be a nonorientable Klein surface of even algebraic genus* $p\geq 2$ *and let G be a supersoluble group of automorphisms of X. Then* $|G|\leq 4p$. *This bound is attained for every even p and the corresponding Klein surface has p boundary components and so it is unique up to topological type.*

Proof. Let G be a supersoluble group of automorphisms of a nonorientable Klein surface of even algebraic genus p. Then $G=\Lambda/\Gamma$ for some NEC group Λ and a nonorientable surface group Γ of algebraic genus p. Let $|G|=N$. First we shall see that $\mu(\Lambda)\geq\pi(N-4)/2N$. We can assume that $\mu(\Lambda)<\pi/2$ since otherwise there is nothing to do. So by 4.1.1 Λ can be assumed to have one of the signatures specified there. Groups with signatures

$$(0;+;[-];\{(2,2,3,l)\}),\ (0;+;[3];\{(2,2)\}),\ (0;+;[2,3];\{(-)\}),$$

can be eliminated using considerations similar to those used in the proof of theorem 4.3.14. So Λ can be assumed to have signature

$$(0;+;[-];\{(2,2,2,l)\}).$$

Clearly 4 divides $|G|$ and l is even. Now $\mu(\Lambda)=\pi(l-2)/2l$ and so $|G|=$

$=4l(p-1)/(l-2)$. Let $l=2k'$. Then $l-2=2(k'-1)$ and so we see that if k' is odd then 4 does not divide $|G|$. Thus k' is even and so 8 divides $|G|$ and Λ can be assumed to have signature

$$(0;+;[-];\{(2,2,2,4k)\})$$

and $|G|=8k(p-1)/(2k-1)$.

By 4.1.2 G is generated by three elements a,b and c of order 2 such that ab and ac have orders 2 and 4k respectively and generate the whole group G. Let q be the order of bc.

First we shall see that 2 divides q and 4 does not. In fact let $k=2^{\alpha}m$ and $q=2^{\beta}n$, where m and n are odd. By 4.3.1 there exists a normal subgroup H of G of odd order such that G/H is a 2-group. Clearly the images $\tilde{a}, \tilde{b}$ and $\tilde{c}$ of a,b and c respectively are elements of order 2. Moreover $\tilde{a}\tilde{b}$, $\tilde{a}\tilde{c}$ and $\tilde{b}\tilde{c}$ have orders 2, $2^{\alpha+2}$ and 2^{β} and $\tilde{a}\tilde{b}$, $\tilde{a}\tilde{c}$ generate G/H. By 4.3.8 $\beta \leq 1$. Clearly $\beta \neq 0$ since otherwise G/H would be a cyclic group of order 2 (generated by $\tilde{a}\tilde{b}$), whilst $\tilde{a}\tilde{c}$ is an element of order ≥ 4. So $\beta=1$.

Now we shall see that $q=2$. Assume to get a contradiction that this is not the case and let us assume that G has been chosen with the smallest possible order. Since 4 does not divide q, G cannot be a 2-group. Let p' be the greatest odd prime dividing $|G|$. Then by 4.3.1(c) there exists a normal subgroup L of G of order p'. From the minimality of G we deduce that $(bc)^2 \in L$ and so $q=2p'$. Thus $L=<(bc)^2>$. Since L is normal in G, $a(bc)^2a$ is a power of $(bc)^2$. Hence we can apply 4.3.11 to deduce that $(bc)^2=1$, a contradiction. We see that $q=2$ and so from 4.3.11 G is a dihedral group of order $N=8k$ and in this case $\mu(\Lambda)=\pi(N-4)/2N$.

Now by the Hurwitz Riemann formula

$$N=|G|=\mu(\Gamma)/\mu(\Lambda)\leq 2\pi(p-1)/\pi((N-4)/2N)=4N(p-1)/(N-4)$$

and so $N\leq 4p$. This completes the first part of the proof.

Now given an even integer $p\geq 2$ let $D_{2p}=<x,y|x^2,y^2,(xy)^{2p}>$. Choosing $a=x$, $b=(xy)^p$ and $c=y$ we see from 4.1.2 that D_{2p} can be presented as Λ/Γ, where Λ is an NEC group with signature

$$(0;+;[-];\{(2,2,2,2p)\})$$

and Γ is a nonorientable surface group of algebraic genus p. Finally bc has order 2 and so Γ has p empty period-cycles. Hereby $X=D/\Gamma$ is the Klein surface we have looked for. From the proof of the first part it follows that this surface is unique up to the topological type. This completes the proof.

4.4. NILPOTENT GROUPS OF AUTOMORPHISMS OF KLEIN SURFACES

A finite nilpotent group is a direct product of its Sylow subgroups. As a result an M^*-group cannot be nilpotent. So if $G=\Lambda/\Gamma$ is a nilpotent surface kernel factor group of an NEC group Λ then $\mu(\Lambda)>\pi/6$. By 4.1.1 $\mu(\Lambda)\geq\pi/4$ and the bound is attained only for an NEC group Λ with signature $(0;+;[-];\{(2,2,2,4)\})$. By 4.1.2 G is generated by elements of order 2 and so since G is nilpotent it is a 2-group. Thus using the Hurwitz Riemann formula we obtain at once the following

Corollary 4.4.1. *A nilpotent group of automorphisms of a Klein surface of algebraic genus* p *has at most* 8(p-1) *elements. Moreover this bound may be attained only for* p-1 *being a power of* 2.

Our next goal is to describe the species of Klein surfaces with nilpotent group of automorphisms of maximal possible order. Our first result in this direction is the following

Theorem 4.4.2. *A Klein surface* X *of algebraic genus* $p\geq 3$ *admitting a nilpotent group of order* 8(p-1) *as a group of automorphisms is orientable.*
Proof. Let $X=H/\Gamma$, where Γ is a nonorientable surface group of algebraic genus p and let $G=\Lambda/\Gamma$ be a nilpotent group of automorphisms of X of order 8(p-1). Then Λ has signature $(0;+;[-];\{(2,2,2,4)\})$ and so by lemma 4.1.2 G is generated by three elements a,b and c of order 2 obeying nontrivially the relations

$$(ab)^2=(ac)^4=1$$

and in addition ab and ac generate the whole group G. Since bc=(ab)(ac) is the product of two elements of order 2 and 4 respectively we see that $bc\neq 1$ and so by 4.3.8, bc has order 2. But then G is a dihedral group of order 8, *i.e.* p=2.

Theorem 4.4.3. *(1) There are only two topological types of Klein surfaces of algebraic genus* p=2 *with a nilpotent group of automorphisms of order 8: a torus with one hole and a real projective plane with two holes.*

(2) Let X *be a Klein surface of algebraic genus* $p\geq 3$ *with* k *boundary components having a nilpotent group of automorphisms of order* 8(p-1). *Then*

(i) X *is orientable.*

(ii) $p-1=2^n$, *for some integer* n.

(iii) $k=2^m$ *for some integer* m *such that* $1\leq m\leq\begin{cases}2 & \text{if } n=1,\\ n & \text{if } n\neq 1.\end{cases}$

Conversely, given integers n *and* m *such that* $1 \leq m \leq \begin{cases} 2 & \text{if } n=1, \\ n-1 & \text{if } n \neq 1 \end{cases}$ *there exists an orientable Klein surface of algebraic genus* $p=2^n+1$ *with* $k=2^m$ *boundary components and with nilpotent group of automorphisms of order* 8(p-1).

Proof. Let $X=H/\Gamma$ be a Klein surface of algebraic genus $p \geq 2$ with a nilpotent group of automorphisms G of order 8(p-1). Then $G \cong \Lambda/\Gamma$, where Λ is an NEC group with signature

$$(0;+;[-];\{(2,2,2,4)\}).$$

By 4.1.2 G is generated by three elements a,b and c of order 2 such that ab and ac have orders 2 and 4 respectively.

First let p=2. Then G is a group of order 8. Moreover we see that a and c generate in G a dihedral subgroup of order 8 and so G is dihedral itself and $b \in \langle a,c \rangle$. If b=a then ab=1, and if b=c or b=aca then ab has order 4. So $b=(ac)^2$ or b=cac. In the first case bc has order 2 and ab, ac generate the group G whilst in the second one bc has order 4 and clearly the group generated by ab and ac is a cyclic group of order 4. So by 4.1.2 $G=D_4$ acts on two Klein surfaces: a real projective plane with 2 boundary components and on a torus with 1 boundary component. This completes the proof of (1).

Now we will prove the first part of (2). X is orientable by the previous theorem, $p-1=2^n$ for some $n \geq 1$, by 4.4.1, and by 4.1.2 $k=2^m$ is such a power also. First we will show that $m \neq 0$. In fact if this were not so then, using 4.1.2, there would exist a group G of order 2^{n+3} generated by three elements a,b and c of order 2 such that ab, ac and bc would be elements of order 2, 4 and $q=2^{n+2}$ respectively. So G would be a dihedral group generated by b and c. But then $a=(bc)^{q/2}$ or $a=(bc)^{\alpha}b$ for some α in range $0 \leq \alpha \leq q-1$. The first case is clearly impossible since ac has order 4 whilst in the second case $\alpha=q/2$ because ab has order 2. But then $ac=(bc)^{\alpha+1}$ would have order q since $\alpha+1$ is odd. So q=4 and thus n=0, a contradiction. On the other hand if m>n then there would exist a group G generated by elements a,b,c of order 2 such that ab, ac and bc have orders 2,4 and $q=2^{n-m+2}<2^2$. So q=2 and thus G is a group of order ≤ 16. Hence p=3 *i.e.* n=1 and n-m+2=1, *i.e.* m=2. This completes the proof of the first part of (2).

Now let $n \geq 2$ and let m be an integer such that $1 \leq m < n$. We shall see that there exists an orientable Klein surface of algebraic genus $p=2^n+1$ with $k=2^m$ boundary components and with a nilpotent group of automorphisms of order 8(p-1).

For, let r=n-m+2. Form the semidirect product G of the cyclic group $H=\langle w|w^{2^r}\rangle$ by the Klein four group $K=\langle x,y|x^2,y^2,[x,y]\rangle$ subject to the following action of K on H:

$$w^x = w^{2^{r-1}+1}, \quad w^y = w^{-1}.$$

Then a=x, b=y and c=yw have order 2, they generate G and ab, ac, bc have orders 2, 4 and 2^r respectively. Clearly G is a nilpotent group of order 2^{r+2}.

Now since n>m, r>2 and so there exists an NEC group Λ_1 with signature

$$(0;+;[-];\{(2,4,2^r)\}).$$

Consider the homomorphism θ from Λ onto G induced by the assignment:

$$\theta(c_0)=\theta(c_3)=b, \ \theta(c_1)=a, \ \theta(c_2)=c, \ \theta(e)=1.$$

By 2.2.4 $\Gamma_1=\ker\theta$ has no proper periods and by 2.3.3 it has no period-cycles. So Γ_1 is a surface group without period-cycles. Now using similar arguments to those used in the proof of (iii) in 4.3.15(2) we complete the proof. For, notice first that there exists a surface group Γ_2 such that:

(i) Γ_2 is a surface group without period-cycles,

(ii) Γ_2 is normal in Λ_1,

(iii) $|\Gamma_1/\Gamma_2|=2^s$, where $s\geq m-1$.

In fact it is sufficient to choose $\Gamma_2=\Gamma_l^{(2)}$, the lth 2-Frattini subgroup for l large enough.

Let $\mathfrak{G}=\Lambda_1/\Gamma_2$, $H=\Gamma_1/\Gamma_2$. Then $\mathfrak{G}/H\cong G$. Now $\mathfrak{G}$ is a 2-group and so there exists a normal series $\{M_i\}$ of $\mathfrak{G}$ with factors of order 2. Let $H_i=M_i\cap H$. Then it can be easily checked that there exists i such that H_i has order 2^{s-m+1}. Then $G_1=\mathfrak{G}/H_i$ is a group of order 2^{n+3}.

Let $\psi:\Lambda_1\longrightarrow\mathfrak{G}=\Lambda_1/\Gamma_2$ be the canonical epimorphism and let Γ_3 denote the inverse image of H_i. Then Γ_3 is a surface group without period-cycles as a subgroup of Γ_1 of finite index. Moreover using 2.2.4 and 2.3.3 once more we see that G_1 is generated by three elements a_1,b_1 and c_1 of order 2 such that a_1b_1, a_1c_1 and b_1c_1 have orders 2, 4 and 2^r respectively. By 4.3.8 a_1b_1 and a_1c_1 generate a proper subgroup of G_1. Therefore using 4.1.2 we obtain that G_1 acts on an orientable Klein surface of algebraic genus $p=2^n+1$ having $k=2^m$ boundary components.

To finish the proof we need to show the existence of orientable surfaces of algebraic genus p=3 with 2 and 4 boundary components and a nilpotent group of automorphisms of order 16. For, let $G=D_4\times Z_2=\langle x,y|x^2,y^2,(xy)^4\rangle\times\langle z|z^2\rangle$ and consider two triples (a,b,c)=(x,xz,yz) and (a′,b′,c′)=(x,z,y) of generators of G of order 2. Then #(ab)=#(a′b′)=2, #(ac)=#(a′c′)=#(bc)=4 and #(b′c′)=2 and so the assertion follows from 4.1.2. This completes the proof.

For nonorientable surfaces we get:

Theorem 4.4.4. *Let* G *be a nilpotent group of automorphisms of a nonorientable* *bordered Klein surface of algebraic genus* $p \geq 3$. *Then* $|G| \leq 4p$. *Moreover this bound is attained if and only if* p *is a power of* 2. *In such a case* $G = D_{2p}$ *and the corresponding Klein surface has* p *boundary components.*

Proof. Let $G = \Lambda/\Gamma$, where Λ is an NEC group and Γ is a nonorientable surface group of algebraic genus p, *i.e.* $\mu(\Gamma) = 2\pi(p-1)$. Assume that $|G| = N$. First we will show that $\mu(\Lambda) \geq \pi(N-4)/2N$. If $\mu(\Lambda) \geq \pi/2$, then there is nothing to do. So we can assume that $\mu(\Lambda) < \pi/2$ and thus that Λ has one of the signatures listed in 4.1.1. Since Γ has a nonempty set of period-cycles and it is nonorientable, the period-cycle of Λ is nonempty by 2.1.3.

Notice also that since G is a product of its Sylow subgroups all periods in this period-cycle are powers of 2; otherwise Γ would have a nontrivial period-cycle by 2.3.3 or a proper period by 2.2.4.

If Λ has signature $(0;+;[3];\{(2,2)\})$ then $x_1 = e^{-1}$ and since G is a direct product of its Sylow subgroups the images in G of e and c_i commute for all i. In particular the relation $ec_0e^{-1} = c_2$ in Λ implies $c_0 = c_2$ in the factor group. Now since Γ is a bordered surface subgroup of Λ, a canonical reflection of Λ belongs to Γ. If c_0 belongs to Γ then c_1 is the only orientation reversing proper generator of Λ and it is easy to see that it and the other proper generators of Λ cannot produce a nonorientable word in Γ. In fact using the relation $x_1 = e^{-1}$ in Λ and the fact that the images of x_1 and c_i in G commute, one can choose such a word w in the form $w = x_1^k c_1$. But then c_1 would be equal to x_1^k in the quotient, a contradiction since the image of x_1 has order 3 whilst the image of c_1 has order 2. Similarly if c_1 belongs to Γ then the remaining proper generators of Λ cannot produce a nonorientable word in Γ.

Thus we can assume that Λ has signature

(4.4.4.1) $$(0;+;[-];\{(2,2,2,k)\}).$$

We will show that the only nilpotent nonorientable surface kernel factor of Λ is the dihedral group of order 2k. In fact let G be a nonorientable surface kernel factor of Λ. Then by 4.1.2 G is generated by three elements a, b and c of order 2 such that ab and ac have orders 2 and $k \geq 4$ respectively and ab, bc generate the whole group G. But then by 4.3.8 bc has order 2 and so G is a dihedral group of order $2k = N$.

Now by the Hurwitz Riemann formula

$$N = |G| = \mu(\Gamma)/\mu(\Lambda) \leq 2\pi(p-1)/\pi((N-4)/2N) = 4N(p-1)/(N-4).$$

So $N \leq 4p$. This completes the first part of the proof. Also, if $N = 4p$ we get $G = D_{2p}$. Since G is nilpotent, p must be a power of 2 and $X = H/\Gamma$ has, by 4.1.2, p boundary components.

Conversely if $p \geq 3$ is a power of 2, let Λ be an NEC group with signature $(0;+;[-];\{(2,2,2,2p)\})$ and let $G = \langle x,y \mid x^2, y^2, (xy)^{2p} \rangle$ be the dihedral group of order 4p. We define

$$\theta(c_0)=x,\ \theta(c_1)=1,\ \theta(c_2)=(xy)^p,\ \theta(c_3)=y,\ \theta(c_4)=x,\ \theta(e)=1.$$

It is clear that this assignment induces a homomorphism from Λ onto G and $\Gamma=\ker\theta$ is a surface group. Moreover $w=(c_0c_3)^pc_2$ is a nonorientable word in Γ. Thus Γ is nonorientable. By the Hurwitz Riemann formula, the algebraic genus of Γ is p. By 4.1.2 Γ has p empty period-cycles. Thus $X=H/\Gamma$ is a nonorientable Klein surface of algebraic genus p with p boundary components admitting the dihedral group of order 4p as a group of automorphisms.

Nilpotent groups of automorphisms of Klein surfaces of maximal possible order turn out to be 2-groups by 4.4.1 and 4.4.4. The remaining part of the chapter is devoted to the study of p-groups (p an odd prime) of automorphisms of Klein surfaces. (Up to now we denoted the algebraic genus by p. In order to follow this convention we denote in last two theorems of this section a prime number by q.) We start with the nonorientable case. What is a little strange, this case turns out to be easier than the orientable one.

Theorem 4.4.5. *Let* G *be a* q*-group* (q *an odd prime*) *of automorphisms of a nonorientable Klein surface of algebraic genus* $p\geq 2$. *Then* $|G|\leq q(p-1)/(q-1)$.

The necessary and sufficient condition for the existence of a nonorientable Klein surface of algebraic genus $p\geq 2$ *with* k *boundary components and* q*-group of automorphisms of order* $q(p-1)/(q-1)$ *is:*

(a) $p=(q-1)q^n+1$ *for some* $n\geq 0$,

(b) $k=\begin{cases} 1 \text{ or } q & \text{when } n=0,\\ q^r & \text{for some } r \text{ in range } 0\leq r\leq n, \text{ when } n>0.\end{cases}$

Proof. Let G be a group in question. Then as always we represent G as a factor Λ/Γ, where Λ is an NEC group and Γ is a nonorientable surface group with $\mu(\Gamma)=2\pi(p-1)$. We will show that $\mu(\Lambda)\geq 2\pi(q-1)/q$.

Since Γ is a bordered surface group, $\sigma(\Lambda)$ has some period-cycle. Since $[\Lambda:\Gamma]=|G|$ is odd, Λ is nonorientable by 2.1.2. Thus in particular the orbit genus of Λ is greater than or equal to 1. Moreover all period-cycles of Λ are empty, by 2.3.1. We can assume that Λ has a proper period, since otherwise $\mu(\Lambda)$ would be a multiple of 2π. Moreover all proper periods are powers of q, since in the other case they would produce proper periods in Γ by 2.2.3. But then $\mu(\Lambda)\geq 2\pi(q-1)/q$, as desired.

Now by the Hurwitz Riemann formula we obtain that $|G|=\mu(\Gamma)/\mu(\Lambda)\leq$ $\leq 2\pi(p-1)/2\pi((q-1)/q)=q(p-1)/(q-1)$. This completes the first part of the proof. Now we shall prove the second part.

The bound $2\pi(q-1)/q$ is attained only for an NEC group Λ with signature

(4.4.5.1) $\qquad(1;-;[q];\{(-)\}).$

Of course, for some $n\geq 0$, $q^{n+1}=|G|=2\pi(p-1)/2\pi((q-1)/q)$ *i.e.* $p=(q-1)q^n+1$.

Moreover, from 2.4.2, part (iii), $k=q^r$ for some r in range $0 \leq r \leq n+1$. Even more, let us call a,b,u the images in $G=\Lambda/\Gamma$ of the canonical generators d, x and e of Λ. If $u=1$ we get $a^2b=1$. Hence G is a cyclic group generated by a and, since $x^q=1$, G has order q. Thus, if $n>0$, we get $u \neq 1$ and using again 2.4.2, $k \neq q^{n+1}$. This proves the necessity.

To prove the sufficiency assume first that $n=0$. Then $G=\langle w|w^q \rangle$. We define two epimorphisms $\theta_1,\theta_2:\Lambda \longrightarrow G$ as follows:

$$\theta_1(c)=1,\ \theta_1(x)=w,\ \theta_1(e)=w,\ \theta_1(d)=w^{-1},$$
$$\theta_2(c)=1,\ \theta_2(x)=w,\ \theta_2(e)=1,\ \theta_2(d)=w^{(q-1)/2}.$$

Then by 2.1.2 and 2.3.1 $\ker\theta_1$ and $\ker\theta_2$ are nonorientable surface groups with 1 and q empty period-cycles, respectively.

Now assume that $n \geq 1$ and r is an integer in range $0 \leq r \leq n$. Let $s=n-r+1$. All reduces to find a group G of order q^{n+1} generated by two elements a and b of orders q and q^s respectively. Once it will be done, let u be an element of G verifying $u^2=(ab)^{-1}$ (such an element exists because G has odd order). Now if Λ is an NEC group with signature (4.4.5.1), then the kernel Γ^* of the epimorphism $\eta:\Lambda \longrightarrow G$ defined by

$$\eta(c)=1,\ \eta(x)=a,\ \eta(e)=b,\ \eta(d)=u$$

is, by 2.1.2, 2.2.3 and 2.3.1, a nonorientable surface group with q^r period-cycles and so $X=H/\Gamma^*$ is the surface we are looking for.

To construct G we start with a cyclic group G_1 of order q^s. Assume first that $(q,s) \neq (3,1)$, and let w be the generator of G_1 and Λ_s be a fuchsian group with signature $\sigma_s=(0;+;[q,q^s,q^s];\{-\})$. We define $\theta:\Lambda_s \longrightarrow G_1$ as follows:

$$\theta(x_1)=w^{q^{s-1}},\ \theta(x_2)=w,\ \theta(x_3)=w^{-(q^{s-1}+1)}.$$

Then $\ker\theta=\Gamma$ is a fuchsian surface group.

We claim that there exists a normal subgroup Γ_1 of Λ_s such that

(i) Γ_1 is a fuchsian surface group,

(ii) $|\Gamma/\Gamma_1|=q^t$, where $t \geq r$.

In fact it is sufficient to take $\Gamma_1=\Gamma_l^{(q)}$, the lth q-Frattini subgroup of Γ for l large enough.

Let $\mathfrak{G}=\Lambda_s/\Gamma_1$ and $H=\Gamma/\Gamma_1$. Then $\mathfrak{G}/H \cong G_1$ and in particular $|\mathfrak{G}|=q^{s+t}$. Now $\mathfrak{G}$ has a normal series $\{M_i\}_{0 \leq i \leq s+t}$ with $M_0=\mathfrak{G}$, $M_{s+t}=1$ and cyclic factors M_i/M_{i+1} of order q. Let $H_i=H \cap M_i$, $i=0,\dots,s+t$. Then it is easy to check that for some i_0, $G=\mathfrak{G}/H_{i_0}$ is a group of order q^{n+1}. We claim that this is the group we have looked for. In fact let $\psi:\Lambda_s \longrightarrow \mathfrak{G}$ be the natural epimorphism. Denote the inverse image of H_{i_0} in Λ_s by Γ_2. Then Γ_2 is a fuchsian surface group as a subgroup of Γ. Moreover $\Lambda_s/\Gamma_2 \cong \psi(\Lambda_s)/\psi(\Gamma_2) \cong \mathfrak{G}/H_{i_0}=G$. Thus the images a and b of x_1 and x_2 have orders q and q^s respectively and generate G as desired.

Notice that all above does not hold true for $q=3$ and $s=1$, because in such a case $\mu(\sigma_s)=0$. So, in order to complete the proof it remains to show that given a positive integer n there exists a 3-group of order 3^n generated by two elements of order 3. For, notice that this is obvious for $n=1$ and 2. For $n=3$ one can easily check that the order of a group G with presentation $<a,b|a^3,b^3,(ab)^3,(ab^{-1})^3>$ is 27. Finally for $n=4$ consider the group of order 81 from the Burnside list with presentation

$$<a,b,c|a^9,b^3,c^3,[a,b],b^{-1}[a,c],[c^{-1},b]a^3>.$$

Then $b=a^{-1}c^{-1}ac$ and so b is superfluous. Moreover it can be easily checked that $(ac)^3=1$. So G is a surface kernel factor of a fuchsian group Λ with signature $(0;+;[3,3,9];\{-\})$ and in the same way as in the case $(q,s)\neq(3,1)$, changing Λ_s above by Λ, we show the existence of such a factor of order 3^n for arbitrary $n\geq 5$.

Our final goal in this section is to prove a similar result concerning orientable Klein surfaces. We will need a definition.

Given $n\geq 1$ let $N(n,q)=N(n)$ be the smallest integer such that there exists a group of order $q^{N(n)}$ generated by two elements of order q whose product has order q^n.

Remark 4.4.6. The number N(n) exists for every $n\geq 1$. In fact it is obvious that $N(1)=1$. If $n>1$, then there is a fuchsian group Γ with signature $(0;+;[q,q,q^n];\{-\})$ and one can show using theorem 2.2.4 that $\Gamma_l^{(q)}$ is a fuchsian surface group for l large enough (*e.g.* $l=n$) and so $\Gamma/\Gamma_l^{(q)}$ is a finite group generated by two elements (images of x_1 and x_2) of order q whose product has order q^n.

The following theorem gives a bound for the order of q-groups of automorphisms of orientable Klein surfaces as well as it classifies species of the corresponding surfaces up to the knowledge of the sequence $\{N(n)\}_{n\in\mathbb{N}}$.

Theorem 4.4.7. *Let* G *be a* q*-group* (q *an odd prime*) *of automorphisms of an orientable Klein surface of algebraic genus* $p\geq 2$. *Then* $|G|\leq q(p-1)/(q-2)$.

The necessary and sufficient condition for the existence of a nonorientable Klein surface of algebraic genus $p\geq 2$ *with* k *boundary components and* q*-group of automorphisms of order* $q(p-1)/(q-2)$ *is:*

(a) $p=(q-2)q^n+1$ *for some* $n\geq 0$,

(b) $k=q^{n-s+1}$ *for some* s *such that* $N(s)\leq n+1$.

Proof. The proof is very similar to the proof of the previous theorem. Let G be a group in question. Then G can be represented as a factor Λ/Γ where Λ is an NEC group and Γ is an orientable surface group with $\mu(\Gamma)=2\pi(p-1)$. Similarly as in the previous theorem we can show that $\mu(\Lambda)\geq 2\pi(q-2)/q$ and this time the bound in question is attained only for a group Λ with signature $(0;+;[q,q];\{(-)\})$. So in particular the first part follows.

To prove the second part first notice that for some $n\geq 0$ $q^{n+1}=|G|= =2\pi(p-1)/2\pi((q-2)/q)$. So $p=(q-2)q^n+1$. Denote by a,b,u the images in $G=\Lambda/\Gamma$ of the canonical generators x_1, x_2 and e. Then a and b have order q. Let $\#(ab)=\#(u)=q^s$ for some s. By 2.4.2 $k=q^{n+1-s}$, where clearly $n+1\geq N(s)$. This proves the necessity.

Now given $n\geq 0$ and s such that $N(s)\leq n+1$ let G_1 be a q-group of order $q^{N(s)}$ generated by two elements of order q whose product has order q^s. As in the previous theorem we deduce the existence of a group G of order q^{n+1} generated by two elements a and b of order q whose product has order q^s. Let Λ be an NEC group with signature $(0;+;[q,q];\{(-)\})$. Then the assignment defined by:

$$\theta(x_1)=a,\ \theta(x_2)=b,\ \theta(e)=(ab)^{-1},\ \theta(c)=1$$

induces an epimorphism $\theta:\Lambda\longrightarrow G$ having an orientable surface group Γ as the kernel. By the Hurwitz Riemann formula Γ has algebraic genus $p=(q-2)q^n+1$ and by 2.4.2 it has q^{n+1-s} period-cycles. This completes the proof.

4.5. ABELIAN GROUPS OF AUTOMORPHISMS OF KLEIN SURFACES

In this last section we deal with abelian groups. Here we are able to give a complete answer to all questions posed at the beginning of the chapter. These groups are always written additively. We start with the orientable case.

Theorem 4.5.1. *Let* A *be an abelian group of automorphisms of an orientable Klein surface of algebraic genus* $p\geq 2$, $p\neq 5$. *Then* $|A|\leq 2(p+1)$. *There exists an orientable Klein surface of algebraic genus* $p=5$ *admitting* $Z_2\oplus Z_2\oplus Z_2\oplus Z_2$ *as a group of automorphisms, and this is the unique case in which the inequality* $|A|\leq 2(p+1)$ *does not hold true.*

Proof. Let $A\cong\Lambda/\Gamma$, where Γ is a bordered and orientable surface group and Λ is an NEC group with signature

(4.5.1.1) $\qquad (g;\pm;[m_1,\ldots,m_r];\{(n_{i1},\ldots,n_{is_i})|1\leq i\leq k\}).$

First we will show that $\mu(\Lambda)\geq\pi$ unless $k=1$ and $g=0$. Let $|A|=N$.

By the results of section 3 of chapter 2, $k\geq 1$. If $k\geq 3$ then $\mu(\Lambda)\geq 2\pi$ whilst if $k=2$ then either a period-cycle is nonempty or $r\geq 1$ or $g\geq 1$. But then $\mu(\Lambda)\geq\pi$. So let $k=1$ and call $s_1=s$. If $g>1$ then $\mu(\Lambda)\geq 2\pi$. If $g=1$ then signσ="-", since otherwise $\mu(\Lambda)\geq 2\pi$, but then $r>0$ or $s\geq 2$, and so $\mu(\Lambda)\geq\pi$.

We must analyze now what happens for $k=1$, $g=0$. Then since A is abelian, Λ has signature

$$(0;+;[m_1,\ldots,m_r];\{(2,\overset{s}{\ldots},2)\}),$$

with $s\neq 1$ by 2.4.1.

Assume first that $|A|=N>16$. We will show that $\mu(\Lambda)\geq\pi(N-4)/N$. For, denote by θ the corresponding epimorphism $\Lambda\longrightarrow A$ with $\ker\theta=\Gamma$. Clearly $\theta(c_i)=0$ for some $0\leq i\leq s$, since otherwise Γ has not period-cycles, by results of section 3 in chapter 2. In what follows we will use this observation without referring explicitely to it.

Now assume that $r=0$. Then $s\geq 5$. If $s\geq 6$ then $\mu(\Lambda)\geq\pi$, whilst if $s=5$ then A would be generated by at most 4 elements of order 2 and therefore it would be a group of order ≤ 16. So let $r>0$.

The area $\mu(\Lambda)\geq\pi$ unless either $s=3$, $r=1$ and $m_1\leq 3$ or $s=2$, $r=1$ or $s=0$, $r=2$. But in the first case A would be generated by two elements of order 2 and one of order ≤ 3. So $|A|\leq 12$. Hence let $s=2$ and $r=1$, *i.e.* let Λ have signature of the form

$$(0;+;[m];\{(2,2)\}).$$

Then A is generated by two elements of orders 2 and m respectively. Hence $m=N$ or $m=N/2$. If $m=N$ then $\mu(\Lambda)=\pi(N-2)/N>\pi(N-4)/N$, whilst if $m=N/2$ *i.e.* Λ has signature

(4.5.1.2) $$(0;+;[N/2];\{(2,2)\}),$$

then $\mu(\Lambda)=\pi(N-4)/N$.

Finally, if $s=0$ and $r=2$, Λ has signature

$$\sigma(\Lambda)=(0;+;[k,l];\{(-)\}),\ \max\{k,l\}>2.$$

So A is generated by two elements of orders k and l. But then N divides kl and in particular $N/k\leq l$. So

$$\mu(\Lambda)\geq 2\pi(-1+(1-1/k)+(1-k/N))=2\pi(N(k-1)-k^2)/kN.$$

It is easy to see that the last value is strictly smaller than $\pi(N-4)/N$ if and only if $N<2k$ *i.e.* for $k=N$. But then, since $l\geq 2$,

$$\mu(\Lambda)\geq 2\pi(-1+(1-1/2)+(1-1/N))=\pi(N-2)/N>\pi(N-4)/N.$$

The bound $\mu(\Lambda)=\pi(N-4)/N$ is attained, in case $s=0$ for $l=2$, $k=N/2$, *i.e.* for a group with signature

(4.5.1.3) $$(0;+;[2,N/2];\{(-)\}).$$

Now let $X=H/\Gamma$ be an orientable Klein surface of algebraic genus p and let A be an abelian group of automorphisms of X of order >16. Then $A=\Lambda/\Gamma$ for some NEC group Λ. Then

$$N=|A|=\mu(\Gamma)/\mu(\Lambda)\leq 2\pi(p-1)/\pi((N-4)/N)=2N(p-1)/(N-4).$$

Therefore $N\leq 2(p+1)$.

Now consider abelian groups of order ≤ 16. If A is cyclic then it follows from 3.2.18.(4) that for an orientable Klein surface of algebraic genus p admitting A as a group of automorphisms the inequality $|A| \leq 2(p+1)$ holds. Obviously the inequality also holds when $|A| \leq 6$. So we must deal with non cyclic groups verifying $8 \leq |A| \leq 16$.

Let us consider the group $A = \mathbb{Z}_2 \oplus \mathbb{Z}_4$, for example, and let $A = \Lambda/\Gamma$, where Λ is an NEC group and Γ is an orientable surface group of algebraic genus p. We will show that $\mu(\Lambda) \geq \pi/2$. This clearly holds when $k \geq 2$ or $k=1$ and $g \geq 1$. So we can assume that $k=1$ and $g=0$. Since reflections in Λ provide in A only elements of order 2, Λ has proper periods and one of them is equal to 4. If $r \geq 2$ then $\mu(\Lambda) \geq \pi/2$. If $r=1$ then the period-cycle is nonempty. But then also $\mu(\Lambda) \geq \pi/2$. So

$$8 = |A| = \mu(\Gamma)/\mu(\Lambda) \leq 2\pi(p-1)/(\pi/2) = 4(p-1).$$

Therefore $p \geq 3$ and as a consequence $|A| = 8 \leq 2(p+1)$.

Similar arguments show that the bound in question holds for the genera of surfaces admitting all abelian groups but $A = \mathbb{Z}_2 \oplus \mathbb{Z}_2 \oplus \mathbb{Z}_2 \oplus \mathbb{Z}_2$ as a group of automorphisms. Let $A = \Lambda/\Gamma$, where Γ is a bordered orientable surface group. Since A has only elements of order 2 all proper periods and periods in period-cycles are equal to 2. Therefore $\mu(\Lambda)$ is a multiple of $\pi/2$. In particular $\mu(\Lambda) \geq \pi/2$. Consider an NEC group Λ with signature

$$(0;+;[-];\{(2,2,2,2,2)\})$$

with area $\pi/2$ and a homomorphism $\theta:\Lambda \longrightarrow A$ induced by the assignment:

$$\theta(c_0) = (1,0,0,0),\ \theta(c_1) = (0,0,0,0),\ \theta(c_2) = (0,1,0,0),$$
$$\theta(c_3) = (0,0,1,0),\ \theta(c_4) = (0,0,0,1),\ \theta(c_5) = (1,0,0,0),$$
$$\theta(e) = (0,0,0,0).$$

Clearly $\Gamma = \ker\theta$ is an orientable surface group. By the Hurwitz Riemann formula $16 = |A| = \mu(\Gamma)/\mu(\Lambda) = 2\pi(p-1)/(\pi/2) = 4(p-1)$. So $p=5$. Thus we see that there exists an orientable Klein surface of algebraic genus $p=5$ admitting an abelian group A of order 16 as a group of automorphisms. As we remarked $\pi/2$ is the lower bound for the area of Λ. Therefore A cannot act as a group of automorphisms of a surface of genus $p<5$.

In next theorem we shall see that the bound obtained is the best one.

Theorem 4.5.2. *Given an integer* $p \geq 2$ *there exists an orientable Klein surface* X *of algebraic genus* p *admitting an abelian group* A *of order* $2(p+1)$ *as a group of automorphisms. Necessarily* $A = \mathbb{Z}_2 \oplus \mathbb{Z}_{p+1}$ *and the surface* X *may have* 1, 2 *or* p+1 *boundary components. The last possibility always occurs, whilst the first actually occurs when* p *is even and the second when* p *is odd.*

Proof. We showed in the previous theorem that if A is an abelian group of order $N = 2(p+1)$ and $A = \Lambda/\Gamma$, where Λ is an NEC group and Γ is an orientable surface group of algebraic genus p, then $\mu(\Lambda) \geq \pi(N-4)/N$ and the bound may be only attained for NEC groups with signature

$$(0;+;[2,p+1];\{(-)\}),\ \text{or}\ (0;+;[p+1];\{(2,2)\}),$$

see (4.5.1.2) and (4.5.1.3).

Let Λ be an NEC group with the first signature and assume that A is an orientable surface kernel factor of Λ. Then A is clearly generated by two elements of orders 2 and p+1 respectively. So $A=\mathbb{Z}_2\oplus\mathbb{Z}_{p+1}$. On the other hand consider the homomorphism $\theta:\Lambda\longrightarrow A=\mathbb{Z}_2\oplus\mathbb{Z}_{p+1}$ given by

$$\theta(c)=(0,0),\ \theta(e)=(-1,-1),\ \theta(x_1)=(1,0),\ \theta(x_2)=(0,1).$$

Then $\Gamma=\ker\theta$ is an orientable surface group. Since $\mu(\Lambda)=\pi(N-4)/N$, $X=H/\Gamma$ is a Klein surface of algebraic genus p having A as a group of automorphisms. Now $\theta(e)$ has order 2(p+1) if p is even and p+1 if p is odd. So by 2.3.3 X has 1 or 2 boundary components according as p is even or odd respectively.

Now let Λ be an NEC group with the second signature. Also here $A=\mathbb{Z}_2\oplus\mathbb{Z}_{p+1}$ is the only candidate for an orientable surface kernel factor group of Λ. We will show that such a factor actually exists. For, consider the homomorphism $\theta:\Lambda\longrightarrow A$ induced by the assignment

$$\theta(c_0)=(1,0),\ \theta(c_1)=(0,0),\ \theta(c_2)=(1,0),\ \theta(x)=(0,1),\ \theta(e)=(0,-1).$$

Clearly $\Gamma=\ker\theta$ is an orientable surface group. So $X=H/\Gamma$ is an orientable Klein surface of genus p having, by 2.3.3, p+1 boundary components.

Dealing with nonorientable surfaces we obtain

Theorem 4.5.3. *Let* A *be an abelian group of automorphisms of a nonorientable Klein surface of algebraic genus* $p\geq 2$, $p\neq 3$. *Then* $|A|\leq 2p$. *There exists a nonorientable Klein surface of algebraic genus* p=3 *admitting* $\mathbb{Z}_2\oplus\mathbb{Z}_2\oplus\mathbb{Z}_2$ *as a group of automorphisms and this is the unique case in which the inequality* $|A|\leq 2p$ *does not hold.*

Proof. Let $A=\Lambda/\Gamma$ be a nonorientable surface kernel factor and let θ be the corresponding homomorphism. As in the proof of 4.5.1 we argue that $\mu(\Lambda)\geq\pi$ unless k=1 and the orbit genus g of Λ is 0. So let Λ has signature

$$(0;+;[m_1,\dots,m_r];\{(2,\overset{s}{\dots},2)\}),$$

where periods in the period-cycle are equal to 2 since A is abelian.

Assume first that $|A|=N>8$. We will show that $\mu(\Lambda)\geq\pi(N-2)/N$. Since Λ and Γ have different sign, N must be even, by 2.1.2.

If s=0 then, since Γ has a nonempty set of period-cycles, the corresponding reflection of Λ belongs to Γ. Thus there are no nonorientable word in Γ and so Γ would be orientable by 2.1.2. As a result $s\geq 2$.

If r=0 then $s\geq 5$, since $\mu(\Lambda)>0$. If $s\geq 6$ then $\mu(\Lambda)\geq\pi$ whilst if s=5 then, since Λ is generated by reflections, at least one of them being in Γ, and a nonorientable word belongs to Γ, A would be generated by three elements of order 2 and then $|A|\leq 8$. So let r>0.

If $s\geq 4$ or if s=3 and $r\geq 2$ then $\mu(\Lambda)\geq\pi$. If s=3, r=1 and $m_1=2$ then $|A|\leq 8$, whilst if $m_1 \geq 4$ then $\mu(\Lambda) \geq \pi$. We claim that $|A| \leq 6$ when $m_1=3$. In fact by 2.1.2 a nonorientable word w would belong to Γ. Since Γ has period-cycles, a canonical reflection of Λ would belong to Γ. If one of c_0, c_3 belongs to Γ then the second one also does. Since Γ is a surface group $c_1,c_2 \notin \Gamma$. So $w=w(x_1,c_1,c_2)$. But since w is nonorientable, either c_1 or c_2 appears in Γ an odd number of times. But then $\theta(c_1)$ or $\theta(c_2)$ is redundant and so A is an abelian group generated by two elements of orders 2 and 3 respectively, *i.e.* $|A| \leq 6$. In a similar way we arrive to this conclusion when c_1 or c_2 belongs to Γ.

If $s=2$ then $r \geq 1$, since otherwise $\mu(\Lambda)<0$, but if $r \geq 2$ then $\mu(\Lambda) \geq \pi$. So let $s=2$ and $r=1$ *i.e.* let Λ have signature $(0;+;[m];\{(2,2)\})$. Then $\theta(c_i)=0$ for some i, say for $i=1$. Since Γ is a nonorientable surface group, a nonorientable word belongs to Γ by 2.1.2. But $\theta(c_0)=\theta(c_2)$, $\theta(e)=-\theta(x)$ and so this word can be chosen to be $w=x^k c_0$. But then A would be a cyclic group generated by the image of x and therefore $m=N$. As a result $\mu(\Lambda)=\pi(N-2)/N$. The bound is attained only for an NEC group Λ with signature

$$(0;+;[N];\{(2,2)\}). \tag{4.5.3.1}$$

Now let $X=H/\Gamma$ be a nonorientable Klein surface of algebraic genus $p \geq 2$ and let $A=\Lambda/\Gamma$ be an abelian subgroup of Aut(X) of order $N>8$. By the previous part $\mu(\Lambda) \geq \pi(N-2)/N$ and so by the Hurwitz Riemann formula

$$N=|A|=\mu(\Gamma)/\mu(\Lambda) \leq 2\pi(p-1)/\pi((N-2)/N)=2N(p-1)/(N-2).$$

Hence $N \leq 2p$.

Now consider abelian groups of order ≤ 8. The inequality clearly holds when $|A| \leq 4$. By 3.2.18.1. it also holds when A is cyclic. So it only remains to deal with the groups $A=Z_2 \oplus Z_4$ and $A=Z_2 \oplus Z_2 \oplus Z_2$.

Consider the first group and let $A=\Lambda/\Gamma$ be a nonorientable surface kernel factor. We will show that $\mu(\Lambda) \geq \pi/2$. Since reflections in Λ provide in A only elements of order 2, Λ has a proper period equal to 4. Moreover $s \geq 2$ since otherwise Γ would be orientable. But then $\mu(\Lambda) \geq \pi/2$ and the equality holds only for an NEC group Λ with the signature $(0;+;[4];\{(2,2)\})$. Assume that $\theta:\Lambda \longrightarrow A$ is an epimorphism whose kernel is a nonorientable surface group Γ. Since Γ has period-cycles, $\theta(c_i)=0$ for some $i=0,1,2$. If one of $\theta(c_0)$, $\theta(c_2)$ is equal to 0 then the second one also is equal to 0. Assume first that this is the case. Then since $\theta(e)=-\theta(x)$, A is generated by the images of x and c_1. But since Γ is nonorientable, a nonorientable word in Λ belongs to Γ. As before this word can be chosen to be $w=x^k c_1$. As a result $\theta(c_1)=-k\theta(x)$ and so A is a cyclic group of order 4. Similarly we argue that A is a cyclic group when $\theta(c_1)=0$. Thus $\mu(\Lambda)$ must be strictly greater than $\pi/2$ and so by the Hurwitz Riemann formula $p>3$. Therefore $|A|=8 \leq 2p$.

Now let $A=Z_2 \oplus Z_2 \oplus Z_2$ and let $A=\Lambda/\Gamma$, where Γ is a bordered nonorientable surface group. Since A has only elements of order 2, all proper periods and periods in period-cycles are equal to 2. Therefore $\mu(\Lambda)$ is a multiple of $\pi/2$. If $\mu(\Lambda)\neq\pi/2$ then $\mu(\Lambda)\geq\pi>\pi(N-2)/N$ and so $|A|<2p$. If $\mu(\Lambda)=\pi/2$, the algebraic genus p of $X=H/\Gamma$ verifies $8=4(p-1)$, *i.e.* $p=3$. So, it only remains to realize A as a group of automorphisms of a nonorientable surface of algebraic genus 3. Indeed, let Λ be an NEC group with signature

$$(0;+;[-];\{(2,2,2,2,2,2)\}).$$

Then $\mu(\Lambda)=\pi/2$ and take the homomorphism $\theta:\Lambda\longrightarrow A$ induced by the assignment

$$\theta(c_0)=(1,0,0),\ \theta(c_1)=(0,0,0),\ \theta(c_2)=(0,1,0),\ \theta(c_3)=(1,1,0),$$
$$\theta(c_4)=(0,0,1),\ \theta(c_5)=(1,0,0),\ \theta(e)=(0,0,0).$$

Clearly $\Gamma=\ker\theta$ is a surface group. Moreover $w=c_0c_2c_3$ is a nonorientable word in Γ. So Γ is nonorientable. By the Hurwitz Riemann formula $\mu(\Gamma)=4\pi$. Therefore $X=H/\Gamma$ is a nonorientable Klein surface of algebraic genus $p=3$ admitting the group A as a group of automorphisms. This completes the proof.

The counterpart of the last theorem is

Theorem 4.5.4. *Given an integer* $p\geq2$ *there exists a nonorientable Klein surface* X *of algebraic genus* p *with an abelian group of automorphisms of order* 2p. *Such a surface has necessarily* p *boundary components and an abelian group for which this bound is attained must be cyclic.*

Proof. We showed in the previous theorem that if A is an abelian group of order $N=2p$ and $A=\Lambda/\Gamma$, where Λ is an NEC group and Γ is a nonorientable surface group of algebraic genus p, then by 4.5.3.1, $\sigma(\Lambda)=(0;+;[2p];\{(2,2)\})$. Then the image $\tilde{x}$ of x in A has order 2p by 2.2.3 and so A is cyclic. Moreover by 2.3.3, $X=H/\Gamma$ has p boundary components. Finally let $\theta:A\longrightarrow\mathbb{Z}_{2p}$ be the epimorphism induced by

$$\theta(c_0)=\theta(c_2)=p,\ \theta(c_1)=0,\ \theta(x)=1 \text{ and } \theta(e)=2p-1.$$

Clearly $\Gamma=\ker\theta$ is a surface group and $w=c_0x^p$ is a nonorientable word in Γ. So Γ is nonorientable by 2.1.3. As a result A is a group of automorphisms of $X=H/\Gamma$, a nonorientable Klein surface of algebraic genus p which, by 2.3.3, has p boundary components.

4.6. NOTES

Groups of automorphisms of bordered Klein surfaces started to be investigated by May in **[95]**. Groups of big enough order were constructed by May in **[97]**. Theorems 4.2.1 and 4.2.2 were proved by May in **[95]** using a different technique. The approach employed in the proof of 4.2.5 comes from Macbeath **[83]**. Families of M^*-groups presented in 4.2.13 and 4.2.14 were found by May in **[100]**. Although we follow the general line proposed by May, we consider our proofs essentially simplified and more direct. Corollary 4.2.15, that is an immediate consequence of two previous theorems, was proved by May in **[100]**, and 4.2.18 was proved in **[99]**. Among results concerning M^*-groups, first characterized by May **[98]**, not presented in this chapter, it is worth to list a result of May **[99]** saying that for a given prime q $(q\neq 5)$ there is no Klein surface with maximal symmetry having algebraic genus $p=q+1$ as well as that there is no such a surface with $k=q$ boundary components. It is also worth to notice that there are infinitely many simple M^*-groups **[43,46,55,119]**. Theorems 4.3.2 and 4.3.7 were proved by May in **[102]**. Lemma 4.3.8 was proved in **[27]**. Its importance lies in the fact that it shows that the approach suggested in **[101]** for finding nonorientable Klein surfaces of algebraic genus $p\geq 3$ with nilpotent group of automorphisms of order 8(p-1) fails. Theorem 4.4.1 was proved in May **[101]**; our proof is new. Theorem 4.4.2 proved in **[27]** gives a negative answer to the conjecture posed by May in **[101]**. Our theorem 4.4.3 together with an example of May of a bordered Klein surface of algebraic genus $p=2^n+1$ with $k=2^n$ boundary components and a group of automorphisms of order 8(p-1), presented in **[101]**, gives the complete classification of the species of Klein surfaces with a nilpotent group of automorphisms of largest possible order. In **[63,70,71,104,122,123]** bounds were found for the order of a group of automorphisms of bordered orientable Klein surfaces having, either topological genus $g\leq 5$ or $k\leq 3$ boundary components. In **[33,34,35,37,39,40,56,57,59,63,69, 79,83,87,105,110,114,131,132]** the reader can find results of similar type concerning Riemann surfaces. Other problems of finding lower bounds for the order of the largest group of automorphisms that a Klein surface of certain topological type can admit, have been studied in **[1,89,121]**. For the general results concerning group theory that we used here we refer the reader to **[93,109,120]**.

CHAPTER - 5
The automorphism group of compact Klein surfaces with one boundary component

A bordered compact Klein surface S of algebraic genus ≥2 can be represented as a quotient H/Γ for some bordered surface group Γ. Having S so represented, a finite group G is a subgroup of Aut(S) if and only if $G=\Lambda/\Gamma$ for some NEC group Λ. It is extremely difficult to calculate Aut(S) for a given S. Even if one tries only to determine when a group can be realized as the full group of automorphisms Aut(S′) for some S′ with the same topological type as S, a major difficulty appears, since $\mathrm{Aut}(S')=\Lambda'/\Gamma'$ where Γ' is isomorphic to Γ and Λ' is the normalizer of Γ' in $\Omega=\mathrm{Aut}(H)$, and it is a rather difficult problem to determine it. We solve this question for the class of surfaces with one boundary component in Section 2. The key point in the solution is that almost all NEC signatures are maximal in some sense, and given such a maximal signature τ, there exists a maximal NEC group Λ with signature τ. In particular, $\Lambda/\Gamma=\mathrm{Aut}(H/\Gamma)$ for every surface NEC group Γ contained in Λ as a normal subgroup. This is the main result, Theorem 5.1.2, in the first Section. It will be also useful in Ch.6, and opens the door to a vast program: fixed a topological class of surfaces, determine those finite groups that can be realized as the full group of automorphisms of surfaces in the given class.

As a consequence of 5.1.2 we also prove in Corollary 5.1.4 that each finite group is realized as the full group of automorphisms of some surface, both in the orientable and nonorientable cases.

5.1. MAXIMAL NEC GROUPS

Definition 1. An NEC group is said to be *maximal* if there does not exist another NEC group containing it properly.

Definition 2. An NEC signature τ is said to be *maximal* if for every NEC group Λ' containing an NEC group Λ with signature τ, the equality $d(\Lambda)=d(\Lambda')$ (dimensions of Teichmüller spaces) implies $\Lambda=\Lambda'$.

Remarks 5.1.1. (1) Let τ be the signature of a proper NEC group. If τ^+ is

maximal, then so does τ. Indeed, assume on the contrary the existence of an NEC group Λ' containing properly an NEC group Λ with signature τ, and $d(\Lambda)=d(\Lambda')$. Since Λ is a proper NEC group, Λ'^+ contains properly Λ^+ and, by 0.3.2 (i), $d(\Lambda'^+)=2d(\Lambda')=2d(\Lambda)=d(\Lambda^+)$. Hence $\tau^+=\sigma(\Lambda^+)$ is not maximal.
(2) All fuchsian signatures that fail to be maximal are listed in the right hand side columns of 0.3.5 and 0.3.6.

The following theorem establishes the close relation between the notions of maximal signature and maximal NEC group.

Theorem 5.1.2. *Given a maximal NEC signature τ, there exists a maximal NEC group Γ with $\sigma(\Gamma)=\tau$.*

Proof. Let $\mathcal{A}=\{\tau_i|i\in I\}$ be the set of signatures of NEC groups containing some NEC group with signature τ. For every $i\in I$ we fix an NEC group Λ_i with signature τ_i. We also fix an NEC group Λ with signature τ. With the notations in 0.3.7 let us consider, for each $i\in I$, the set $T_i=T(\Lambda,\Lambda_i)$ of all elements $[r]\in T(\Lambda)$ such that, for some representative r of $[r]$, there exists an NEC group Λ_r with signature τ_i containing $r(\Lambda)$. Let $T=\bigcup_{i\in I} T_i$. With these notations all reduces to show that $T\neq T(\Lambda)$. In such a case, for $[r]\in T(\Lambda)\setminus T$, the group $r(\Lambda)=\Gamma$ is maximal and $\sigma(\Gamma)=\tau$. The last is obvious. For the first, assume that Γ is properly contained in an NEC group Λ'. Then $\tau'=\sigma(\Lambda')\in\mathcal{A}$, *i.e.* $\tau'=\tau_i$ for some $i\in I$ and so $[r]\in T_i$, absurd. Since the set of all signatures is countable, so does I and, consequently, it is enough to prove that $\dim T_i<d(\Lambda)=\dim T(\Lambda)$, for every $i\in I$. But, by 0.3.7,

$$T_i=\bigcup_{[\alpha]\in M(\Lambda)} T(\alpha)[\bigcup_{[j]\in I(\Lambda,\Lambda_i)} \operatorname{Im}T(j)]$$

and $M(\Lambda)$ is countable, whilst $I(\Lambda,\Lambda')$ is also countable because Λ is finitely generated and Λ_i is countable. Hence, since $T(\alpha)$ is an isometry, and for every $[j]\in I(\Lambda,\Lambda_i)$ the space $\operatorname{Im}T(j)$ has dimension $d(\Lambda_i)=\dim T(\Lambda_i)$, it is enough to check that $d(\Lambda_i)<d(\Lambda)$ for every $i\in I$. This is obviously true since τ is maximal.

Corollary 5.1.3. *Let Λ be an NEC group containing a surface NEC group Γ as a normal subgroup. Assume that $\sigma(\Lambda^+)$ is maximal. Then the topological surface H/Γ can be endowed with an structure of Klein surface such that* $\mathrm{Aut}(H/\Gamma)=\Lambda/\Gamma$.

Proof. By 5.1.1, $\tau=\sigma(\Lambda)$ is maximal. Then, by 5.1.2 there exists a maximal NEC group Λ' and an isomorphism $\phi:\Lambda\longrightarrow\Lambda'$. In particular, Λ' is the normalizer of $\Gamma'=\phi(\Gamma)$ in Ω and so, the Klein surface $S=H/\Gamma'$, with the same topological type as H/Γ, verifies $\mathrm{Aut}(S)=\Lambda'/\Gamma'\cong\Lambda/\Gamma$.

As a first consequence of 5.1.2 we obtain

Corollary 5.1.4. *Let G be a finite group. There exist compact and bordered Klein surfaces X_1 and X_2 such that $\mathrm{Aut}(X_1)=\mathrm{Aut}(X_2)=G$, X_1 is orientable and X_2 is nonorientable. Moreover, X_1 and X_2 can be chosen having $|G|$ boundary components.*

Proof. Let $u_1,\dots,u_g$ be a system of generators of G. Without loss of generality we can assume $g \geq 2$. Let

$$\tau_1=(g;+;[-];\{(-)\}) \quad\text{and}\quad \tau_2=(2g+1;-;[-];\{(-)\}).$$

Using 2.2.5 to compute τ_1^+ and τ_2^+ we observe they are maximal. By 5.1.1 and 5.1.2 there exist maximal NEC groups Λ_1 and Λ_2 with signatures τ_1 and τ_2. Define epimorphisms $\theta_1:\Lambda_1 \longrightarrow G$ and $\theta_2:\Lambda_2 \longrightarrow G$ by

$$\theta_1(a_i)=\theta_1(b_i)=u_i,\ 1\leq i\leq g;\ \theta_1(e)=\theta_1(c_0)=1;$$

$$\theta_2(d_{2l-1})=u_l,\ \theta_2(d_{2l})=u_l^{-1},\ 1\leq l\leq g;\ \theta_2(d_{2g+1})=\theta_2(e)=\theta_2(c_0)=1.$$

If $k=|G|$, then $\ker\theta_1$ and $\ker\theta_2$ are surface groups with k period-cycles, algebraic genus $p=k(2g-1)+1\geq 2$ and sign of $\sigma(\ker\theta_1)$ is "+", sign of $\sigma(\ker\theta_2)$ is "-". Then, if $X_1=H/\ker\theta_1$ and $X_2=H/\ker\theta_2$, both are compact Klein surfaces of algebraic genus ≥ 2, with $|G|$ boundary components and, since Λ_1 and Λ_2 are maximal, $\mathrm{Aut}(X_1)=\mathrm{Aut}(X_2)=G$.

We obtain now a partial but useful converse of 5.1.2.

Theorem 5.1.5. *Let τ be one of the following signatures:*

(i) $(0;+;[-];\{(-),(-),(-)\})$
(ii) $(0;+;[m];\{(-),(-)\})$
(iii) $(0;+;[m_1,m_2];\{(-)\})$
(iv) $(0;+;[m];\{(2,2)\})$
(v) $(1;-;[-];\{(-),(-)\})$
(vi) $(1;-;[m];\{(-)\})$
(vii) $(2;-;[-];\{(-)\})$

If Λ is an NEC group with signature τ and it contains a bordered surface group Γ as a normal subgroup with abelian factor, then there exists an NEC group Γ'' containing Λ as a subgroup of index 2 and Γ as a normal subgroup.

Proof. None of the above signatures is maximal; the corresponding pairs (σ'',τ) can be found in 2.4.7 and notice that in each case $[\sigma'':\tau]=2$. The proof consists in the following: by 2.4.5.(1), for each σ'' there exists an NEC group Γ'' with signature σ'' containing Λ. We express the canonical generators of Λ in terms of those of Γ''. Using this, we will be able to show that there exists $a\in\Gamma''\backslash\Lambda$ such that $a\gamma' a^{-1}=\gamma'^{-1} \bmod [\Lambda,\Lambda]$ for every canonical generator γ' of Λ. We claim that then $a\gamma a^{-1}\in\Gamma$ for every element γ of Γ. Indeed let us represent such γ as $\gamma(w_1,\dots,w_s)$, where $w_1,\dots,w_s$ are canonical generators of Λ. Then, since $G=\Lambda/\Gamma$ is abelian, for the canonical epimorphism $\theta:\Lambda \longrightarrow G$, we have

$$\theta(a\gamma(w_1,\ldots,w_s)a^{-1})=\theta(\gamma(w_1^{-1},\ldots,w_s^{-1}))=\theta(\gamma(w_1,\ldots,w_s)^{-1})=1$$

and so $a\gamma a^{-1}\in\Gamma$ as desired. Thus Γ is a normal subgroup of Γ'' and the proof will be complete.

(i) Let $\sigma''=(0;+;[-];\{(2,2,2,2,2,2)\})$. Then by 2.4.7 $\tau\triangleleft\sigma''$ and so there exists an NEC group Γ'' with signature σ'' containing Λ as a subgroup of index 2. Now let $\beta:\Gamma''\longrightarrow\Gamma''/\Lambda=\mathbb{Z}_2=\langle y\rangle$ be the canonical epimorphism. Using results of chapter 2, it is easy to check that β is forced to be defined by either

$$\beta(c_i'')=y^{i+1},\ 0\le i\le 6,\ \beta(e'')=1$$

or

$$\beta(c_i'')=y^i,\ 0\le i\le 6,\ \beta(e'')=1.$$

In the first case the canonical generators of Λ are expressed in terms of Γ'' as follows:

$$c_{i0}'=c_{2i-1}'',\ e_i'=c_{2i-2}''c_{2i}'',\ 1\le i\le 3$$

and in the second one by

$$c_{i0}'=c_{2i}'',\ e_i'=c_{2i-1}''c_{2i+1}'',\ 1\le i\le 3$$

with all indices taken mod 6. In fact, in the first case we must only check that the above elements satisfy the relations in Λ. First

$$(e_i')^{-1}c_{i0}'e_i'c_{i0}'=c_{2i}''c_{2i-2}''c_{2i-1}''c_{2i-2}''c_{2i}''c_{2i-1}''=(c_{2i}''c_{2i-1}'')^2=1,$$

since c_{2i-1}'' commutes with c_{2i-2}'' by the relations in Γ''. The remaining relations are obvious. Analogously we argue in the second case.

Consider now the first case (the second is similar). Choose $a=c_0''$. Then

$$ac_{10}'a=c_0''c_1''c_0''=c_1''=c_{10}'^{-1}$$

$$ac_{20}'a=c_0''c_3''c_0''=(c_0''c_2'')(c_2''c_3''c_2'')(c_2''c_0'')=e_1'c_{20}'e_1'^{-1}=c_{20}'^{-1}\mathrm{mod}[\Lambda,\Lambda]$$

$$ac_{30}'a=c_0''c_5''c_0''=(c_0''c_2'')(c_2''c_4'')c_5''(c_4''c_2'')(c_2''c_0'')=(e_1'e_2')c_{30}'(e_1'e_2')^{-1}=c_{30}'^{-1}\mathrm{mod}[\Lambda,\Lambda]$$

$$ae_1'a=c_0''c_0''c_2''c_0''=e_1'^{-1}$$

$$ae_2'a=c_0''c_2''c_4''c_0''=(c_0''c_2'')(c_4''c_2'')(c_2''c_0'')=e_1'(e_2')^{-1}(e_1')^{-1}=(e_2')^{-1}\ \mathrm{mod}[\Lambda,\Lambda]$$

$$ae_3'a=c_0''c_4''c_6''c_0''=(c_0''c_2'')(c_2''c_4'')(c_6''c_4'')(c_4''c_2'')(c_2''c_0'')=(e_1'e_2')(e_3')^{-1}(e_1'e_2')^{-1}=(e_3')^{-1}\ \mathrm{mod}[\Lambda,\Lambda].$$

In the same way we handle cases (ii) - (vii). We just give the corresponding data σ'', β, a, and the relations between the generators of Λ and Γ''. The computations are left to the reader.

(ii) $\sigma''=(0;+;[-];\{(2,2,2,2,2m)\})$; $\beta(c_j'')=y^{j+1}$, $0\le j\le 4$, $\beta(e'')=1$, $c_{i0}'=c_{2i-1}''$, $e_i'=c_{2i-2}''c_{2i}''$, $1\le i\le 2$, $x'=c_4''c_0''$, $a=c_0''$.

(iii) $\sigma''=(0;+;[-];\{(2,2,m_1,m_2)\})$; $\beta(c_j'')=y$, $j\ne 1$, $\beta(c_1'')=1$, $\beta(e'')=1$ $c_0'=c_1''$, $e'=c_0''c_2''$, $x_1'=c_2''c_3''$, $x_2'=c_3''c_0''$, $a=c_0''$.

(iv) $\sigma''=(0;+;[-];\{(2,2,2,m)\})$; $\beta(c_0'')=\beta(c_3'')=y$, $\beta(c_1'')=\beta(c_2'')=1$, $\beta(e'')=1$ $c_i'=c_{i+1}''$, $0\le i\le 1$, $c_2'=c_3''c_1''c_3''$, $e'=c_0''c_3''$, $x'=c_3''c_0''$, $a=c_0''$.

(v) $\sigma''=(0;+;[2];\{(2,2,2,2)\})$; $\beta(e'')=y$, $\beta(c_j'')=y^{j+1}, 0\le j\le 4$, or $\beta(c''_j)=y^j$, $0\le j\le 4$

$d'=x''c''_0$, $e'_1=c''_0c''_2$, $c'_{10}=c''_1$, $e'_2=c''_2c''_4$, $c'_{20}=c''_3$ in the first case

$d'=x''c''_1$, $e'_1=c''_1c''_3$, $c'_{10}=c''_2$, $e'_2=c''_3c''_5$, $c'_{20}=c''_4$ in the second one, $a=c''_1$.

(vi) $\sigma''=(0;+;[2];\{(2,2,m)\})$; $\beta(c''_1)=1$, $\beta(x'')=\beta(e'')=\beta(c''_i)=y$, $i\neq 1$

$d'=c''_0x''$, $x'=c''_3c''_2$, $e'=c''_2c''_0$, $c'_0=c''_1$, $a=c''_0$.

(vii) $\sigma''=(0;+;[2,2];\{(2,2)\})$; $\beta(c''_0)=\beta(c''_2)=\beta(x''_1)=\beta(x''_2)=y$, $\beta(c''_1)=1$, $\beta(e'')=1$,

$d'_1=c''_2x''_1$, $d'_2=x''_2c''_0$, $e'=c''_0c''_2$, $c'_0=c''_1$, $a=c''_0$.

Remarks 5.1.6. (1) Observe that each extension

$$1\longrightarrow\Lambda\longrightarrow\Gamma''\overset{\beta}{\longrightarrow}\mathbb{Z}_2\longrightarrow 1$$

considered in the proof splits, *i.e.*, β has a right inverse. Therefore Γ'' is a semidirect product $\Lambda{:}\mathbb{Z}_2$. Moreover, let $\mathbb{Z}_2$ act on $G=\Lambda/\Gamma$ by $ygy^{-1}=g^{-1}$. Then the epimorphism $\theta:\Lambda\longrightarrow G$ preserves this action and as a consequence there is a group epimorphism $\eta:\Gamma''\longrightarrow\tilde{G}=G{:}\mathbb{Z}_2$ defined coordinatewise, *i.e.*, $\eta(\gamma')=\theta(\gamma')$, if $\gamma'\in\Lambda$, $\eta(y)=y$, with $\ker\eta=\ker\theta$.

(2) Observe also that in case $\Gamma=\ker\theta$ has only one period-cycle then G is cyclic unless Λ has signature (iv) with m=2. In this exceptional case $G=\mathbb{Z}_2\oplus\mathbb{Z}_2$.

5.2. THE AUTOMORPHISM GROUP

Along this section, *surface* means compact Klein surface of algebraic genus $p\ge 2$ with one boundary component. Let X be a surface. Since finite groups of homeomorphisms of a circle are cyclic or dihedral, the same holds true for Aut(X). Note also that $p(X)=\alpha g(X)$ and so orientable surfaces have even algebraic genus.

We fix the following notations. Given $p\ge 2$,

$$\sigma_p^{\pm}=(p/\alpha;\pm;[-];\{(-)\}),\quad \alpha=\begin{cases}2 & \text{if sign } \sigma_p = "+", \text{ (p even)}\\ 1 & \text{if sign } \sigma_p = "-"\end{cases}$$

$\mathcal{K}_+(p)$, resp. $\mathcal{K}_-(p)$, is the set of orientable, resp. nonorientable, surfaces of algebraic genus p.

Our goal in this section is to compute, for each $p\ge 2$, the sets $G_+(p)$ of non trivial groups G for which there exists $X\in\mathcal{K}_+(p)$ with Aut(X)=G, and $G_-(p)$ defined in the obvious way. Let us call $Z_+(p)$ (respectively $D_+(p)$) the cyclic (respectively dihedral) groups in $G_+(p)$. Then $G_+(p)=Z_+(p)\cup D_+(p)$, and analogously $G_-(p)=Z_-(p)\cup D_-(p)$, with obvious meaning. For the sake of convenience, we introduce the sets $O_+(p)$ (respectively $O_-(p)$) of integers $N\ge 2$ such that $\mathbb{Z}_N$ is a subgroup, not necessarily proper, of Aut(X) for some $X\in\mathcal{K}_+(p)$ (respectively $X\in\mathcal{K}_-(p)$).

Moreover, for the sake of notational simplicity, we shall write $N\in Z_\pm(p)$,

$N \in D_{\pm}(p)$ instead of $Z_N \in Z_{\pm}(p)$, $D_N \in D_{\pm}(p)$.

Remarks 5.2.1. (1) Obviously both $Z_{\pm}(p)$ and $D_{\pm}(p)$ are contained in $O_{\pm}(p)$.
(2) With the terminology introduced in Chapter 3, $N \in O_{\pm}(p)$ if and only if there exists some signature τ such that $(\sigma_p^{\pm},\tau)$ is an N-pair. Moreover, if this signature can be chosen with τ^+ maximal then, by 5.1.1 and 5.1.2, $N \in Z_{\pm}(p)$. On the other hand, if every such τ appears in the list of 5.1.5, then $N \notin Z_{\pm}(p)$.

We are going to determine $Z_{\pm}(p)$. To do that, and for later convenience, we shall also compute $O_{\pm}(p)$. First of all we state the following immediate consequence of 2.4.2 and 2.4.4.

Lemma 5.2.2. *Let Λ be an NEC group containing a surface group Γ with a unique period-cycle as a normal subgroup of index greater than two. Then, if Λ/Γ is cyclic, all period-cycles of Λ are empty.*

In Chapter 3, Section 1, we found necessary and sufficient conditions for (σ,τ) to be an N-pair. The proof of the following theorem essentially consists in applying theorems proven there, in the case when σ has only one period-cycle.

Theorem 5.2.3. *(1)* $2 \in Z_-(p)$; $2 \in Z_+(p)$ *if and only if* p *is even. In what follows,* N *is a natural number,* $N \geq 3$.

(2) $N \in O_+(p)$ *if and only if* p *is even and* $p-1 = N\left[2g'-1+\sum_{i=1}^{r}(1-1/m_i)\right]$ *for some (non-negative integers)* g', r, $m_1,\dots,m_r$, *for which* $l.c.m.(m_1,\dots,m_r)=N$. *Moreover,* $N \in Z_+(p)$ *if and only if we can choose either* $g' \neq 0$ *or* $r \geq 3$.

(3) Let N *be odd. Then* $N \in O_-(p)$ *if and only if* $p-1 = N\left[g'-1+\sum_{i=1}^{r}(1-1/m_i)\right]$, *for some* $g' \geq 1$, *and* $m_1,\dots,m_r$ *dividing* N. *Moreover,* $N \in Z_-(p)$ *if and only if we can choose* r *and* g' *with* $r+g' \geq 3$.

(4) Let N *be even. Then* $N \in O_-(p)$ *if and only if either*

$p-1 = N\left[2g'+k'-2+\sum_{i=1}^{r}(1-1/m_i)\right]$, *for some* $g' \geq 0$, $k' \geq 2$, *and* $m_1,\dots,m_r$ *dividing* N, *or*

$p-1 = N\left[g'+k'-2+\sum_{i=1}^{r}(1-1/m_i)\right]$, *for some* $g' \geq 1$, $k' \geq 1$, *and* $m_1,\dots,m_r$ *dividing* N,

such that $\varepsilon = \sum_{i=1}^{r} N/m_i$ *is odd if* $k'=1$.

Moreover, $N\in Z_-(p)$ *if and only if the numbers above may be so chosen that* $r+k'\geq 4$ *when* $g=0$ *in the first case, or* $r+k'\geq 3$ *when* $g'=1$ *in the second case.*

Proof. (1) Take $\tau=(0;+;[2,2];\{(2,2)\})$ if $p=2$, and $\tau=(0;+;[\overset{p-1}{2,\ldots,2}];\{(-),(-)\})$ if $p>2$. From 3.1.6, (σ_p^-,τ) is a 2-pair, and τ^+ is maximal, by 2.2.5, 0.3.5 and 0.3.6. By (2) in remark 5.2.1, $2\in Z_-(p)$.

Since orientable surfaces have even algebraic genus, the necessity of condition "p is even" is clear here and also in the part (2) of the theorem. For the sufficiency, it is enough to choose $\tau=(0;+;[\overset{p+1}{2,\ldots,2}];\{(-)\})$. Then (σ_p^+,τ) is a 2-pair by 3.1.5, and τ^+ is maximal, because $p+1\geq 3$. Now apply 5.2.1 (2).

(2) Assume that $N\in O_+(p)$ is odd. Then there exists an N-pair (σ_p^+,τ) verifying conditions in 3.1.2. There, (3.1) implies $k'=1$, $l_1=N$ and so, by (2) and (3.2) in 3.1.2,

$$p-1=N\left[2g'-1+\sum_{i=1}^{r}(1-1/m_i)\right],\ l.c.m.(m_1,\ldots,m_r)=N.$$

Suppose also that $N\in Z_+(p)$ and $g'=0$. Then $\tau=(0;+;[m_1,\ldots,m_r];\{(-)\})$ and $r\geq 2$ because $\mu(\tau)>0$. Since τ cannot be in the list of 5.1.5, (cf. remark 5.2.1) $r\geq 3$.

Conversely, let us take $\tau=(g';+;[m_1,\ldots,m_r];\{(-)\})$. By 3.1.2, (σ_p^+,τ) is an N-pair, *i.e.*, $N\in O_+(p)$. Moreover, the restriction $r\geq 3$ if $g'=0$ implies that τ^+ is maximal. Thus $N\in Z_+(p)$.

If $N\in O_+(p)$ is even, and (σ_p^+,τ) is an N-pair, then conditions in either 3.1.5 or 3.1.9 must be satisfied. In the second case, by the last lemma, we deduce, using (3.1), that $s'=k'$, $p'=1$, $l_1=N$, which are not compatible with (3.2) there. Hence, conditions in 3.1.5 hold true. By 5.2.2 and (3.1) in 3.1.5, $p'=1$ and $l_1=N$. Thus by (3.4) $k'=p'=1$ and so the equality $p-1=N\left[2g'-1+\sum_{i=1}^{r}(1-1/m_i)\right]$ is nothing else that $\mu(\sigma_p^+)=N\mu(\tau)$, and in addition, from (3.2) and (3.3) in 3.1.5, $N=l.c.m.(m_1,\ldots,m_r)$. Condition $r\geq 3$ for $g'=0$ when $N\in Z_+(p)$, and the converse, are proven in the same way that in case of odd N.

(3) If $N\in O_-(p)$ is odd then, by 3.1.3, there exists $\tau=(g';-;[m_1,\ldots,m_r];\{(-)\})$ such that (σ_p^-,τ) is an N-pair. Note that $g'\geq 1$. Then the desired equality is (2) in 3.1.3. Moreover, in case $N\in Z_-(p)$, τ is neither (vi) nor (vii) in 5.1.5. Also, $\mu(\tau)\neq 0$ implies $(r,g')\neq(0,1)$. Thus $r+g'\geq 3$. For the converse notice that for τ above, (σ_p^-,τ) is an N-pair and the imposed restrictions ensure that τ^+ is maximal.

(4) Assume that $N\in O_-(p)$ is even and take an N-pair (σ_p^-,τ). Then τ verifies

conditions in either 3.1.6 or 3.1.8, where, by 5.2.2, $s'=k'$. In both cases $p'=1$, $l_1=N$, by (3.1). In particular $k'\geq 1$ and in the first case, $p'<k'$ implies $k'\geq 2$. In the second one, $g'\geq 1$, because sign(τ)="-". Now the equalities in the statement are nothing else but $\mu(\sigma_p^-)=N\mu(\tau)$. We prove now that ε is odd when $k'=1$. In this case $p'=k'$, $l_1=N$ and so, by (3.3) in 3.1.8, there exist integers $\alpha_1,\ldots,\alpha_r,\beta$ such that $g.c.d.(\alpha_i,m_i)=g.c.d.(\beta,N)=1$, $S=\sum_{i=1}^{r}\alpha_i N/m_i+\beta$ is even. In particular, β is odd whilst α_i is odd when N/m_i is. Thus each $(\alpha_i-1)N/m_i$ is even and so $\varepsilon=S-\beta-\sum_{i=1}^{r}(\alpha_i-1)N/m_i$ is odd.

Moreover, if either $r+k'\leq 3$ with $g'=0$ in the first case or $r+k'\leq 2$ with $g'=1$ in the second one, then τ would be one of the signatures (i), (ii), (v) or (vi) in 5.1.5, and so, by 5.2.1, $N\notin Z_-(p)$.

Let us prove that these conditions are also sufficient. We take the signature $\tau=(g';\pm;[m_1,\ldots,m_r];\{\overbrace{(-),\ldots,(-)}^{k'}\})$ with sign(τ)="+" if the first condition in the statement holds and sign(τ)="-" otherwise. Applying 3.1.6 and 3.1.8 with $p'=1$, $l_1=N$, we deduce that (σ_p^-,τ) is an N-pair, *i.e.*, $N\in O_-(p)$. Note that $\tau^+=(\alpha g'+k'-1;+;[m_1,m_1,\ldots,m_r,m_r])$, which is maximal unless either $\alpha g'+k'-1=1$, $r=1$, or $\alpha g'+k'-1=2$, $r=0$, or $\alpha g'+k'-1=0$, $r=2$. Since $r+k'\geq 4$ when $\alpha=2$, $g'=0$, and $r+k'\geq 3$ when $\alpha=g'=1$, and additionally $k'\geq\alpha$, the unique case to be studied is $\alpha=1$, $g'=2$, $k'=1$, $r=0$. But then $k'=1$ and $\varepsilon=0$, absurd. This completes the proof.

Remark 5.2.4. For prime N it is easy to decide whether $N\in O_\pm(p)$ or not. Of course we can assume $N\neq 2$, and so $m_1=\ldots=m_r=N$ in statements (2) and (3) of the last theorem. Hence, writing $r-1=t$, we get

(i) $N\in O_+(p)$ if and only if $N(2g'+t)=p+t$ for some $g'\geq 0$, $t\geq 0$.

(ii) $N\in O_-(p)$ if and only if $N(g'+t)=p+t$ for some $g'\geq 1$, $t\geq -1$.

Moreover, from 3.2.13, $N\leq p+1$ if $N\in O_+(p)$ and $N\leq p$ if $N\in O_-(p)$.

The dihedral group D_N is presented from now on as $D_N=\langle x,y|x^2,y^2,(xy)^N\rangle$. The next theorem, together with theorem 5.2.3, determines $D_\pm(p)$ and so $G_\pm(p)$ as proposed.

Theorem 5.2.5. *(1)* $N\in D_-(p)$ *if and only if* $N\in O_-(p)$.

(2) $2\in D_+(p)$ *if and only if* p *is even.*

(3) $3\notin D_+(2)$.

(4) Assume $N\geq 3$, $(p,N)\neq(2,3)$. *Then* $N\in D_+(p)$ *if and only if* $N\in O_+(p)$.

Proof. (1) It is enough to prove the "if part". As in the other parts of the theorem, the strategy consists in starting with an N-pair (σ_p^-,σ') and finding a maximal signature τ with $[\tau:\sigma']=2$ together with an epimorphism from a group with signature τ onto D_N whose kernel has signature σ_p^-.

Assume $N\in O_-(p)$ is odd. From (3) in 5.2.3, $p-1=N\left[g'-1+\sum_{i=1}^{r}(1-1/m_i)\right]$ for some divisors $m_1,\ldots,m_r$ of N and some $g'\geq 1$. Let Λ be a maximal NEC group with signature $\tau=(0;+;[\overset{g'}{2,\ldots,2}];\{(2,2,m_1,\ldots,m_r)\})$, whose existence is given by 5.1.1 and 5.1.2, because τ^+ is maximal. We write $b_j=\sum_{i=1}^{j}N/m_j$, for $j=1,\ldots,r$, $b=b_r-1$ and $a=\begin{cases} b/2 & \text{if } b \text{ is even} \\ (N+b)/2 & \text{if } b \text{ is odd.}\end{cases}$

Now define $\theta:\Lambda\longrightarrow D_N$ by

$$\theta(c_0')=x,\ \theta(c_1')=1,\ \theta(c_2')=y,\ \theta(c_{j+2}')=y(xy)^{b_j},\ 1\leq j\leq r, \text{ and}$$

$$\begin{cases} \theta(e')=\theta(x_i')=y(xy)^a,\ 1\leq i\leq g', & \text{if } g' \text{ is odd} \\ \theta(e')=(xy)^{a+1},\ \theta(x_1')=x,\ \theta(x_2')=x(yx)^{a+1},\ \theta(x_i')=y,\ 3\leq i\leq g', & \text{if } g' \text{ is even.}\end{cases}$$

It induces an epimorphism and

$$w=\begin{cases} e'c_2'\,(c_0'c_2')^a & \text{if } g' \text{ is odd} \\ x_2'c_0'\,(c_2'c_0')^{a+1} & \text{if } g' \text{ is even}\end{cases}$$

is a nonorientable word in $\ker\theta$. From 2.1.3, 2.2.4, 2.3.3 and 2.3.6, $\sigma(\ker\theta)=$ $=\sigma_p^-=(p;-;[-];\{(-)\})$. Hence, $\mathrm{Aut}(H/\ker\theta)=\Lambda/\ker\theta=D_N$ and so $N\in D_-(p)$.

Now let N be even. For N=2 the signature $\tau=(0;+;[-];\{(\overset{p+3}{2,\ldots,2})\})$ is maximal since τ^+ is so. By 5.1.2 there exists a maximal NEC group Λ with signature τ. Consider the epimorphism $\theta:\Lambda\longrightarrow D_2$ given by

$$\theta(e')=1,\ \theta(c_0')=x,\ \theta(c_1')=1,\ \theta(c_2')=y,\ \theta(c_{j+2}')=y(xy)^j,\ 1\leq j\leq p-1,$$
$$\theta(c_{p+2}')=xy,\ \theta(c_{p+3}')=x.$$

Also now $c_0'c_2'c_{p+2}'$ is a nonorientable word in $\ker\theta$, and $\sigma(\ker\theta)=\sigma_p^-$. Thus $2\in D_-(p)$.

If $2<N\in O_-(p)$, conditions in part (4) of 5.2.3 are satisfied and we choose the maximal signature $\tau=(0;+;[\overset{r'}{2,\ldots,2}];\{(2,2,m_1,\ldots,m_r,2,2)\})$, with

$$r'=\begin{cases} 2g'+k'-2 & \text{if the first condition holds true,} \\ g'+k'-2 & \text{otherwise.}\end{cases}$$

Let Λ be a maximal NEC group with signature τ. With the notation introduced in the case of odd N we define $\theta:\Lambda\longrightarrow D_N$ as follows:

$$\theta(e')=x^{1-r'},\ \theta(c_0')=x,\ \theta(c_1')=1,\ \theta(c_2')=y,\ \theta(c_{j+2}')=y(xy)^{b_j},\ 1\leq j\leq r,$$
$$\theta(c_{r+3}')=(xy)^{N/2},\ \theta(c_{r+4}')=x,\ \theta(x_i')=x,\ 1\leq i\leq r'.$$

Since $w=(c'_0c'_2)^{N/2}c'_{r+3}$ is a nonorientable word in the kernel, $N\in D_-(p)$.

From now on, $\sigma_p=(p/2;+;[-];\{(-)\})=\sigma_p^+$.

(2) We have only to prove that $2\in D_+(p)$ for even p. Let Λ be a maximal NEC group with signature $\tau=(0;+;[-];\{(\overset{p+3}{2,\ldots,2})\})$ and $\theta:\Lambda\longrightarrow D_2$ be defined as

$$\theta(e')=1,\ \theta(c'_0)=x,\ \theta(c'_1)=1,\ \theta(c'_2)=y,\ \theta(c'_{j+2})=y(xy)^j,\ 1\le j\le p,\ \theta(c'_{p+3})=x.$$

By 2.2.4 $\ker\theta$ has no proper periods and by 2.3.3 it has only one (empty) period-cycle. So since $\bar{\mu}(\tau)=(p-1)/4$, by 2.1.3 it is enough to check that $\ker\theta$ does not contain nonorientable words. But if this were not so then, since D_2 is abelian, this word could be chosen to be $w=c_0^{\varepsilon_0}c_2^{\varepsilon_2}\ldots c_{p+3}^{\varepsilon_{p+3}}$, where $\varepsilon=$ $=\varepsilon_0+\varepsilon_2+\ldots+\varepsilon_{p+3}$ is odd. Then $x^{\varepsilon_0+\varepsilon_3+\ldots+\varepsilon_{p+3}}=y^{\varepsilon_2+\varepsilon_4+\ldots+\varepsilon_{p+2}}$, and so, both $\delta_1=\varepsilon_0+\varepsilon_3+\ldots+\varepsilon_{p+3}$ and $\delta_2=\varepsilon_2+\varepsilon_4+\ldots+\varepsilon_{p+2}$ are even. Thus also $\varepsilon=\delta_1+\delta_2$ is even, a contradiction. Thus, by the maximality of Λ, $\mathrm{Aut}(H/\ker\theta)=\Lambda/\ker\theta=D_2$ and we conclude that $2\in D_+(p)$.

(3) Assume that $3\in D_+(2)$. Then there exist NEC groups Γ and Λ with $\Lambda/\Gamma\cong D_3=$ $=\mathrm{Aut}(H/\Gamma)$, and $\sigma(\Gamma)=\sigma_2^+$. We compute $\tau=\sigma(\Lambda)$. Since $\mathbb{Z}_3$ is a subgroup of D_3, there exists an NEC group Γ_1 with $\Gamma\triangleleft\Gamma_1\triangleleft\Lambda$ and $\Gamma_1/\Gamma=\mathbb{Z}_3$. From 5.2.3 (2) we get $\sigma'=\sigma(\Gamma_1)=(0;+;[3,3];\{(-)\})$. Now (σ',τ) is a normal pair of index 2 and so, by 2.4.7, τ has one of the following forms: $(0;+;[3];\{(2,2)\})$; $(0;+;[2,3];\{(-)\})$ or $(0;+;[-];\{(2,2,3,3)\})$. Suppose that τ is the third one. Then by 2.4.7 we can choose a group Γ'' containing Λ as a subgroup of index 2, with signature $\sigma''=(0;+;[-];\{(2,2,2,3)\})$. Now it is enough to construct, for any epimorphism $\theta:\Lambda\longrightarrow D_3$ with kernel Γ, a commutative square with η surjective

$$\begin{array}{ccc} \Gamma'' & \xrightarrow{\ \eta\ } & D_3\times\mathbb{Z}_2 \\ \uparrow i & & \uparrow j \\ \Lambda & \xrightarrow{\ \theta\ } & D_3 \end{array} \tag{5.2.5.1}$$

since in such a case $\ker\eta=\ker\theta=\Gamma$ and $D_6=D_3\times\mathbb{Z}_2$ is a group of automorphisms of H/Γ which implies $\mathrm{Aut}(H/\Gamma)\neq D_3$.

For, let us see first how a homomorphism θ must look like. Since Γ is a surface group with one period-cycle, $\theta(c'_1)=1$, and $\theta(c'_0)$, $\theta(c'_2)$ are elements of order 2 whose product has order 3. So without loss of generality we can assume $\theta(c'_0)=x$, $\theta(c'_2)=y$. But then, since $e'=1$, $\theta(c'_4)=x$. Finally $\theta(c'_3)$ must be an element of order 2 such that $\theta(c'_2)\theta(c'_3)$ and $\theta(c'_3)\theta(c'_4)$ are elements of order 3. Therefore $\theta(c'_3)=xyx$.

Now let us see how Λ is embedded in Γ''. For, consider the canonical epimorphism $\beta:\Gamma''\longrightarrow\Gamma''/\Lambda=\mathbb{Z}_2=\langle z\rangle$. Then using results of chapter 2 we easily agree that β is forced to be

either $\qquad\beta(c''_1)=y,\ \beta(c''_i)=1,$ for $i\neq 1$, $\beta(e'')=1,$

or $\beta(c''_2)=y,\ \beta(c''_i)=1$, for $i\neq 2$, $\beta(e'')=1$.

Consider the first case (the second is similar). Arguing as in the proof of 5.1.5(i), the canonical generators of Λ can be expreseed by

$$i(c'_0)=i(c'_4)=c''_1c''_3c''_1,\ i(c'_1)=c''_2,\ i(c'_2)=c''_3,\ i(c'_3)=c''_0.$$

Let $j(x)=x$, $j(y)=y$. Then with the above data the reader can easily check that the assignment

$$\eta(c''_0)=\eta(c''_4)=xyx,\ \eta(c''_1)=xyxz,\ \eta(c''_2)=1,\ \eta(c''_3)=y$$

induces an epimorphism $\Gamma''\longrightarrow D_3\times\mathbb{Z}_2$ verifying 5.2.5.1.

Consider now the first signature. Then we use the same construction as above. Indeed the reader can check that the following are the corresponding data:

$$\theta(x')=xy,\ \theta(e')=(xy)^2,\ \theta(c'_0)=x,\ \theta(c'_1)=1, \theta(c'_2)=y,$$
$$\beta(c''_0)=z,\ \beta(c''_1)=\beta(c''_2)=1,\ \beta(c''_3)=\beta(c''_4)=z,$$
$$i(x')=c''_3c''_0,\ i(e')=c''_0c''_3,\ i(c'_0)=c''_1,\ i(c'_1)=c''_2,\ i(c'_2)=c''_3c''_1c''_3,$$
$$\eta(c''_0)=\eta(c''_4)=zx,\ \eta(c''_1)=x,\ \eta(c''_2)=1,\ \eta(c''_3)=zxyx.$$

Finally, if we take the second signature, then $\theta(x'_1)$ and $\theta(x'_2)$ are generators of orders 2 and 3 respectively and so $\theta(e')$ has order 2. Then by 2.3.3, $\Gamma=\ker\theta$ has 3 period-cycles, a contradiction. We are finished.

(4) Clearly $D_+(p)\subseteq O_+(p)$, and using (2) it suffices to prove that $N\in D_+(p)$ provided $3\leq N\in O_+(p)$, $(p,N)\neq(2,3)$. Of course, conditions in part (2) of 5.2.3 are satisfied. We claim:

(5.2.5.2) If $N=2^sA$ with $s\neq 0$ and odd A, and $m_1,\dots,m_t$ are those of m_i divisible by 2^s, then t is odd.

In fact, $\sum_{i=1}^{t} N/m_i=N(2g'-1+r)+1-p-\sum_{i=t+1}^{r} N/m_i$, and the right hand side of this equality is odd. Hereby t is also odd.

In particular, we see that the set $\{m_1,\dots,m_r,N\}$ verifies condition (2) in lemma 3.1.1 with $M=N$, and so there exist $\xi_1,\dots,\xi_r,\lambda$ in $\mathbb{Z}_N$ with orders $m_1,\dots,m_r,N$ respectively such that $\xi_1+\dots+\xi_r+\lambda=0$. Let $b_j\in\mathbb{Z}$, $1\leq j\leq r$, and $u\in\mathbb{Z}$, be representatives of $\sum_{i=1}^{j}\xi_i$ and λ, respectively. Take an NEC group Λ with signature $\tau=(g';\pm;[-];\{(2,2,m_1,\dots,m_r)\})$, the sign$(\tau)=$"+" only for $g'=0$.

Let $\theta:\Lambda\longrightarrow D_N=\langle x,y|x^2,y^2,(xy)^N\rangle$ be the homomorphism induced by

$$\theta(c'_0)=\theta(c'_{r+2})=(xy)^{u-1}x,\ \theta(c'_1)=1,\ \theta(c'_2)=y,$$

$$\theta(c'_{j+2})=y(xy)^{b_j},\ 1\leq j\leq r\text{-}1,\ \theta(e')=1,\ \theta(d'_i)=y,\ 1\leq i\leq g' \text{ if } g'>0.$$

The numbers b_j are so chosen that θ preserves periods, *i.e.*, $\theta(c'_0c'_1)$, $\theta(c'_1c'_2)$ are elements of order 2 and $\theta(c'_{i+2}c'_{i+3})$ has order m_{i+1} for $i=0,\dots,r$-1. So by 2.2.4 $\Gamma=\ker\theta$ has no proper periods. Now θ is surjective, because $y\in\operatorname{im}\theta$ and

$(xy)^u=\theta(c_0'c_2')$ has order N. The last also shows that Γ has one (empty) period-cycle. To see that $\sigma(\Gamma)=\sigma_p$ it remains to show that Γ is orientable. Indeed an element w(x,y) representing the unit in D_N has even length. On the other hand the image of each proper generator of Λ has odd length, by the definition of θ. We see that all proper generators of Λ represent orientation reversing elements. So in particular every word in Γ must be orientable. Therefore Γ is orientable by the second part of 2.1.3. Hence $X=H/\Gamma\in\mathcal{K}_+(p)$ and D_N is a subgroup of Aut(X). All reduces to see that Λ can be chosen the normalizer of Γ in Aut(H). If this were not so then in particular τ and τ^+ would not be maximal and since $r>0$ we have from 0.3.5 and 0.3.6 that $g'=0$, $r=2$, $m_1=m_2$. We also conclude that if $\tilde{\sigma}$ is a fuchsian signature such that $\tau^+<\tilde{\sigma}$, then $[\tilde{\sigma}:\tau^+]=2$.

Now $m_1=m_2=N$ since $N=l.c.m.(m_1,m_2)$. In particular by 5.2.5.2, N is odd. Now let σ'' be a signature such that $\sigma''^+=\tilde{\sigma}$ and $\tau<\sigma''$. Then since $[\sigma'':\tau]=[\tilde{\sigma}:\tau^+]=2$ we deduce from 2.4.7 that $\sigma''=(0;+;[-];\{(2,2,2,N)\})$.

Concluding we have showed that if $N\in O_+(p)$, $(p,N)\neq(2,3)$ and $N\notin D_+(p)$ then (in the previous notations) $\tau=(0;+;[-];\{(2,2,N,N)\})$ and for any NEC group Λ with signature τ and for any epimorphism $\theta:\Lambda\longrightarrow D_N$ whose kernel has signature σ_p^+, there exists an NEC group Γ'' with signature σ'' and a commutative diagram

$$\begin{array}{ccc} \Gamma'' & \xrightarrow{\ \eta\ } & D_{2N} \\ \uparrow i & & \uparrow j \\ \Lambda & \xrightarrow{\ \theta\ } & D_N \end{array} \tag{5.2.5.3}$$

where η is surjective.

In particular this holds for the epimorphism θ defined by

$$\theta(c_0')=\theta(c_4')=x,\ \theta(c_1')=1,\ \theta(c_2')=y,\ \theta(c_3')=xyx.$$

Let us see how the remaining homomorphisms in 5.2.5.3 must look like. Clearly we do not lose generality assuming that $D_{2N}=D_N\times\mathbb{Z}_2=\langle x,y\rangle\times\langle z\rangle$, and $j(x)=x$, $j(y)=y$. Denote by β the canonical projection $\Gamma''\longrightarrow\Gamma''/\Lambda=\mathbb{Z}_2=\langle u|u^2\rangle$. As in the case (3), β has either the form

$$\beta(c_1'')=u,\ \beta(c_i'')=1 \text{ for } i\neq 1,\ \beta(e'')=1,$$

or

$$\beta(c_2'')=u,\ \beta(c_i'')=1 \text{ for } i\neq 2,\ \beta(e'')=1.$$

Consider the first case. Then the canonical generators of Λ can be expressed by those of Γ'' as follows:

$$i(c_0')=i(c_4')=c_1''c_3''c_1'',\ i(c_1')=c_2'',\ i(c_2')=c_3'',\ i(c_3')=c_0''.$$

Finally since η makes (5.2.5.3) commutative

$$\eta(c_0'')=\eta(c_4'')=\eta i(c_3')=j\theta(c_3')=xyx,\ \eta(c_2'')=\eta i(c_1')=j\theta(c_1')=1,$$
$$\eta(c_3'')=\eta i(c_2')=j\theta(c_2')=y$$

and $\eta(c_1'')$ is an element obeying

$$x=\theta(c_0')=j\theta(c_0')=\eta i(c_0')=\eta(c_1''c_3''c_1'')=\eta(c_1'')\eta(c_3'')\eta(c_1'')=\eta(c_1'')y\eta(c_1'').$$

Since η is onto, $\eta(c''_1)=az$ for some element a of D_N of order 2. But as N is odd, $a=(xy)^b x$ for some b. Therefore

$$x=(xy)^b xzy(xy)^b xz=(xy)^{2b+1}x$$

and so $b=(N-1)/2$. Hence

$$\eta(c''_0)\eta(c''_1)=xyx(xy)^{(N-1)/2}xz=(yx)^{(N-3)/2}z$$

is an element of order 2. Therefore N=3. But then $12=|D_{2N}|=\mu(\sigma_p)/\mu(\sigma'')=$ $=12(p-1)$, *i.e.*, $(p,N)=(2,3)$ against the hypothesis. To the same contradiction we arrive starting with the other homomorphism β.

Remark 5.2.6. From 4.2.1, $2N\leq 12(p-1)$ (of course in case of cyclic and dihedral groups this bound can be essentially improved, see for instance 3.2.9 and 3.2.18). Hence the above theorems show that problems of finding $Z_{\pm}(p)$ and $D_{\pm}(p)$ are questions of finitistic nature. Of course one would like to have a numerical condition describing these sets without the existential quantifier. This seems however to be a rather difficult problem.

We finish this chapter with an example of computing explicitly $G_+(16)$ in order to ilustrate how our results work.

Example 5.2.7. From 3.2.18 $N\leq 2(p+1)$ or $N\leq 2p$ according to $N\in O_+(p)$ or $N\in O_-(p)$ respectively. In our case

(5.2.7.1) If $N\in O_+(16)$, then $N\leq 34$. Moreover, if N is prime, $N\leq 17$ (see 5.2.4).

Let P denote the set of odd primes, Q the set of odd integers.

(5.2.7.2) $$O_+(16)\cap P=\{3,5,17\}.$$

In fact, if $N\in O_+(16)$ is prime, then, by 5.2.4, $N(2g'+t)=16+t$ for some $g'\geq 0$ and $t\geq 0$. If $t\geq 5$ we get $N\leq 21/5$ and for $t=4$ (resp. 2, 1) we get $N=5$ (resp. 3, 17) with $g'=0$, 2, 0. There are no solutions when $t=0$ or 3.

Let $N\in O_+(16)$; by (2) in 5.2.3 we have $15 = N\,(2g'-1+\sum_{i=1}^{r}(1-1/m_i))$, with $l.c.m.(m_1,\dots,m_r)=N$.

(5.2.7.3) If $N \geq 15$, then $g'=0$. Moreover, if N is odd, then $r=2$.

The first assertion is obvious. Thus $r\geq 2$ and $1<\sum_{i=1}^{r}(1-1/m_i)\leq 2$. However if N is odd and $r\geq 3$ then $\sum_{i=1}^{r}(1-1/m_i)\geq 2(1-1/3)+1-1/5>2$, a contradiction.

(5.2.7.4) $$Z_+(16)\cap Q=\{3,5,9\}.$$

The inclusion "$\subseteq$" is an immediate consequence of (2) in 5.2.3, and two previous claims. The converse follows from 5.2.3 applied for N = 3,5,9 with the values: $g'=0$, each $m_i=N$ and $r=9,5,3$ respectively.

Now we prove

(5.2.7.5) $$O_+(16)\cap Q=\{3,5,9,17\}.$$

Note that $15=17(1-1/17-1/17)$. So "$\supseteq$" follows from (2) in 5.2.3 and the last claim. Conversely assume that $N\in O_+(16)\cap Q$, and $N\neq 3,5,9,17$. By 5.2.7.2, $N\geq 15$ and by 5.2.7.3, $15=N(1-1/m_1-1/m_2)$, where *l.c.m.*$(m_1,m_2)=N$. Clearly $N\neq 15$ and as we remarked at the beginning $N\leq 34$. So by 5.2.7.2, $N=25$ or 27 (notice that 21 and 33 do not belong to $O_+(16)$ because 7 and 11 respectively do not). Then necessarily $m_1=N$ and so $N-16=N/m_2$ divides N, absurd.

For even numbers we get

(5.2.7.6) $$Z_+(16)\cap 2N=\{2,4,6,8,10,12,16\}.$$

To prove "$\subseteq$" notice first that $14\notin Z_+(16)$ because $7\notin O_+(16)$. Also, if $N>16$ belongs to $Z_+(16)$ we know, by 5.2.7.3 and 5.2.3 that

(*) $$15=N(-1+\sum_{i=1}^{r}(1-1/m_i)),\ r\geq 3,\ l.c.m.(m_1,\dots,m_r)=N.$$

Thus $\sum_{i=1}^{r}(1-1/m_i)\geq 1-1/2+1-1/2+1-1/8$, *i.e.*, $16<N\leq \frac{120}{7}<18$, a contradiction.

For the converse, observe that $2\in Z_+(16)$ by 5.2.3. To see that the other numbers belong to the left hand side of 5.2.7.3, it is sufficient to check that the following data are solutions of (*):

$$\begin{array}{ll} r=7,\ m_i=2,\ m_j=4,\ 1\leq i\leq 2,\ 3\leq j\leq 7, & \text{if } N=4\\ r=6,\ m_i=2,\ m_j=3,\ 1\leq i\leq 3,\ 4\leq j\leq 6, & \text{if } N=6\\ r=5,\ m_i=2,\ m_5=8,\ 1\leq i\leq 4, & \text{if } N=8\\ r=3,\ m_1=N,\ m_2=m_3=\begin{cases}5\\3\\2\end{cases} & \begin{array}{l}\text{if } N=10\\ \text{if } N=12\\ \text{if } N=16\end{array} \end{array}$$

Finally we get

(5.2.7.7) $$O_+(16)\cap 2N=(Z_+(16)\cap 2N)\cup\{18,20,24,32,34\}.$$

Now $22,26,28,30\notin O_+(16)$ since by 5.2.7.2, $7,11,13\notin O_+(16)$ and by 5.2.7.5, 15 does not belong to $O_+(16)$. So inclusion "$\subseteq$" is evident in virtue of 5.2.7.1. For the other we write $15=N(2g'-1+1-1/m_1+1-1/m_2)$, where *l.c.m.*$(m_1,m_2)=N$, with $g'=0$, $m_1=N$ (except $m_1=17$ for $N=34$) and $m_2=9$ (for $N=18$), 5 (for $N=20$), 3 (for $N=24$), 2 (for $N=32$ or 34).

Finally, combining 5.2.7.7 and 5.2.5 and denoting

$$I=\{2,3,4,5,6,8,9,10,12,16\},\ J=I\cup\{17,18,20,24,32,34\}$$

we obtain

(5.2.7.8) $$G_+(16)=\{Z_N|N\in I;\ D_N|N\in J\}.$$

5.3 NOTES

Although proved in a different way, proposition 5.1.5 was stated in **[16]**.

The proof of Corollary 5.1.4 is inspired by the one given by Greenberg (who also introduced the notion of maximal signature) [52,53] for the analogous result for Riemann surfaces, and it appeared in [17].

Very few results are known about the families of groups Aut(X), where X runs over Klein surfaces of given topological type. However we must quote here the results obtained in [16] for planar Klein surfaces.

CHAPTER - 6
The automorphism group of hyperelliptic compact Klein surfaces with boundary

We obtain here a complete solution of the problem of determine what groups can be realized as the full group of automorphisms Aut(X) of a compact hyperelliptic Klein surface X. To succeed we use an extremely useful property of Aut(X) in this case; namely, the existence of a unique central element ϕ of order 2 such that X/ϕ has algebraic genus zero and one boundary component.

We divide this chapter into three sections. In section 1 we give a characterization of hyperelliptic Klein surfaces in terms of NEC groups. A group of automorphisms of X comes from a certain NEC group Λ and in section 2 we obtain a finite set of candidates for the signature of Λ as a function of the topological type of X. We shall determine in the last section all groups appearing as Aut(X) for each topological type. Along this chapter X is a compact bordered Klein surface of algebraic genus ≥ 2, topological genus g, and $X=H/\Gamma$ for some bordered surface NEC group Γ.

6.1. HYPERELLIPTIC KLEIN SURFACES

Definition 1. A compact and bordered Klein surface X is said to be *hyperelliptic* (shortly *HKS*) if and only if there exists an involution ϕ of X (an automorphism of order 2) such that X/ϕ has algebraic genus 0. The automorphism ϕ will be called a *hyperellipticity automorphism.*

In order to characterize hyperelliptic Klein surfaces in terms of NEC groups, we shall need the following technical lemmas.

Lemma 6.1.1. *A surface* X *is an HKS if and only if there exists an NEC group* Γ_1 *of algebraic genus zero containing* Γ *as a subgroup of index* 2.

Proof. For the "if" part let us suppose that a group Γ_1 in question exists. Let ϕ be the generator of Γ_1/Γ. Then $X/\phi=H/\Gamma_1$ has algebraic genus zero and therefore X is hyperelliptic.

Now let X be hyperelliptic and let ϕ be the corresponding hyperellipticity automorphism. Then by the last remark in chapter 1, $<\phi>\cong\Gamma_1/\Gamma$ for some NEC group Γ_1. Now $X/\phi=H/\Gamma_1$ has algebraic genus zero and so Γ_1 is a group we were looking for.

A group Γ_1 satisfying the lemma above will be called a *group of hyperellipticity of* X. We fix this notation in what follows.

Lemma 6.1.2. *(a) If* X *is an HKS then all proper periods and all periods of the period-cycles of* Γ_1 *are equal to two.*

(b) If X *is orientable then either* Γ_1 *has empty set of proper periods or all its period-cycles are empty.*

(c) Let X *be an orientable HKS with* k *boundary components. Then either* $g=0$, *and the signature of* Γ_1 *is* $\sigma(\Gamma_1)=(0;+;[-];\{(2,\overset{2k}{\ldots},2)\})$, *or* $k<3$ *and then* Γ_1 *has signature* $(0;+;[2,\overset{2g+k}{\ldots},2];\{(-)\})$. *Notice that since* X *has algebraic genus* ≥ 2, *both situations cannot occur simultaneously.*

Proof. (a) is an immediate consequence of 2.4.4.

(b) Assume to a contradiction that the signature of Γ_1 has both a proper period $m_1=2$ and some nonempty period-cycle C_i. Let $\theta:\Gamma_1\longrightarrow\Gamma_1/\Gamma\cong Z_2=\langle x|x^2\rangle$ be the canonical projection. Then, since Γ has no proper periods, by 2.2.4, $\theta(x_1)\neq 1$, and since Γ has only empty period-cycles, by 2.3.2, there is a reflection c_{ij} such that $\theta(c_{ij})\neq 1$. Thus $x_1c_{ij}\in\Gamma$ is a nonorientable word and so by 2.1.3 Γ is nonorientable, a contradiction.

(c) Let Γ_1 be a group of hyperellipticity, with orbit genus g' and $k'>0$ period-cycles. Then, since $0=\alpha g'+k'-1$ we deduce that $k'=1$ and $g'=0$. By the previous part, either the period-cycle is empty or Γ_1 has no proper periods. But then by the Hurwitz Riemann formula Γ_1 has signature

$$(0;+;[2,\overset{2g+k}{\ldots},2];\{(-)\}) \text{ or } (0;+;[-];\{(2,\overset{4g+2k}{\ldots},2)\}).$$

In the first case $k\leq 2$ by part (2) of 2.4.4 and in the second one $k=2g+k$ by the same argument and so $g=0$.

Theorem 6.1.3. *A surface* X *of topological genus* g *and* k *boundary components is an HKS if and only if there exists an NEC group* Γ_1 *with* $[\Gamma_1:\Gamma]=2$, *and whose signature is:*

(i) $(0;+;[-];\{(2,\overset{2k}{\ldots},2)\})$ *if* $g=0$.

(ii) $(0;+;[2,\overset{2g+k}{\ldots},2];\{(-)\})$ $(k<3)$ *if* $g\neq 0$ *and* X *is orientable.*

(iii) $(0;+;[2,\overset{g}{\ldots},2];\{(2,\overset{2k}{\ldots},2)\})$ *if* X *is nonorientable.*

Proof. A group Γ_1 with a signature listed in (i)-(iii) has algebraic genus zero. Thus X is hyperelliptic by 6.1.1. Now we prove the converse.

If X is orientable then $\sigma(\Gamma_1)$ is either (i) or (ii) by (c) in the previous lemma. Let us suppose that X is nonorientable. Let g',k' be the orbit genus and the number of period-cycles of Γ_1 respectively. Then, the equality $0=\alpha g'+k'-1$ implies $g'=0$, $k'=1$ and so by the first part of 6.1.2 the signature of Γ_1 is $(0;+;[2,\overset{u}{\ldots},2];\{(2,\overset{w}{\ldots},2)\})$.

Since Γ has k empty period-cycles and no proper period, we have $w=2k$ by 2.4.4. By the Hurwitz Riemann formula $u=g$.

We will show that the group of hyperellipticity is unique. For, we shall need the following

Lemma 6.1.4. *Let Γ be an NEC group with signature* $(p;+;[-];\{-\})$ *and let Γ_1 and Γ_2 be NEC groups with signature* $(0;+;[2,\overset{2p+2}{\ldots},2];\{-\})$ *containing Γ. Then $\Gamma_1=\Gamma_2$.*

Proof. From the Hurwitz Riemann formula, $[\Gamma_i:\Gamma]=2$. Assume $\Gamma_1\neq\Gamma_2$ and let ϕ_1 and ϕ_2 be the generators of $G_1=\Gamma_1/\Gamma$ and $G_2=\Gamma_2/\Gamma$ respectively. Let G be the dihedral subgroup of Aut(H/Γ) generated by ϕ_1 and ϕ_2, say of order $2N>2$. By the last remark of chapter 1, there exists an NEC group Γ' such that $G=\Gamma'/\Gamma$. By 2.4.2 and 2.4.4 $\sigma(\Gamma')$ has the form $(g';+;[m_1,\ldots,m_r];\{-\})$. The numbers $m_1,\ldots,m_r$ are associated to the points in H/Γ with nonempty stabilizer with respect to G.

Let us denote by $F(G,\Gamma)$ the number of pairs (f,q) with $1_G\neq f\in G$, $q\in H/\Gamma$ and $f(q)=q$. As a fundamental region for Γ can be obtained by a union of 2N copies of a fundamental region for Γ', and the classes modΓ of the canonical hyperbolic generators of Γ' have orders $m_1,\ldots,m_r$, we obtain

$$F(G,\Gamma)=\sum_{i=1}^{r}(m_i-1)2N/m_i=2N\sum_{i=1}^{r}(1-1/m_i). \tag{6.1.4.1}$$

On the other hand, the dihedral group G can be written as a union

$$G=H_0\cup H_1\cup H_2\cup\ldots\cup H_N,$$

where H_0 is the cyclic subgroup of G of order N and $H_1,\ldots,H_N$ are subgroups of order two such that $H_i\cap H_j=\{1_G\}$ for each $i\neq j$, $0\leq i,j\leq N$. Consequently

$$F(G,\Gamma)=\sum_{j=0}^{N}F(H_j,\Gamma)\geq\sum_{j=1}^{N}F(H_j,\Gamma).$$

But each $H_j=\Gamma_1/\Gamma$ for $1\leq j\leq N$, and so, by using 6.1.4.1, we get

$$F(H_j,\Gamma)=2\sum_{i=1}^{2p+2}(1-1/2)=2p+2.$$

Hence $2N\sum_{i=1}^{r}(1-1/m_i)\geq N(2p+2)$. However, $\bar{\mu}(\Gamma')=2g'-2+\sum_{i=1}^{r}(1-1/m_i)$, and by the Hurwitz Riemann formula, we get

$$2p-2=\bar{\mu}(\Gamma)=2N(2g'-2)+2N\sum_{i=1}^{r}(1-1/m_i)\geq -4N+N(2p+2)\geq N(2p-2),$$

a contradiction.

Theorem 6.1.5. *Let X be an HKS, and let ϕ be a corresponding automorphism of the hyperellipticity of X. Then ϕ is unique and central in the full group of automorphisms of X.*

Proof. Let us assume that X is orientable with k boundary components. Let $p=2g+k-1$ be its algebraic genus and suppose that $\phi_1=\Gamma_1/\Gamma$ and $\phi_2=\Gamma_2/\Gamma$ are two distinct automorphisms of the hyperellipticity of X. By 2.2.5 and 6.1.3 $\sigma(\Gamma^+)=(p;+;[-];\{-\})$ and $\sigma(\Gamma_1^+)=\sigma(\Gamma_2^+)=(0;+;[2,\overset{2p+2}{\ldots},2];\{-\})$. Hence, by the last lemma, $\Gamma_1^+=\Gamma_2^+$ and the signatures of Γ_1 and Γ_2 are of type (i) or (ii) in 6.1.3. In both cases, since Γ is a surface group, there exists an element $v\in\Gamma_1^+\setminus\Gamma$. In fact in the first case it is enough to take a product of two consecutive reflections in Γ_1 and in the second one a canonical elliptic generator. Hence $\Gamma_1=\Gamma\cup v\Gamma$. Now $v\in\Gamma_1^+=\Gamma_2^+\subseteq\Gamma_2$ and so also $\Gamma_2=\Gamma\cup v\Gamma$. Hence $\Gamma_1=\Gamma_2$ and therefore $\phi_1=\phi_2$. The proof in the nonorientable case is similar and we omit it.

Let us check now that ϕ is central in Aut(X). Given $\psi\in$Aut(X) we consider the involution $\Phi=\psi\phi\psi^{-1}$. Let us denote by Γ_1 and Γ_2 NEC groups for which $<\phi>=\Gamma_1/\Gamma$ and $<\Phi>=\Gamma_2/\Gamma$. Both are conjugate as subgroups of $\Omega=$Aut(H), via a representative of ψ modΓ. In particular they are isomorphic and so $X/\Phi=H/\Gamma_1$ has algebraic genus zero, *i.e.* Φ is a hyperellipticity automorphism of X. From the uniqueness part, $\phi=\Phi$ and so $\psi\phi=\phi\psi$.

6.2. THE SIGNATURES ASSOCIATED TO THE AUTOMORPHISM GROUP

Let X be an HKS, and let G be a group of automorphisms of X containing the automorphism of the hyperellipticity. Then $G=\Gamma'/\Gamma$, for some NEC group Γ', and $\Gamma\triangleleft\Gamma_1\triangleleft\Gamma'$. In the previous section we showed that Γ_1/Γ is a central subgroup of Γ'/Γ and in 6.1.3 we described possible signatures of Γ_1 as a function of the topological type of X. In this section we shall determine a finite set of candidates for the signature of Γ'. As a consequence of our results we obtain all candidates for the group of automorphisms of a given HKS.

Proposition 6.2.1. *In the conditions above, let* $N=[\Gamma':\Gamma_1]$. *Then the signature of Γ' has one of the following forms:*

(i) $(0;+;[N,2,\overset{r}{\ldots},2];\{(2,\overset{s}{\ldots},2)\})$,

(ii) $(0;+;[2N,2,\overset{r}{\ldots},2];\{(2,\overset{s}{\ldots},2)\})$,

(iii) $(0;+;[2,\overset{r}{\ldots},2];\{(N/2,2,\overset{s}{\ldots},2)\})$, N *even*,

(iv) $(0;+;[2,\overset{r}{\ldots},2];\{(N,2,\overset{s}{\ldots},2)\})$, N *even*.

Proof. Recall that by 6.1.3 Γ_1 has signature

$$(0;+;[2,\overset{l}{\ldots},2];\{(2,\overset{l'}{\ldots},2)\}) \text{ for some } l,l'.$$

Let Γ' have signature of general form

$$(g';\pm;[m_1,\ldots,m_r];\{(n_{i1},\ldots,n_{is_i})|\ i=1,\ldots,k'\}),$$

and let $\theta:\Gamma' \longrightarrow \Gamma'/\Gamma_1$ be the canonical epimorphism. By results of section 3 in Chapter 2, each period-cycle of Γ_1 comes from a period-cycle of Γ' not preserved by θ. Thus, θ preserves all period-cycles except one, say $C_1 = (n_{11},\ldots,n_{1s_1})$.

Consider two cases:

Case 1. All reflections corresponding to C_1 belong to Γ_1. Then, since Γ_1 has only one period-cycle, this period-cycle is, by theorems 2.3.1 and 2.3.2

$$(n_{11},\ldots,n_{1s_1},\overset{N}{\ldots},n_{11},\ldots,n_{1s_1}).$$

So in particular $n_{11}=\ldots=n_{1s_1}=2$. By theorems 2.2.3 and 2.2.4 the set of proper periods of Γ_1 has the form

$$[(m_i/p_i)^{N/p_i}\ (1\le i\le r),\ (n_{ij}/q_{ij})^{N/2q_{ij}}\ (2\le i\le k',\ 1\le j\le s_i)]$$

for some integers $1\le p_i\le m_i\le N$, $1\le q_{ij}\le n_{ij}\le N/2$. Thus

$$\mu(\Gamma_1)=-1+\sum_{i=1}^{r} N/p_i(1-p_i/m_i)+\sum_{i=2}^{k'}\left[\sum_{j=1}^{s_i} N/2q_{ij}(1-q_{ij}/n_{ij})\right]+N/2\sum_{j=1}^{s_1}(1-1/n_{1j}).$$

It is clear that for every $p_i\ge 1$, $q_{ij}\ge 1$

$$N/p_i(1-p_i/m_i)\le N(1-1/m_i)\ \text{and}\ N/2q_{ij}(1-q_{ij}/n_{ij})\le N/2(1-1/n_{ij}).$$

So looking at $\mu(\Gamma')$ and using the Hurwitz Riemann formula we obtain at once that $-1\ge N(\alpha g'+k'-2)$. Thus $k'=1$, $g'=0$. Now it is easy to check that

$$0=N\mu(\Gamma')-\mu(\Gamma_1)=1-N+N\sum_{i=1}^{r}(1-1/p_i)$$

and the unique solution of this equation is:

(6.2.1.1) *"all p_i but one, say p_{i_0}, are equal to* 1 *and* $p_{i_0}=N$".

Since $m_i/p_i=1$ or 2, we have $m_i=2$ for $i\ne i_0$ and $m_{i_0}=N$ or $2N$. Hence using remark 0.2.6.4 we arrive to one of the first two signatures.

Case 2. Not all reflections corresponding to C_1 belong to Γ_1. Then N is even and by theorem 2.3.3, there exists just one pair (u,v) such that $1\le u\le v\le s_1$,

$$c_{10},\ldots,c_{1u-1},c_{1v+1},\ldots,c_{1s_1}\notin\Gamma_1,\ c_{1u},\ldots,c_{1v}\in\Gamma_1$$

and the period-cycle of Γ_1 has the form:

$$(n_{1,v+1}/2,n_{1,v},\ldots,n_{1,u+1},n_{1,u}/2,n_{1,u+1},\ldots,n_{1,v})^{N/2}$$

for some $1\le u\le v\le s_1$. So in particular $n_{1,u+1}=\ldots=n_{1,v}=2$. By theorem 2.2.4 the set of proper periods of Γ_1 has the form:

$$[(m_i/p_i)^{N/p_i}(1\le i\le r),\ (n_{ij}/q_{ij})^{N/2q_{ij}}\ (2\le i\le k',\ 1\le j\le s_i),\ (n_{1t}/q_{1t})^{N/2q_{1t}}\ (1\le t\le u-1,\ v+1\le t\le s_1)]$$

for some $1\le p_i\le m_i\le N$, $1\le q_{ij}\le n_{ij}\le N/2$. So

$$\mu(\Gamma_1)=-1+\sum_{i=1}^{r} N/p_i(1-p_i/m_i)+\sum_{i=2}^{k'}\left[\sum_{j=1}^{s_i} N/2q_{ij}(1-q_{ij}/n_{ij})\right]+$$

$$+\sum_{t\in\{1,\dots,u-1,v+2,\dots,s_1\}} N/2q_{1t}(1-q_{1t}/n_{1t})+N/4(1-2/n_{1u})+N/4(1-2/n_{1,v+1})+N/2\sum_{t=u+1}^{v}(1-1/n_{1t}).$$

Now using the same arguments as in the previous case and the obvious relations

$$N/4(1-2/n_{1h})\leq N/2(1-1/n_{1h})$$

for $h=u,v+1$, we obtain again that $-1\geq N(\alpha g'+k'-2)$. So $k'=1$ and $g'=0$. Moreover it is easy to check that

$$0=N\mu(\Gamma')-\mu(\Gamma_1)=1-N/2+N\sum_{i=1}^{r}(1-1/p_i)+N/2\sum_{t\in\{1,\dots,u-1,v+2,\dots,s_1\}}(1-1/q_{1t})$$

and the unique solution of this equation is:

(6.2.1.2) *"all* p_i *are equal to* 1, *and all* q_{1t} *but one, say* q_{1t_0}, *are equal to* 1 *and* $q_{1t_0}=N/2$".

So $m_1=\dots=m_r=2$, $n_{1t}=2$ for $t\neq t_0$ and $n_{1t_0}=N/2$ or N. Moreover since $n_{1,v+1}/2=1$ or 2 and $n_{1u}/2=1$ or 2, we have that $n_{1,v+1}=2$ or 4 and similarly for n_{1u}.

We claim that actually none of $n_{1u},n_{1,v+1}$ is equal to 4. In fact assume that $n_{1,v+1}=4$. Recall that $c_{1v}\in\Gamma_1$ and $c_{1,v+1}\notin\Gamma_1$. Clearly $c_{1v}\notin\Gamma$ since otherwise $(c_{1v}c_{1,v+1})^2\in\Gamma$ would be an orientation preserving element of order 2, a contradiction because Γ is a surface group. So $c_{1v}\Gamma$ is a generator of Γ_1/Γ. By 6.1.5 it is central in Γ'/Γ. But then $(c_{1v}c_{1,v+1})^2\Gamma=((c_{1v}\Gamma)(c_{1,v+1}\Gamma))^2=$ $=(c_{1v}^2\Gamma)(c_{1,v+1}^2\Gamma)=\Gamma$, *i.e.* $(c_{1v}c_{1,v+1})^2\in\Gamma$, a contradiction again. The case $n_{1u}=4$ can be ruled out in the same way. Therefore using remark 0.2.6.4 we arrive to one of the remaining two signatures.

Remark 6.2.2. The integers p_i and q_{ij} employed in the proof of the proposition above are so chosen (see the formulation of theorems of chapter 2 employed in the proof) that p_i is the order of $\Gamma_1x_i\in\Gamma'/\Gamma_1$, and q_{ij} is the order of $\Gamma_1c_{i,j-1}c_{ij}\in\Gamma'/\Gamma_1$. So in particular from 6.2.1.1 and 6.2.1.2 it follows that in case of the two first signatures Γ'/Γ_1 is a cyclic group of order N generated by Γ_1x_1 and in the case of the remaining signatures Γ'/Γ_1 is a dihedral group $D_{N/2}$ generated by Γ_1c_{10}, Γ_1c_{11}.

Remark 6.2.3. From the proof of the proposition it also follows that:

(1) If Γ_1 has signature $(0;+;[-];\{(2,\overset{2k}{\dots},2)\})$, then Γ' may have only a signature of type (i), (iii) with $s\neq 0$, $r=0$.

(2) If Γ_1 has signature $(0;+;[2,\overset{2g+k}{\dots},2];\{(-)\})$, then Γ' may have either

a signature of type (i), (ii) with s=0, or (iii), (iv) with arbitrary r and s.

(3) If Γ_1 has signature $(0;+;[2,\overset{g}{\ldots},2],\{(2,\overset{2k}{\ldots},2)\})$, $g>0$, $k>0$, then Γ' may have a signature of type (i),...,(iv) with $s\neq 0$.

By definition 1 the group of automorphisms of an HKS has even order. The previous theorem and results of chapter 2 enable us to reach much more precise information.

Theorem 6.2.4. *Let* X *be an HKS and let* G *be the group of its automorphisms, say of order* 2N. *Then* $G=Z_{2N}$, D_N, $Z_2\times Z_N$ or $Z_2\times D_{N/2}$. *In the second and fourth cases* N *is necessarily even.*

Proof. By theorem 6.1.5 there exist NEC groups Γ_1 and Γ' such that $G=\Gamma'/\Gamma$ and $\Gamma_1/\Gamma\cong Z_2$ is a central subgroup of G. Of course, $[\Gamma':\Gamma_1]=N$. Theorems 6.1.3 and 6.2.1 describe possible signatures of Γ_1 and Γ' respectively. The proof splits naturally into four cases.

Case 1. Γ' has a proper period equal to 2N. Then, since Γ has no proper periods, x_1 induces in Γ'/Γ, by theorem 2.2.4, an element of order 2N. So G is cyclic.

Case 2. Γ' has a period-cycle with a period equal to N. Then N is even by 6.2.1. Let c_0,c_1 be the canonical reflections corresponding to this period. Then, if Γ contained c_0 and c_1, it would have a nonempty period-cycle by theorems 2.3.2 and 2.3.3, whilst if only one of them or a nontrivial power of their product belongs to Γ then, by theorem 2.2.4, Γ has a proper period. So c_0,c_1 induce in Γ'/Γ two elements of order two whose product has order N. Therefore $G=D_N$.

Case 3. Γ' has a proper period equal to N. Then arguing as in case 1 we prove that x_1 induces in Γ'/Γ an element of order N. Let $a\in\Gamma_1\backslash\Gamma$. We shall show that the subgroup $<\Gamma a>\cong Z_2$ meets $<\Gamma x_1>\cong Z_N$ in the identity. Assume to a contradiction that $\Gamma a\in<\Gamma x_1>$. Then N is even and $\Gamma a=\Gamma x_1^{N/2}$. So in particular $\Gamma_1=\Gamma_1 a=\Gamma_1 x_1^{N/2}$, a contradiction (see remark 6.2.2). Therefore since Γa is a central element of G we get $G\cong Z_2\times Z_N$.

Case 4. Γ' has a period-cycle with a period equal to N/2. Let c_0,c_1 be the canonical reflections corresponding to this period. Then, as we remarked in 6.2.2, $\Gamma_1 c_0,\Gamma_1 c_1$ generate the group $\Gamma'/\Gamma_1\cong D_{N/2}$. So in particular Γc_0, Γc_1 generate a dihedral subgroup $D_{N/2}$ of G. Let Γa generate Γ_1/Γ. We shall see that $<\Gamma a>\cap<\Gamma c_0,\Gamma c_1>=\{1\}$. If this were not the case then, either $\Gamma a=\Gamma(c_0c_1)^k c_0$ for some k, or N/2 is even and $\Gamma a=\Gamma(c_0c_1)^{N/4}$. So in particular either

$$(c_0c_1)^k c_0\in\Gamma_1 \text{ or } (c_0c_1)^{N/4}\in\Gamma_1.$$

The second case contradicts 6.2.1.2 whilst in the first one, since Γ_1 is a normal subgroup of Γ', either c_0 or c_1 would belong to Γ_1 what is also impossible. Now, since Γa is a central element of G, we obtain $G \cong Z_2 \times D_{N/2}$.

6.3. THE AUTOMORPHISM GROUP

Let X be an HKS of algebraic genus $p \geq 2$. Then X can be represented as a quotient H/Γ where Γ is a bordered surface group *i.e.* an NEC group with signature $(g;\pm;[-];\{(-),\overset{k}{\ldots},(-)\})$ $(p=\alpha g+k-1,\ k\geq 1)$. Let G be the group of automorphisms of X, say of order 2N. Then $G=\Gamma'/\Gamma$ for some NEC group Γ' containing a group Γ_1 with the signature described in 6.1.3 such that $\Gamma \triangleleft \Gamma_1 \triangleleft \Gamma'$ and Γ_1/Γ is a central subgroup of Γ'/Γ. Proposition 6.2.1 describes possible signatures that Γ' may have whilst theorem 6.2.4 gives candidates for G.

In the remainder of this chapter θ will denote the canonical epimorphism from Γ' onto G, z will be the central element, $\pi: G \longrightarrow G/<z>$ the natural epimorphism and $\theta' = \pi\theta$.

By theorem 6.1.3 the study of groups of automorphisms of HKS splits naturally into 3 parts that we shall analyze in three consecutive theorems.

Theorem 6.3.1. *Let* X *be an HKS of algebraic genus* $p \geq 2$ *and topological genus* 0 *having* k *boundary components. Then* Aut(X) *is* $Z_N \times Z_2$ *or* $D_{N/2} \times Z_2$ *for some integer* N. *Furthermore*

(i) *There exists such an HKS having* $Z_N \times Z_2$ *as the group of automorphisms if and only if* N *is a proper divisor of* k.

(ii) *There exists such an HKS having* $D_{N/2} \times Z_2$ *as the group of automorphisms if and only if* N *is an even divisor of* 2k.

Proof. First notice that $k \geq 3$. Also G=Aut(X) has even order 2N. The signature of Γ is $(0;+;[-];\{(-),\overset{k}{\ldots},(-)\})$. By theorem 6.1.3, Γ_1 has signature $(0;+;[-];\{(2,\overset{2k}{\ldots},2)\})$ and, by 6.2.1 and 6.2.3, Γ' has one of the following signatures:

$$\tau_1=(0;+;[N];\{(2,\overset{s}{\ldots},2)\}),\ \text{where } s=2k/N,$$

$$\tau_2=(0;+;[-];\{(N/2,2,\overset{s}{\ldots},2)\}),\ \text{where } s=(2k/N)+2, \text{ and N is even.}$$

(The number s can be calculated by the Hurwitz Riemann formula). Thus the first part is a consequence of the proof of theorem 6.2.4 (cases 3 and 4).

(i) Assume first that $G=Z_N \times Z_2 = <x,y \mid x^N, y^2, [x,y]>$ is the group of automorphisms of X. Then Γ' has signature τ_1. Let $z \in G$ be a central element of G such that $G/<z> \cong Z_N$ and let $\theta: \Gamma' \longrightarrow G$ be an epimorphism such that $\ker\theta = \Gamma$ and $\ker\theta' = \Gamma_1$. Now by 2.3.2 and 2.3.3

$$\theta'(c_0)=\theta'(c_1)=\ldots=\theta'(c_s)=1.$$

As a result $\theta(c_i)=z^{\varepsilon_i}$, where $\varepsilon_i=0$ or 1 for $i=0,1,\ldots,s$.

Now since Γ has no proper periods, there is no $0\le i<s$ such that $\theta(c_i)=\theta(c_{i+1})=z$, by 2.2.4. Moreover since Γ has only empty period-cycles there is also no $0\le i<s$ such that $\theta(c_i)=\theta(c_{i+1})=1$, by 2.3.3. On the other hand $\theta(c_s)=\theta(e)\theta(c_0)\theta(e)^{-1}=\theta(c_0)$. So we see that s must be even and in particular N divides k. If k=N then the signature of Γ' is $(0;+;[N];\{(2,2)\})$ and therefore by 5.1.5, $\Gamma'/\Gamma\neq\mathrm{Aut}(X)$.

Conversely assume that N is a proper divisor of k. Since $s=2k/N>2$, $\tau_1^+=(0;+;[N,N,2,\overset{s}{\ldots},2];\{-\})$ is maximal, from 0.3.5 and 0.3.6. Hence by 5.1.1 and 5.1.2 there exists a maximal NEC group Γ' with signature τ_1. Consider the epimorphism $\theta:\Gamma'\longrightarrow G$ defined as follows: $\theta(x_1)=x$, $\theta(e)=x^{-1}$, $\theta(c_i)=y^{i+1}$, $i=0,\ldots,s$. Choose $z=y$ as the central element. Then $\theta':\Gamma'\longrightarrow G'=\langle x'|x'^N\rangle$ is given by $\theta'(x_1)=x'$, $\theta'(e)=x'^{-1}$, $\theta'(c_i)=1$, $0\le i\le s$. Now using results of Chapter 2, the reader can easily check that $\ker\theta\cong\Gamma$ and $\ker\theta'\cong\Gamma_1$. Hence $X=H/\ker\theta$ is hyperelliptic and since Γ' is maximal, $\mathrm{Aut}(X)=\Gamma'/\ker\theta=G$.

(ii) Now let $G=D_{N/2}\times\mathbb{Z}_2=\langle x,y|x^2,y^2,(xy)^{N/2}\rangle\times\langle w|w^2\rangle$. Then Γ' has signature τ_2. In particular we see that if an HKS in question exists then N divides 2k and N is even.

Conversely let N be an even divisor of 2k and let $s=2k/N+2$. Again $\tau_2^+=(0;+;[N/2,2,\overset{s}{\ldots},2];\{-\})$ is maximal and so there exists a maximal NEC group Γ' with $\sigma(\Gamma')=\tau_2$. Consider the homomorphism $\theta:\Gamma'\longrightarrow G$ defined as follows:

$$\theta(c_0)=\theta(c_{s+1})=x,\ \theta(c_1)=y,\ \theta(c_i)=w^{i+1},\ i=2,\ldots,s,\ \theta(e)=1.$$

Then, by 2.3.3, $\ker\theta$ has $((s-2)/2)N=k$ empty period-cycles and by 2.2.4 it has no proper periods. So $\ker\theta=\Gamma$. Moreover $\ker\theta'$ has a period-cycle $(2,\overset{s'}{\ldots},2)$, where $s'=(s-2)N=2k$ and it has no proper periods. So also $\ker\theta'=\Gamma_1$, and $X=H/\ker\theta$ is a hyperelliptic surface with $\mathrm{Aut}(X)=\Gamma'/\ker\theta=G$.

Theorem 6.3.2. *Let* X *be an orientable HKS of algebraic genus* $p\ge2$, *topological genus* $g\neq0$ *and* k *boundary components. Then* Aut(X) *is one of the groups:* $\mathbb{Z}_{2N}$, $\mathbb{Z}_N\times\mathbb{Z}_2$, D_N, $D_{N/2}\times\mathbb{Z}_2$ *for some integer* N. *Furthermore*

(i) *There exists such a surface with* $\mathbb{Z}_{2N}$ *as the group of automorphisms if and only if* N *is a proper divisor of* p.

(ii) *There exists such a surface with* $\mathbb{Z}_N\times\mathbb{Z}_2$ *as the group of automorphisms if and only if* N *is a proper divisor of* p+1.

(iii) *There exists such a surface with* D_N *as the group of automorphisms if and only if* N *is an even divisor of* 2p.

(iv) *There exists such a surface with* $D_{N/2}\times\mathbb{Z}_2$ *as the group of automorphisms if and only if* N *is an even divisor of* 2(p+1).

Proof. The first part is a direct consequence of theorem 6.2.4. We shall prove the second part.

The group Γ has signature $(g;+;[-];\{(-),\overset{k}{\ldots},(-)\})$ and so by theorem 6.1.3 Γ_1 has signature $(0;+;[2,\overset{2g+k}{\ldots},2];\{(-)\})$. By remark 6.2.3 Γ' may have one of the following signatures:

$\tau_1=(0;+;[2N,2,\overset{r}{\ldots},2];\{(-)\}),\quad r=p/N.$

$\tau_2=(0;+;[N,2,\overset{r}{\ldots},2];\{(-)\}),\quad r=(p+1)/N.$

$\tau_3=(0;+;[2,\overset{r}{\ldots},2];\{(N,2,\overset{s}{\ldots},2)\}),\quad s=(2p/N)+2-2r,\quad N$ even.

$\tau_4=(0;+;[2,\overset{r}{\ldots},2];\{(N/2,2,\overset{s}{\ldots},2)\}),\ s=(2(p+1)/N)+2-2r,$ N even.

The relations between the numbers r and s are calculated by the Hurwitz Riemann formula.

(i) Assume $\Gamma'/\Gamma=\mathbb{Z}_{2N}=\mathrm{Aut}(X)$. By the proof of theorem 6.2.4 Γ' has signature τ_1 and so in particular N divides p. If $p=N$, $\Gamma'/\Gamma\neq\mathrm{Aut}(X)$ by 5.1.5. Conversely, assume that p is a proper divisor of N. Then $r>1$ and so τ_1^+ is maximal. Hence there exists a maximal NEC group Γ' with signature τ_1, by 5.1.1 and 5.1.2. Consider the epimorphism $\theta:\Gamma'\longrightarrow G=\langle x|x^{2N}\rangle$ defined by

$$\theta(x_1)=x,\ \theta(x_2)=\ldots=\theta(x_{r+1})=x^N,\ \theta(e)=x^{rN-1},\ \theta(c)=1.$$

Choose $z=x^N$. Then using results of chapter 2 in the same way as in the proof of the previous theorem one can show that $\ker\theta=\Gamma$ and $\ker\theta'=\Gamma_1$. Consequently, $X=H/\Gamma$ is hyperelliptic and $\mathbb{Z}_{2N}=\mathrm{Aut}(X)$.

The remaining cases can be proved in the same way. We restrict ourselves to define in each case a suitable epimorphism from a maximal NEC group Γ' with signature τ_2, τ_3 or τ_4 respectively onto $G=\mathbb{Z}_N\times\mathbb{Z}_2$, D_N or $D_{N/2}\times\mathbb{Z}_2$, to prove the "if parts". The existence of such a maximal NEC group Γ' is the immediate consequence of 5.1.1 and 5.1.2 since, by 0.3.5 and 0.3.6, τ_2^+, τ_3^+ and τ_4^+ are maximal signatures.

In case (ii) Γ' has signature τ_2 and we define

$$\theta:\Gamma'\longrightarrow\mathbb{Z}_N\times\mathbb{Z}_2=\langle x,y|x^N,y^2,[x,y]\rangle$$

by

$$\theta(x_1)=x,\ \theta(x_2)=\ldots=\theta(x_{r+1})=y,\ \theta(e)=x^{-1}y^r,\ \theta(c)=1.$$

In case (iii) Γ' has signature τ_3 and we define

$$\theta:\Gamma'\longrightarrow D_N=\langle x,y\mid x^2,y^2,(xy)^N\rangle$$

by

$$\theta(x_1)=\ldots=\theta(x_r)=(xy)^{N/2},\ \theta(e)=(xy)^{rN/2},\ \theta(c_0)=x,\ \theta(c_1)=y,\ \theta(c_2)=1,$$
$$\theta(c_i)=x(xy)^{N(s+i+1)/2},\ i=3,\ldots,s+1.$$

Finally in case (iv) Γ' has signature τ_4 and we define

$$\theta:\Gamma'\longrightarrow D_{N/2}\times\mathbb{Z}_2=\langle x,y,z|x^2,y^2,z^2,(xy)^{N/2},[x,z],[y,z]\rangle$$

by

$$\theta(x_1)=\ldots=\theta(x_r)=z,\ \theta(e)=z^r,\ \theta(c_0)=x,\ \theta(c_1)=y,\ \theta(c_2)=1,$$
$$\theta(c_i)=xz^{s+i+1},\ i=3,\ldots,s+1.$$

Theorem 6.3.3. *Let* X *be a nonorientable HKS of topological genus* g *with* k *boundary components and with algebraic genus* $p\geq 2$. *Then* Aut(X) *is one of the groups* Z_{2N}, $Z_N\times Z_2$, D_N, $D_{N/2}\times Z_2$. *Moreover*

(i) *There exists such a surface with* Z_{2N} *as the group of automorphisms if and only if* $N|g\text{-}1$ *and* $N|k$.

(ii) *There exists such a surface with* $Z_N\times Z_2$ *as the group of automorphisms if and only if* $N|g$ *and* $N|k$.

(iii) *There exists such a surface with* D_N *as the group of automorphisms if and only if* $N|2k$, $N|2(g\text{-}1)$ *and* N *is even.*

(iv) *There exists such a surface with* $D_{N/2}\times Z_2$ *as the group of automorphisms if and only if* $N|2k$, $N|2g$ *and* N *is even.*

Proof. The first part follows from 6.2.4. By 6.1.3 the signature of Γ_1 is

$$(0;+;[2,\overset{g}{\ldots},2];\{(2,\overset{2k}{\ldots},2)\})$$

and by remark 6.2.3 Γ' may have one of the following signatures ($s\neq 0$):

$$\tau_1=(0;+;[2N,2,\overset{r}{\ldots},2];\{(2,\overset{s}{\ldots},2)\}),\ s=((2g+2k\text{-}2)/N)\text{-}2r.$$
$$\tau_2=(0;+;[N,2,\overset{r}{\ldots},2];\{(2,\overset{s}{\ldots},2)\}),\ s=((2g+2k)/N)\text{-}2r.$$
$$\tau_3=(0;+;[2,\overset{r}{\ldots},2];\{(N,2,\overset{s}{\ldots},2)\}),\ s=((2g+2k\text{-}2)/N)+2\text{-}2r,\ N \text{ even}.$$
$$\tau_4=(0;+;[2,\overset{r}{\ldots},2];\{(N/2,2,\overset{s}{\ldots},2)\}),\ s=((2g+2k)/N)+2\text{-}2r,\ N \text{ even}.$$

As in the proof of 6.3.1 (i) we argue that s is even for τ_1 and τ_2. In the proof of 6.2.1 we arrived to signatures τ_1 and τ_2 just when all $c_i\in\Gamma_1$. So, in case (i) by 2.3.1 and 2.3.2, $2k=sN$. Since s is even N divides k. But then also N divides g-1. Now assume that $N|k$, $N|g\text{-}1$. Choose $r=(g\text{-}1)/N$, $s=2k/N$, and $t=2(g+k\text{-}1)/N$. Then $\tau_1^+=(0;+;[2N,2,\overset{t}{\ldots},2];\{\text{-}\})$ is maximal, because $t\geq 4$. Thus there exists a maximal NEC group Γ' with signature τ_1 and we define an epimorphism from Γ' onto $Z_{2N}=\langle x|x^{2N}\rangle$ as follows:

$$\theta(x_1)=x,\ \theta(x_2)=\ldots=\theta(x_{r+1})=x^N,\ \theta(e)=x^{rN-1},\ \theta(c_i)=x^{iN},\ i=0,\ldots,s.$$

As before we argue that $\ker\theta=\Gamma$, and that for $z=x^N$, $\ker\theta'=\Gamma_1$. Consequently, $X=H/\Gamma$ is hyperelliptic and, Γ' being maximal, $\mathrm{Aut}(X)=\Gamma'/\Gamma=Z_{2N}$.

Remaining three cases can be proved in the same way. We restrict ourselves to define suitable epimorphisms for the "if parts". As always, Γ' can be chosen maximal by 5.1.1 and 5.1.2, because τ_2^+, τ_3^+ and τ_4^+ are maximal signatures, from 0.3.5 and 0.3.6.

In case (ii) we choose $r=g/N$, $s=2k/N$. The epimorphism $\theta:\Gamma'\longrightarrow Z_N\times Z_2= \langle x,y\mid x^N,y^2,[x,y]\rangle$ we define is given by:

$$\theta(x_1)=x,\ \theta(x_2)=\ldots=\theta(x_r)=y,\ \theta(e)=x^{-1}y^{r-1},\ \theta(c_i)=y^{i+1},\ i=0,\ldots,s.$$

In case (iii) we take arbitrary r in range $0\leq r\leq (g\text{-}1)/N$ and decompose the corresponding s into the sum s_1+s_2, where $s_1=(2(g\text{-}1)/N)\text{-}2r$, and $s_2=(2k/N)+2$. Then we define $\theta:\Gamma'\longrightarrow D_N=\langle x,y|x^2,y^2,(xy)^N\rangle$ as

$$\theta(x_1)=\ldots=\theta(x_r)=(xy)^{N/2},\ \theta(e)=(xy)^{rN/2},\ \theta(c_0)=\theta(c_{s+1})=x,$$

$$\theta(c_i)=\begin{cases} y(xy)^{N(i+1)/2}, & i=1,2,\ldots,s_1+1,\\ (xy)^{N(i+1)/2}, & i=s_1+2,\ldots,s_1+s_2.\end{cases}$$

The case (iv) is similar. We choose again arbitrary r in range $0\leq r\leq g/2$ and decompose the corresponding s into the sum s_1+s_2, where $s_1=2g/N-2r$, and $s_2=(2k/N)+2$. Then if $D_{N/2}\times\mathbb{Z}_2=\langle x,y,z|x^2,y^2,z^2,(xy)^{N/2},[x,z],[y,z]\rangle$, we define $\theta:\Gamma'\longrightarrow D_{N/2}\times\mathbb{Z}_2$ by

$$\theta(x_1)=\ldots=\theta(x_r)=z,\ \theta(e)=z^r,\ \theta(c_0)=\theta(c_{s+1})=x,$$

$$\theta(c_i)=\begin{cases} yz^{i+1}, & i=1,2,\ldots,s_1+1,\\ z^{i+1}, & i=s_1+2,\ldots,s_1+s_2.\end{cases}$$

6.4. NOTES

The characterization of hyperelliptic compact Klein surfaces given in 6.1.3 was established in [21]. Its counterpart for Riemann surfaces was obtained by Maclachlan [91], who expressed in terms of fuchsian groups the existence of a canonical involution. The combinatorial proof of 6.1.4 is inspired in Accola [2].

Partial results on automorphisms groups of hyperelliptic compact Klein surfaces with nonempty boundary were obtained, for surfaces of genus 2 and 3 in [26] and [22], respectively, and for groups with big enough order, in [18], [19] and [20].

Finally we remark that results on compact hyperelliptic Riemann surfaces of low genus are due to Wiman [127], Kuribayashi,A. [76], Kuribayashi,I. [77], and Lønsted-Kleiman [81], and for groups with big enough order, they appear in [25]. A characterization of those groups that can be realized as subgroups of Aut(S) for hyperelliptic compact Riemann surfaces S, is given by Brandt-Stichtenoth in [9] -see also Tsuji [122]-.

APPENDIX
Compact Klein surfaces and real algebraic curves.

The main goal in this appendix is to prove that categories $\mathcal{F}_{\mathbb{R}}$ and $\mathcal{K}$ of algebraic function fields in one variable over $\mathbb{R}$ and compact Klein surfaces are functorially coequivalent. To see that, we shall associate to each field extension $E|\mathbb{R}$ in $\mathcal{F}_{\mathbb{R}}$ a compact Klein surface $S(E|\mathbb{R})$ and we also prove that the field $\mathcal{M}(S)$ of meromorphic functions on a compact Klein surface S is an algebraic function field in one variable over $\mathbb{R}$. Moreover we get the fundamental equalities:

$$\mathcal{M}(S(E|\mathbb{R}))=E;\ S(\mathcal{M}(S)|\mathbb{R})=S.$$

This correspondence has nice properties: S has nonempty boundary if and only if $\mathcal{M}(S)$ is a *real field* (*i.e.*, -1 is not a sum of squares in $\mathcal{M}(S)$). In this case, ∂S is homeomorphic to the compact and irreducible, smooth real algebraic curve C whose field of rational functions is $\mathcal{M}(S)$. Nonorientable surfaces without boundary turn out to be those for which $\mathcal{M}(S)$ is not real, but $\sqrt{-1}\notin\mathcal{M}(S)$.

The tools developed in this book apply to obtain results on the group B(C) of birational isomorphisms of a real irreducible algebraic curve C of (algebraic) genus $p\geq 2$. In fact, the field $\mathbb{R}(C)$ of rational functions on C belongs to $\mathcal{F}_{\mathbb{R}}$ and it is real. So, $S=S(\mathbb{R}(C)|\mathbb{R})$ is a compact bordered Klein surface with algebraic genus p and

$$B(C)=\mathrm{Isom}(\mathbb{R}(C))=\mathrm{Isom}(\mathcal{M}(S))=\mathrm{Aut}(S).$$

All throughout this appendix we fix a compact Klein surface S and a dianalytic atlas $\mathcal{A}=\{(U_i,\phi_i)|i\in I\}$ on S. We shall denote by $\Sigma=\mathbb{C}\cup\{\infty\}$ and $\Delta=\mathbb{C}^+\cup\{\infty\}$ the Riemann sphere and the closed disc with the structures of Klein surfaces given in Chapter 0.

Definition 1. *A meromorphic function on* S *relative to* $\mathcal{A}$ is a family of maps $f_{\mathcal{A}}=(f_i\colon U_i\longrightarrow\Sigma,\ i\in I)$ such that

(1) $f_i\phi_i^{-1}\colon \phi_i(U_i)\longrightarrow\Sigma$ is a meromorphic function.

(2) $f_i(U_i\cap\partial S)\subseteq\mathbb{R}\cup\{\infty\}$.

(3) $f_j|V=\begin{cases} f_i|V & \text{if } \phi_i\phi_j^{-1}|V \text{ is analytic}\\ \bar{f}_i|V & \text{if } \phi_i\phi_j^{-1}|V \text{ is antianalytic}\end{cases}$

for every connected subset V of $U_i \cap U_j$, where $\bar{f}_i$ is the composition of f_i with the complex conjugation.

Under natural operations, the set $\mathcal{M}_{\mathcal{A}}(S)$ of meromorphic functions on S relative to $\mathcal{A}$, is a field.

Remarks A.1. (1) The field $\mathbb{R}$ of real numbers is contained in $\mathcal{M}_{\mathcal{A}}(S)$ under the natural identification

$$c \in \mathbb{R} \longmapsto c_{\mathcal{A}} = (c_i : U_i \longrightarrow \Sigma : x \longmapsto c).$$

(2) It is easily seen that the fields $\mathcal{M}_{\mathcal{A}}(S)$ and $\mathcal{M}_{\mathcal{B}}(S)$ are $\mathbb{R}$-isomorphic for equivalent atlas $\mathcal{A}$ and $\mathcal{B}$ on S. Hence, from now on, we denote by $\mathcal{M}(S)$ the field of meromorphic functions on S with respect to every atlas equivalent to $\mathcal{A}$.

(3) In general, $\mathcal{M}(S)$ does not contain $\mathbb{C}$. More precisely,

$$\sqrt{-1} \in \mathcal{M}(S) \text{ if and only if S is orientable with empty boundary.}$$

In fact, if $x \in \partial S \cap U_i$, the constant function $\sqrt{-1}$ on U_i does not verify condition (2) in Definition 1, whilst in case S is nonorientable, condition (3) fails to be true on a connected subset of $U_i \cap U_j$ for some $i,j \in I$ such that $\phi_i \phi_j^{-1}$ is antianalytic. The converse is obvious since Riemann surfaces admit analytic atlas.

(4) This last argument shows that for a Riemann surface S, each $f \in \mathcal{M}(S)$ induces a morphism of Klein surfaces $\tilde{f}: S \longrightarrow \Sigma$. As a consequence of the compatibility condition (3) in Definition 1, this is not the case for Klein surfaces which are not Riemann surfaces. However,

Proposition A.2. *Each* $f \in \mathcal{M}(S)$ *induces a morphism of Klein surfaces* $\tilde{f}: S \longrightarrow \Delta$. *Moreover, if* $f_1, f_2 \in \mathcal{M}(S)$ *are non-constant with* $f_1 \neq f_2$, *then* $\tilde{f}_1 \neq \tilde{f}_2$.

Before to prove it we need a definition.

Definition 2. Let $f \in \mathcal{M}(S)$ and $x \in U_i$. We define the *order of* f *at* x as

$$\upsilon_x(f) = \text{order at } \phi_i(x) \text{ of } f_i \phi_i^{-1}.$$

The set of *poles* of f is $P(f) = \{x \in S \mid \upsilon_x(f) < 0\}$, and we denote $Z(f) = \{x \in S \mid \upsilon_x(f) > 0\}$ the set of *zeros* of f. Clearly, $Z(f) = P(f^{-1})$ if $f \neq 0$. Since S is compact, P(f), and so Z(f), are finite for $f \neq 0$.

Proof of A.2. For each $i \in I$ we can write

$$f_i \phi_i^{-1} : \phi_i(U_i \setminus P(f)) \longrightarrow \mathbb{C} : x + y\sqrt{-1} \longmapsto u_i(x,y) + v_i(x,y)\sqrt{-1}$$

where $u_i, v_i : \phi_i(U_i \setminus P(f)) \longrightarrow \mathbb{R}$ are analytic functions satisfying Cauchy-Riemann conditions. Put $g_i = u_i \phi_i$, $h_i = v_i \phi_i$. Since $f \in \mathcal{M}(S)$, we get

$$g_i = g_j \text{ and } |h_i| = |h_j| \text{ on } U_i \cap U_j \setminus P(f).$$

Hence we define a morphism $\tilde{f}: S \longrightarrow \Delta: x \longmapsto \begin{cases} g_i(x)+|h_i(x)|\sqrt{-1}, & x\in U_i\setminus P(f) \\ \infty & x\in P(f). \end{cases}$

Moreover, if $f_1,f_2\in\mathcal{M}(S)$ and $\tilde{f}_1=\tilde{f}_2$ on each connected U_i, then $f_1|U_i$ equals either $f_2|U_i$ or $\overline{(f_2|U_i)}$. Since both $(f_1|U_i)\phi_i^{-1}$ and $(f_2|U_i)\phi_i^{-1}$ are non-constant and analytic outside a thin set, we conclude $f_1|U_i=f_2|U_i$ and thus $f_1=f_2$.

Back to Riemann surfaces, it is easily seen that P(f) is nonempty for non-constant $f\in\mathcal{M}(S)$. Hence, the *order of f* defined as

$$o(f)=-\sum_{x\in P(f)} v_x(f)$$

is a positive integer. A crucial result in the theory of Riemann surfaces (see [49], Thms. 8.3, and 16.1) says:

Theorem A.3. *Let S be a compact Riemann surface.*

(1) Given $a\in S$ *there exists* $f\in\mathcal{M}(S)$ *whose unique pole is* a. *In particular* f *is non-constant.*

(2) Given non-constant $f,g\in\mathcal{M}(S)$, *there exists a polynomial* $P\in\mathbb{C}[X,Y]$, $P\neq 0$, *such that* $P(f,g)=0$ *and the degree* $\partial_Y P$ *of* P *with respect to* Y *is less than or equal to* o(f).

(3) As a consequence, the field extension $\mathcal{M}(S)|\mathbb{C}$ *is an algebraic function field in one variable.*

We are interested now to prove

Theorem A.4. *If* S *is a compact Klein surface, then* $\mathcal{M}(S)|\mathbb{R}$ *is an algebraic function field in one variable.*

First we need the following lemma, which shall be also useful later:

Lemma A.5. *Each non-constant morphism* $f: S\longrightarrow S'$ *between Klein surfaces* S *and* S' *induces an* $\mathbb{R}$-*homomorphism* $f^*: \mathcal{M}(S')\longrightarrow\mathcal{M}(S)$ *such that, for every* $g\in\mathcal{M}(S')$, *the morphism* $(f^*(g))^\sim$ *associated to* $f^*(g)$ *is* $\tilde{g}f$. *Moreover,* f^* *is unique with this property.*

Proof. Uniqueness is clear since by A.2 each meromorphic function is determined by the morphism it induces. Let us construct f^*. Given $g\in\mathcal{M}(S')$ we are going to define $h=f^*(g)\in\mathcal{M}(S)$. Let us take two atlas $\mathcal{A}=\{(U_i,\phi_i)|i\in I\}$ and $\mathcal{B}=\{(V_i,\psi_i)|i\in I\}$ on S and S' such that $f(U_i)\subseteq V_i$. We get commutative diagrams

$$\begin{array}{ccccc} U_i & & \xrightarrow{\quad f \quad} & & V_i \\ \big\downarrow \phi_i & & & & \big\downarrow \psi_i \\ \phi_i(U_i) & \xrightarrow{F} & \mathbb{C} & \xrightarrow{\Phi} & \mathbb{C}^+ \end{array}$$

where Φ is the "folding map" and F is an analytic map. Then, if g is given by the family $(g_i : V_i \longrightarrow \Sigma,\ i \in I)$ we define h to be the meromorphic function $h=(h_i=g_i\psi_i^{-1}\Phi F\phi_i\colon U_i \longrightarrow \Sigma)$. It is straightforward to check that $g \longmapsto f^*(g)=h$, is a field $\mathbb{R}$-homomorphism. Moreover, $\tilde{h}=\tilde{g}f$ because $h_i=g_i\psi_i^{-1}\psi_i f=g_i f$.

Corollary A.6. *Given non-constant morphisms* $f_1\colon S \longrightarrow S'$ *and* $f_2\colon S' \longrightarrow S''$, *then* $(f_2f_1)^*=f_1^*f_2^*$.

Proof. It follows at once from A.2 and the uniqueness part in the lemma above.

The proof of A.4 is based on A.3 and the nice relation between $\mathcal{M}(S)$ and $\mathcal{M}(S_c)$, where S_c is the double cover of S. To see that, and also for later purposes (*e.g.* A.19) we need the following

Lemma A.7. *Let* Y *be a compact Riemann surface. Let* α *be an antianalytic involution of* Y *and let* Z *be the quotient of* Y *under the action of* $G=\langle\alpha\rangle$. *Let* $q\colon Y \longrightarrow Z$ *be the canonical projection and* $q^*\colon \mathcal{M}(Z) \longrightarrow \mathcal{M}(Y)$ *the associated* $\mathbb{R}$*-homomorphism. Then:*

(1) $q^*(\mathcal{M}(Z))=\{f \in \mathcal{M}(Y) \mid \alpha^*(f)=f\}$.

(2) If $\alpha^*(\sqrt{-1}) \neq \sqrt{-1}$, *then* $\mathcal{M}(Y)=[q^*(\mathcal{M}(Z))](\sqrt{-1})$.

Proof. (1) Since $q\alpha=q$, we get $q^*=\alpha^*q^*$ and so one inclusion is obvious. Conversely, let $\mathcal{A}=\{(U_i,\phi_i) \mid i \in I\}$ and $\mathcal{B}=\{(V_i,\psi_i) \mid i \in I\}$ be atlas on Y and Z respectively, and $f=(f_i\colon U_i \longrightarrow \Sigma \mid i \in I)$ be a meromorphic function on Y with $\alpha^*(f)=f$. This equality implies that $f_i(y)=f_i(\alpha(y))$ whenever $y \in U_i$, and so there exists a unique well defined map $h_i\colon V_i \longrightarrow \Sigma$ such that $h_i(q|U_i)=f_i$. It is easily checked that $h=(h_i\colon V_i \longrightarrow \Sigma,\ i \in I) \in \mathcal{M}(Z)$ and $\tilde{h}q=\tilde{f}$, *i.e.*, $f=q^*(h)$.

(2) Put $F=q^*(\mathcal{M}(Z))$ and $E=\mathcal{M}(Y)$. From A.1, $F \subseteq F(\sqrt{-1}) \subseteq E$. From (1), $\sqrt{-1} \notin F$ and so it is enough to prove that $[E:F] \leq 2$. To see that, let us take $f \in E \setminus F$. Then if $g \in E$, the elements $a=\dfrac{f\alpha^*(g)-g\alpha^*(f)}{f-\alpha^*(f)}$ and $b=\dfrac{g-\alpha^*(g)}{f-\alpha^*(f)}$ verify $\alpha^*(a)=a$ and $\alpha^*(b)=b$. Hence, by part (1), $a,b \in F$. Also $g=a+bf$ and so $\{1,f\}$ is a basis of E as F-vector space.

Corollary A.8. *Let* S *be a compact Klein surface which is not a Riemann surface. Let* $\pi\colon S_c \longrightarrow S$ *be its double cover and let* $\tau\colon S_c \longrightarrow S_c$ *be the canonical antianalytic involution of* S_c. *Then:*

(1) $\pi^*(\mathcal{M}(S))=\{f\in\mathcal{M}(S_C)\mid\tau^*(f)=f\}$.

(2) $\mathcal{M}(S_C)=[\pi^*(\mathcal{M}(S))](\sqrt{-1})$.

Proof. As we know $S=S_C/<\tau>$. Moreover, by A.1, $\sqrt{-1}\notin\mathcal{M}(S)$, *i.e.*, $\tau^*(\sqrt{-1})\neq\sqrt{-1}$. Now, the result follows from the last lemma.

Along the whole appendix we fix the notation (S_C,π,τ) for the double cover of S constructed in 0.1.12.

Proof of theorem A.4. In case S is a Riemann surface, from A.3 $\mathcal{M}(S)=\mathbb{C}(f,g)$ for some $f\in\mathcal{M}(S)\setminus\mathbb{C}$ and $g\in\mathcal{M}(S)$ algebraic over $\mathbb{C}(f)$. Hence $\mathcal{M}(S)=\mathbb{R}(f)(g,\sqrt{-1})=$ $=\mathbb{R}(f,h)$, using the existence of a primitive element $h\in\mathcal{M}(S)$ of the extension $\mathbb{R}(f)(g,\sqrt{-1})|\mathbb{R}(f)$.

If S is not a Riemann surface we keep the notations above and from A.8, $\mathcal{M}(S_C)=E=F(\sqrt{-1})\supseteq F=\pi^*(\mathcal{M}(S))$. Of course it is enough to prove that $F|\mathbb{R}$ is an algebraic function field. But

$$\text{tr.deg } F|\mathbb{R} = \text{tr.deg } E|\mathbb{R} - \text{tr.deg } E|F = \text{tr.deg } E|\mathbb{R} = \text{tr.deg } E|\mathbb{C} = 1.$$

Thus there exists $f\in F$ transcendental over $\mathbb{R}$. Moreover, by A.3,

$$[F:\mathbb{R}(f)]=[E:\mathbb{C}(f)]\leq o(f).$$

Hence, using again the existence of primitive element we can write $F=\mathbb{R}(f,g)$ for some $g\in F$.

Comment A.9. The correspondence $S\longrightarrow\mathcal{M}(S)$ is a functor from $\mathcal{K}$ into $\mathcal{F}_{\mathbb{R}}$ as a consequence of A.4 and A.6. We shall construct its inverse. For this purpose we fix in what follows an algebraic function field in one variable $E|k$ where $k=\mathbb{R}$ or $\mathbb{C}$.

Definition 3. A *valuation ring* of $E|k$ is a ring V, $k\subseteq V\subseteq E$, $V\neq E$, such that $f^{-1}\in V$ whenever $f\in E\setminus V$. It is easily seen that V is a local and integrally closed domain with quotient field E. The maximal ideal is denoted by $\mathfrak{m}_V=\{f\in V\mid f^{-1}\notin V\}$ and the set of units by $\mathfrak{u}_V=V\setminus\mathfrak{m}_V$. We call $\kappa(V)=V/\mathfrak{m}_V$ the *residual field of* V and the map $\lambda_V:E\longrightarrow\kappa(V)\cup\{\infty\}$ which maps $f\in E$ to $f+\mathfrak{m}_V$ if $f\in V$ and to ∞ if $f\notin V$ is the *place induced by* V. Clearly $\mathfrak{u}_V$ is a subgroup of the multiplicative group $E^*=E\setminus\{0\}$ and the quotient $\Gamma_V=E^*/\mathfrak{u}_V$ is called the *value group* of V. The map $v: E^*\longrightarrow\Gamma_V : h\longmapsto h\mathfrak{u}_V$ is the *valuation* associated to V. The set of valuation rings of $E|k$ is denoted by $S(E|k)$.

Examples A.10. (1) Take $E=k(T)$ where T is an indeterminate. Then $S(E|k)$

consists of $k[T]_{hk[T]}$, $h\in k[T]$ irreducible, and $k[T^{-1}]_{T^{-1}k[T^{-1}]}$. In fact, if $h\in k[T]$ is irreducible, $V=k[T]_{hk[T]}$ is a valuation ring of $E|k$ because $k\subseteq V$, $h^{-1}\in E\setminus V$ and given $u\in E$ we can write $u=FG^{-1}$ for some $F,G\in k[T]$ that factorize

$$F = h^nF_1...F_r,\quad G = h^mG_1...G_s \quad \text{where } h\nmid F_i,G_j.$$

Hence, if $n\geq m$ it is $u\in V$ and otherwise $u^{-1}\in V$. Moreover,

$$\mathfrak{u}_V=\{u=FG^{-1}\in E|n=m\},\quad \mathfrak{m}_V=hV.$$

Also $k[T^{-1}]_{T^{-1}k[T^{-1}]}\in S(E|k)$ since it is the image of $k[T]_{Tk[T]}$ under the k-isomorphism $k[T]\longrightarrow k[T] : T\longmapsto T^{-1}$.

Conversely, let $V\in S(E|k)$. If $T\in V$, then $k[T]\subseteq V$ and so, for $\mathfrak{p}=\mathfrak{m}_V\cap k[T]$ it is $k[T]_{\mathfrak{p}}\subseteq V$. Then $\mathfrak{p}$ is a prime ideal and $\mathfrak{p}\neq(0)$ because $V\neq E$. Thus $\mathfrak{p}=hk[T]$ for some irreducible polynomial $h\in k[T]$. Hence $V'=k[T]_{hk[T]}\subseteq V$ and both are valuation rings with $\mathfrak{m}_V\cap V'=\mathfrak{m}_{V'}$. Now it is immediate that $V=V'$. On the other hand, if $T\notin V$ it is $T^{-1}\in\mathfrak{m}_V\cap k[T^{-1}]$ and arguing as above, $V=k[T^{-1}]_{T^{-1}k[T^{-1}]}$.

We must emphasize two important properties:

(i) V is a *discrete* valuation ring, *i.e.*, $\Gamma_V\cong\mathbb{Z}$. In fact we have seen that $\mathfrak{m}_V=hV$ for some $h\in V$ and so Γ_V is generated by $h\mathfrak{u}_V$. It is torsion-free since $h^n\in\mathfrak{u}_V$ implies $h\in\mathfrak{u}_V$.

(ii) $\kappa(V)|k$ is finite, since $\kappa(V)=k[T]/hk[T]$ and so $[\kappa(V):k]=\deg(h)$.

(2) Let S be a compact Riemann surface and let $x\in S$. The elements $f\in\mathcal{M}(S)$ such that $x\notin P(f)$ form a valuation ring V_x of $\mathcal{M}(S)|\mathbb{C}$. The only non obvious fact is that $V_x\neq\mathcal{M}(S)$. This follows from (1) in A.3.

Comment. The idea now is to endow $S(E|\mathbb{C})$ with a structure of Riemann surface, such that if S is a compact Riemann surface, the map

$$S\longrightarrow S(\mathcal{M}(S)|\mathbb{C}) : x\longmapsto V_x$$

is an isomorphism of Riemann surfaces. After, we shall extend the result to Klein surfaces by changing $\mathbb{C}$ by $\mathbb{R}$. We first introduce some elementary valuation theory.

Let $E|k$ be given; if $V\in S(E|k)$ we can choose $f\in V$ transcendental over k. Put $F=k(f)$. Then

$$W=V\cap F\in S(F|k) \text{ and } \mathfrak{m}_W=\mathfrak{m}_V\cap W.$$

Hence we have two inclusion maps

$$\Gamma_W\hookrightarrow\Gamma_V : x\mathfrak{u}_W\longmapsto x\mathfrak{u}_V;\ \kappa(W)\hookrightarrow\kappa(V) : x+\mathfrak{m}_W\longmapsto x+\mathfrak{m}_V.$$

By standard arguments one can prove

$$[\Gamma_V:\Gamma_W][\kappa(V):\kappa(W)]\leq[E:F].$$

In particular $[\Gamma_V:\Gamma_W]$ and $[\kappa(V):\kappa(W)]$ are finite. Thus

Proposition A.11. *Let* $V \in S(E|k)$. *Then*

(1) V *is a discrete valuation ring. Moreover, if* $h\mathfrak{u}_V$ *generates* Γ_V, *and* $h \in V$, *then* $\mathfrak{m}_V = hV$.

(2) $\bigcap_{n \in \mathbb{N}} \mathfrak{m}_V^n = (0)$.

(3) $[\kappa(V):k]$ *is finite.*

(4) If $k=\mathbb{C}$, *then* $\kappa(V)=\mathbb{C}$. *If* $k=\mathbb{R}$ *then* $\kappa(V)=\mathbb{R}$ *or* $\mathbb{C}$.

Proof. (1) With the notations above, let $e=[\Gamma_V:\Gamma_W]$. We claim that the map

$$\Gamma_V \longrightarrow \Gamma_W : t \longmapsto t^e$$

is an injective group homomorphism. Indeed, if $t=z\mathfrak{u}_V$ with $z \notin \mathfrak{u}_V$, either $z \notin V$ and so $z^e \notin V$ since V is integrally closed, or $z \in \mathfrak{m}_V$, which implies $z^e \in \mathfrak{m}_V$. In both cases $z^e \notin \mathfrak{u}_V$. Since Γ_W is infinite cyclic, see Example A.10(1), the same holds true for Γ_V. The second part and also (2) are obvious. Part (3) is evident since both $[\kappa(V):\kappa(W)]$ and $[\kappa(W):k]$ are finite, the last by Example A.10. Now it is immediate to see (4).

Remarks A.12. (1) Take $h(T)=T^2+1 \in \mathbb{R}[T]$. It is irreducible and so $V=\mathbb{R}[T]_{h\mathbb{R}[T]}$ is a valuation ring of $\mathbb{R}(T)|\mathbb{R}$. In this case $\kappa(V)=\mathbb{R}[T]/(T^2+1)=\mathbb{C}$, although $k=\mathbb{R}$.

(2) Let $\bar{k}$ be the algebraic closure of k in E. Since $k \subseteq V$ and V is integrally closed, we have $\bar{k} \subseteq V$. Hence the map

$$\bar{k} \hookrightarrow \kappa(V) : x \longmapsto x+\mathfrak{m}_V$$

is a field homomorphism. Since $k \subseteq \bar{k} \subseteq \kappa(V)$, and using (3) in proposition above we deduce that $[\kappa(V):\bar{k}]$ is finite. We define the *degree* of V by

$$\deg(V)=[\kappa(V):\bar{k}].$$

Definition 4. Let $v:E^* \longrightarrow \Gamma_V$ be the valuation associated with $V \in S(E|k)$. Let $h \in V$ be such that $h\mathfrak{u}_V$ generates Γ_V, and $\phi_h:\Gamma_V \longrightarrow \mathbb{Z}: h\mathfrak{u}_V \longmapsto 1$. The *order function of* V is the map $\tilde{v}=\phi_h v$, which *does not depend* on h. The following properties are easily checked:

Proposition A.13. *(1)* $V=\{f \in E^* \mid \tilde{v}(f) \geq 0\} \cup \{0\}$.

(2) $\tilde{v}(f)=0$ *if and only if* $f \in \mathfrak{u}_V$; $\tilde{v}(f)=n>0$ *if and only if* $f \in \mathfrak{m}_V^n \setminus \mathfrak{m}_V^{n+1}$.

(3) $\tilde{v}(fg)=\tilde{v}(f)+\tilde{v}(g)$.

(4) $\tilde{v}(f+g) \geq \min\{\tilde{v}(f),\tilde{v}(g)\}$.

(5) $\tilde{v}(f+g)=\min\{\tilde{v}(f),\tilde{v}(g)\}$ *if* $\tilde{v}(f) \neq \tilde{v}(g)$.

The sets of zeros and poles of a given $f \in E^*$ are defined as

$$Z(f)=\{V \in S(E|k) \mid \tilde{v}(f)>0\}; \quad P(f)=Z(f^{-1})=\{V \in S(E|k) \mid \tilde{v}(f)<0\}.$$

We have introduced the language enough to state a basic result concerning valuations on $E|k$. For the proofs the reader can see [36], Chapter I, Theorem

4, and Chapter IV, Lemma 1.

Theorem A.14. *(1) Let us take* $f\in E\backslash\bar{k}$. *Then* $Z(f)$ *is finite and nonempty. Also,*

$$[E:\bar{k}(f)] = \sum_{V\in Z(f)} \tilde{v}(f)\deg(V).$$

(2) Given $V\in S(E|k)$ *there exists* $f\in E$ *such that* $Z(f)=\{V\}$.

As announced before we endow $S(E|\mathbb{C})$ with a structure of compact Riemann surface.

Definition 5. Given $E|k$, $k=\mathbb{R}$ or $\mathbb{C}$, and $V\in S(E|k)$ we know $\kappa(V)\subseteq\mathbb{C}$. Hence, given $f\in E$ we can define

$$\hat{f} : S(E|k) \longrightarrow \Sigma=\mathbb{C}\cup\{\infty\} : V\longmapsto\hat{\lambda}_V(f).$$

We endow $S(E|k)$ with the initial topology for the family of maps $\{\hat{f}|f\in E\}$.

Proposition A.15. $S(E|\mathbb{C})$ *is a compact Riemann surface.*

The proof is given in [36], Chapter VIII. The idea is to consider a copy Σ_f of Σ for each $f\in E$ and to prove that the map

$$S(E|\mathbb{C}) \longrightarrow P=\prod_{f\in E}\Sigma_f : V\longmapsto(\hat{f}(V)|f\in E)$$

is injective and continuous with closed image. In such a way $S(E|\mathbb{C})$ is a Hausdorff, compact, topological space. An analytic atlas on $S(E|\mathbb{C})$ is constructed in the following way. Given $V\in S(E|\mathbb{C})$ its maximal ideal is generated by some element $f_V\in V\backslash\mathbb{C}$. This defines by restriction a homeomorphism

$$\psi_V=\hat{f}_V|U^V : U^V \longrightarrow B(0,\delta),\ \psi_V(V)=0$$

for suitable $\delta>0$, $\delta\in\mathbb{R}$, and some neighbourhood U^V of V. The analyticity of the transition functions $\psi_W\psi_V^{-1}$ is proved by using the implicit function theorem.

Comment. In order to endow $S(E|\mathbb{R})$ with a structure of Klein surface, the strategy is to compare $S(E|\mathbb{R})$ and $S(E(\sqrt{-1})|\mathbb{C})$, applying to this last the proposition above. We begin with the trivial case.

Proposition A.16. *Given* $E|\mathbb{R}$ *with* $\sqrt{-1}\in E$, *then* $S(E|\mathbb{R})=S(E|\mathbb{C})$. *Thus* $S(E|\mathbb{R})$ *is an orientable compact Klein surface with empty boundary.*

Proof. Clearly $E|\mathbb{C}$ is an algebraic function field in one variable. Moreover, if $V\in S(E|\mathbb{R})$ we know by A.12.(2) that $\mathbb{C}=\bar{k}\subseteq V$, *i.e.*, $V\in S(E|\mathbb{C})$.

Lemma A.17. *Let us take* $E|\mathbb{R}$ *with* $\sqrt{-1}\notin E$ *and* $E'=E(\sqrt{-1})$. *Then:*

(1) $S(E|\mathbb{R})$ *is a Hausdorff topological space.*

(2) Each element in E' *can be written, in a unique way, as* $f+\sqrt{-1}\,g$, $f,g\in E$.
(3) If $f,g\in E$ *are not both equal to zero, then* $f^2+g^2\neq 0$.
(4) Given $V\in S(E'|\mathbb{C})$ *and* $\sigma(V)=\{f+\sqrt{-1}\,g\in E'|f-\sqrt{-1}\,g\in V\}$, *it is* $\sigma(V)\in S(E'|\mathbb{C})$ *and* $\mathfrak{m}_{\sigma(V)}=\{f+\sqrt{-1}\,g\in E'|f-\sqrt{-1}\,g\in\mathfrak{m}_V\}$.
(5) Given $V\in S(E'|\mathbb{C})$, *then* $W=V\cap E\in S(E|\mathbb{R})$ *and* $\mathfrak{m}_W=\mathfrak{m}_V\cap E=\mathfrak{m}_V\cap W$. *In particular* $\kappa(W)\subseteq\kappa(V)$.
(6) With the notations in (5), $V=\sigma(V)$ *if and only if* $\kappa(W)=\mathbb{R}$.

Proof. (1) Take $W_1,W_2\in S(E|\mathbb{R})$, $W_1\neq W_2$; *e.g.*, there exists $f\in W_1\setminus W_2$. Then, $\hat{f}(W_2)=\infty\neq\hat{f}(W_1)=a\in\Sigma$. Since the Riemann sphere is a Hausdorff space, there exist open disjoint subsets U_1 and U_2 of Σ, such that $\infty\in U_2$, $a\in U_1$. Then $G_1=(\hat{f})^{-1}(U_1)$ and $G_2=(\hat{f})^{-1}(U_2)$ are open disjoint neighbourhoods of W_1 and W_2 respectively.
(2) It is evident, because $[E':E]=2$.
(3) Suppose $f^2+g^2=0$, $f\neq 0$. Then $u=gf^{-1}\in E$ and $u^2=-1$, *i.e.*, $\sqrt{-1}\in E$.
(4) Of course $\sigma(V)$ is a ring, $\mathbb{C}\subseteq\sigma(V)\subseteq E'$, $\sigma(V)\neq E'$. If $h=f+\sqrt{-1}\,g\in E'\setminus\sigma(V)$, then $f-\sqrt{-1}\,g\notin V$. In particular $(f,g)\neq(0,0)$. Hence $l=f^2+g^2\neq 0$ and $fl^{-1}+\sqrt{-1}\,gl^{-1}=$ $=(f-\sqrt{-1}\,g)^{-1}\in V$. Thus $h^{-1}=fl^{-1}-\sqrt{-1}\,gl^{-1}\in\sigma(V)$. Consequently $\sigma(V)\in S(E'|\mathbb{C})$. The second part is evident.
(5) It is immediate.
(6) Let us suppose $V=\sigma(V)$ but $\kappa(W)\neq\mathbb{R}$. Then, by A.11, $\sqrt{-1}\in\mathbb{C}=\kappa(W)$, *i.e.*, $(f+\mathfrak{m}_W)^2(g+\mathfrak{m}_W)^{-2}=-1$ for some $f,g\in W$, $g\notin\mathfrak{m}_W$. Thus

$$(f+\sqrt{-1}\,g)(f-\sqrt{-1}\,g)=f^2+g^2\in\mathfrak{m}_W\subseteq\mathfrak{m}_V$$

and $\mathfrak{m}_V$ being a prime ideal we can suppose, without loss of generality, $f+\sqrt{-1}\,g\in\mathfrak{m}_V$. Hence $f-\sqrt{-1}\,g\in\mathfrak{m}_{\sigma(V)}=\mathfrak{m}_V$, which implies $2\sqrt{-1}\,g\in\mathfrak{m}_V$, *i.e.*, $g\in\mathfrak{m}_V\cap E=\mathfrak{m}_W$, absurd.

Conversely, let us suppose $V\neq\sigma(V)$. There exist $f,g\in E$ such that $f+\sqrt{-1}\,g\in V$, $f-\sqrt{-1}\,g\notin V$. We can assume $h=fg^{-1}\in W$. Otherwise $gf^{-1}\in W$ and $g-\sqrt{-1}\,f=-\sqrt{-1}\,(f+\sqrt{-1}\,g)\in V$, $g+\sqrt{-1}\,f=\sqrt{-1}\,(f-\sqrt{-1}\,g)\notin V$, which is equivalent to the initial situation.

We claim that $f,g\notin W$. In fact, if $f\in W$ then $g\in V$ and $f-\sqrt{-1}\,g\in V$. Consequently, $f,g\notin W$ and $h=fg^{-1}\in W$. This implies

$$h^2+1=(f^2+g^2)g^{-2}=(f+\sqrt{-1}\,g)(h-\sqrt{-1}\,)g^{-1}\in VV\mathfrak{m}_V\subseteq\mathfrak{m}_V$$

i.e., $u=h+\mathfrak{m}_W\in\kappa(W)$, $u^2=-1$, and so $\kappa(W)\neq\mathbb{R}$.

Proposition A.18. *Let* $E|\mathbb{R}$ *be such that* $\sqrt{-1}\notin E$, *and* $E'=E(\sqrt{-1}\,)$. *Put* $X=S(E|\mathbb{R})$

and $X'=S(E'|\mathbb{C})$. *Then*

(1) The map $\sigma : X' \longrightarrow X'$ *defined above belongs to* $\mathrm{Aut}(X')$ *and* $\sigma^2=1_{X'}$.

(2) Let $p : X' \longrightarrow X : V \longmapsto V \cap E$. *Then* $p\sigma = p$.

(3) For each $W \in X$ *it is* $1 \le \#p^{-1}(W) \le 2$, *and* $\#p^{-1}(W)=1$ *if and only if* $\kappa(W)=\mathbb{R}$. *Moreover if* $V \in X'$ *with* $V \neq \sigma(V)$, *then* $p^{-1}(p(V))=\{V,\sigma(V)\}$, *whilst if* $V=\sigma(V)$ *then* $p^{-1}(p(V))=\{V\}$.

(4) Let us write $G=\{1,\sigma\}$, *and let us denote by* O_V *the orbit of* $V \in X'$ *under the action of* G. *Then the map* $\overline{p} : X'/G \longrightarrow X : O_V \longmapsto p(V)$ *is a homeomorphism.*

(5) X *admits a unique structure of compact Klein surface such that* p *is a morphism.*

(6) $\partial X = \{W \in X \mid \kappa(W)=\mathbb{R}\}$.

Proof. (1) The equality $\sigma^2=1_{X'}$ is evident. Thus σ is bijective. To prove the continuity we take a basic open set $A \subseteq X'$. Then

$$A=(\hat{h}_1)^{-1}(U_1)\cap\ldots\cap(\hat{h}_n)^{-1}(U_n)$$

for certain $h_1,\ldots,h_n \in E'$ and open subsets $U_1,\ldots,U_n$ of Σ. If $h_j=f_j+\sqrt{-1}\,g_j$ with $f_j,g_j \in E$, and $l_j=f_j-\sqrt{-1}\,g_j \in E'$, it follows from (4) in A.17, that

$$\sigma^{-1}(A)=(\hat{l}_1)^{-1}(U_1)\cap\ldots\cap(\hat{l}_n)^{-1}(U_n).$$

Hence, X' being compact and Hausdorff, σ is a homeomorphism. Now, by 0.1.8, all reduces to check that σ is a morphism. Let us take $V_1 \in X'$, $V_2=\sigma(V_1)$. Then $\mathfrak{m}_1=\mathfrak{m}_{V_1}=f_1V_1$, $f_1=g+\sqrt{-1}\,h$, and $\mathfrak{m}_2=\mathfrak{m}_{V_2}=f_2V_2$, $f_2=g-\sqrt{-1}\,h$, by (4) above. Thus, with the notations in A.15, we get a commutative square

$$\begin{array}{ccc} U^{V_1} & \xrightarrow{\ \sigma\ } & U^{V_2} \\ \downarrow \hat{f}_1 & & \downarrow \hat{f}_2 \\ B=B(0\,,\delta) & \xrightarrow{\ 1_B\ } & B(0\,,\delta) \end{array}$$

Consequently σ is a morphism.

(2) It is immediate to see that $p\sigma(V) \subseteq p(V)$ for $V \in X'$. This applied to $V'=\sigma(V)$ gives us $p(V)=p\sigma(V') \subseteq p(V')=p\sigma(V)$.

(3) Using part (2) in A.14, there exists $f \in E$ such that $Z(f)=\{W\}$. Since $f \in E'\setminus\mathbb{C}$, also from A.14 we know that the set $Z'(f)$ of zeros of f in X' is finite and nonempty, and clearly $p^{-1}(W)=Z'(f)$. Now we use part (1) in A.14 to compute $\#Z'(f)$. Note that $[E':E]=[\mathbb{C}(f):\mathbb{R}(f)]=2$, and $\mathbb{R}$ is algebraically closed in E. Hence

$$\tilde{w}(f)[\kappa(W):\mathbb{R}]=\tilde{w}(f)\deg(W)=[E:\mathbb{R}(f)]=[E':\mathbb{C}(f)]=\sum_{V\in Z'(f)}\tilde{v}(f)\deg V=\sum_{V\in p^{-1}(W)}\tilde{v}(f).$$

Also, the equality $\mathfrak{m}_W=\mathfrak{m}_V\cap W$ for $V \in p^{-1}(W)$ implies $\tilde{v}(f)=\tilde{w}(f)$. To prove that, it suffices to see that $\mathfrak{m}_W^n=\mathfrak{m}_V^n\cap W$ for each natural number n. Of course, it is enough to find some $l\in E$ with $\mathfrak{m}_V=lV$. Let us take $f,g\in E$ such that $(f+\sqrt{-1}\,g)V=\mathfrak{m}_V$. We distinguish:

Case 1. If also $f-\sqrt{-1}\,g$ belongs to $\mathfrak{m}_V$, then $f,g\in\mathfrak{m}_V$. Moreover, we can suppose $g\neq0$, $h=fg^{-1}\in V$, and so $f+\sqrt{-1}\,g=g(h+\sqrt{-1}\,)\in gV$. Thus $\mathfrak{m}_V=gV$ and we can choose $l=g$.

Case 2. If $f-\sqrt{-1}\,g\in\mathfrak{u}_V$, then $\mathfrak{m}_V=(f+\sqrt{-1}\,g)(f-\sqrt{-1}\,g)V=(f^2+g^2)V$ and so $l=f^2+g^2$ solves the question. Finally

Case 3. If $f-\sqrt{-1}\,g\notin V$, then both f and g do not belong to V and we can assume that $h=fg^{-1}\in V$. Let us choose now $l=(h^2+1)g\in E$. Clearly

$$l=(f^2+g^2)g^{-1}=(f+\sqrt{-1}\,g)(h-\sqrt{-1}\,),$$

and so, in order to prove the equality $\mathfrak{m}_V=lV$ it is enough to check that $h-\sqrt{-1}\in\mathfrak{u}_V$. Obviously $h-\sqrt{-1}\in V$ and also $h+\sqrt{-1}=(f+\sqrt{-1}\,g)g^{-1}\in\mathfrak{m}_V^2\subseteq\mathfrak{m}_V$, *i.e.*, $h+\mathfrak{m}_V=-\sqrt{-1}+\mathfrak{m}_V$. This implies $(h-\sqrt{-1}\,)+\mathfrak{m}_V=-2\sqrt{-1}+\mathfrak{m}_V\neq0$, and consequently $h-\sqrt{-1}\in\mathfrak{u}_V$ as desired.

Hence $\#p^{-1}(W)=[\kappa(W):\mathbb{R}]=1$ or 2 according to $\kappa(W)=\mathbb{R}$ or $\mathbb{C}$. The second statement is the obvious consequence of this above, (6) in A.17 and part (2).

(4) The map $\bar{p}$ is well defined by (2) and bijective by (3). Hence all reduces to prove that p is continuous. Given a basic open subset in X,

$$A=(\hat{f}_1)^{-1}(U_1)\cap\ldots\cap(\hat{f}_n)^{-1}(U_n),\ f_j\in E,\ U_j \text{ open in } \Sigma,$$

the maps $g_j : X'\longrightarrow\Sigma : V\longmapsto\lambda_{p(V)}(f_j)$ are continuous, and

$$p^{-1}(A)=g_1^{-1}(U_1)\cap\ldots\cap g_n^{-1}(U_n).$$

Thus p is continuous.

(5) This follows at once from (4), theorem 1.1.10 and the equality $\bar{p}\pi=p$, where $\pi:X'\longrightarrow X'/G$ is the canonical projection.

(6) Given $V\in X'$ we denote, as in 1.1.10,

$$G_V=\{\theta\in G|\theta(V)=V\},\ H_V=\{\theta\in G_V|\psi_V\theta\psi_V^{-1} \text{ is analytic}\}.$$

Clearly $\psi_V\sigma\psi_V^{-1} : B(0,\delta)\longrightarrow B(0,\delta) : \xi\longmapsto\bar{\xi}$ is not analytic. Thus, since X' has empty boundary it follows, by the proof of 1.1.10, that

$$\partial(X'/G)=\{O_V|G_V\neq H_V\}=\{O_V|\sigma\in G_V\}=\{O_V|\sigma(V)=V\}.$$

Using the homeomorphism $X'/G\overset{\bar{p}}{\approx}X$ and part (6) in A.17, we get

$$\partial X=\{W\in X|\kappa(W)=\mathbb{R}\}.$$

Comment. Part (5) in the last proposition provides a functor

$$\mathcal{F}_{\mathbb{R}}\longrightarrow\mathcal{K} : E|\mathbb{R}\longmapsto S(E|\mathbb{R}).$$

It is in fact a functor: each $\mathbb{R}$-homomorphism $F : E_1|\mathbb{R}\longrightarrow E_2|\mathbb{R}$ gives us a morphism of Klein surfaces $F^* : S(E_2|\mathbb{R}) \longrightarrow S(E_1|\mathbb{R}) : V \longmapsto F^{-1}(V)$.

Our goal now is to prove that this is the inverse functor of the one constructed in A.4:

$$\mathcal{K} \longrightarrow \mathcal{F}_{\mathbb{R}} : S \longmapsto \mathcal{M}(S)|\mathbb{R}.$$

The first step is:

Proposition A.19. *Let* X = S(E|ℝ) *be the Klein surface associated to* E|ℝ. *Then* $\mathcal{M}(X)=\hat{E}=\{\hat{f}|f\in E\}$, *which is isomorphic to* E *as field extensions of* ℝ.

Proof. We distinguish several possibilities.

Case 1. $\sqrt{-1}\in E$. Then $X=S(E|\mathbb{C})$ and the result is proved in **[36]**, Chapter VIII.

Case 2. $\sqrt{-1}\notin E$. Write $E'=E(\sqrt{-1})$, $X'=S(E'|\mathbb{C})$. Let σ be the antianalytic involution of X' just constructed. Then $X=X'/<\sigma>$ and $p:X'\longrightarrow X$ is the canonical projection. By lemma A.7, $p^*(\mathcal{M}(X))=\{f\in\mathcal{M}(X')|\sigma^*(f)=f\}$. Moreover, $\sigma^*(\sqrt{-1})=-\sqrt{-1}$. In fact, if ε is the field ℝ-automorphism of E' which maps $u+\sqrt{-1}\,v\in E'$ to $u-\sqrt{-1}\,v$, it is clear that the place $\lambda_{\sigma(V)}$ associated to $\sigma(V)\in X'$ is given by $\lambda_{\sigma(V)}=\lambda_V\varepsilon$. Hence, for every $V\in X'$, we get

$$[\sigma^*(\sqrt{-1})]^{\sim}(V)=\sqrt{-1}\,(\sigma(V))=\lambda_{\sigma(V)}(\sqrt{-1})=\lambda_V(-\sqrt{-1})=-\lambda_V(\sqrt{-1})=$$
$$=-(\sqrt{-1})(V)=[(-\sqrt{-1})]^{\sim}(V),\ i.e.,\ \sigma^*(\hat{\sqrt{-1}})=-\sqrt{-1}.$$

Also, from the first case we know that $\mathcal{M}(X')=\hat{E}'$. We can now check the equality $\mathcal{M}(X)=\hat{E}$. In fact given $f\in\hat{E}\subseteq\hat{E}'=\mathcal{M}(X')$, and $V\in X'$, we have

$$(\hat{f}\sigma)(V)=\lambda_{\sigma(V)}(f)=\lambda_{p\sigma(V)}(f)=\lambda_{p(V)}(f)=\lambda_V(f)=\hat{f}(V).$$

Consequently, $[\sigma^*(\hat{f})]^{\sim}=\hat{f}\sigma=\hat{f}$, and by the uniqueness part in A.5, $\sigma^*(\hat{f})=\hat{f}$, *i.e.*, $\hat{f}\in p^*(\mathcal{M}(X))\equiv\mathcal{M}(X)$.

Conversely, if $F\in\mathcal{M}(X)$, then $p^*(F)\in\mathcal{M}(X')=\hat{E}'$, and so $p^*(F)=(g+\sqrt{-1}\,h)^{\wedge}$ for some $g,h\in E$. Since $\sigma^*(\sqrt{-1})=-\sqrt{-1}$, it is

$$\hat{g}+\sqrt{-1}\,\hat{h}=p^*(F)=\sigma^*(p^*(F))=\sigma^*(\hat{g})+\sigma^*(\sqrt{-1})\sigma^*(\hat{h})=\hat{g}-\sqrt{-1}\,\hat{h}$$

and so $\hat{h}=0$, *i.e.*, $p^*(F)=\hat{g}\in\hat{E}$.

Now the main theorem says:

Theorem A.20. *Let* S *be a compact Klein surface and let* $X=S(\mathcal{M}(S)|\mathbb{R})$. *Then* S *and* X *are isomorphic Klein surfaces.*

Proof. We proved in A.4 that $\mathcal{M}(S)|\mathbb{R}$ is an algebraic function field in one variable. Hence, by A.16 and A.18, X is a compact Klein surface.

Case 1. $\sqrt{-1}\in\mathcal{M}(S)$. Then, by (3) in A.1 and A.16, both S and X are compact Riemann surfaces, and we proved in Example A.10,(2) that the map

$$\phi : S \longrightarrow X : x \longmapsto V_x = \{f \in \mathcal{M}(S) | x \notin P(f)\}$$

is well defined. This is the isomorphism we are looking for. The injectivity is clear, by using part (1) in A.3. To see the continuity notice first that

(A.20.1) given $f \in \mathcal{M}(S)$, then $f(x) = \hat{f}(\phi(x))$ for every $x \in S$.

In fact the maximal ideal $\mathfrak{m}_x$ of V_x is given by $\mathfrak{m}_x = \{h \in \mathcal{M}(S) | x \in Z(h)\}$. Thus, if $x \in P(f)$ it is $f(x) = \infty = \hat{f}(\phi(x))$, whilst for $x \notin P(f)$, the meromorphic function $g = f - f(x)$ belongs to $\mathfrak{m}_x$, *i.e.*, $\hat{f}(\phi(x)) = f + \mathfrak{m}_x = f(x) + \mathfrak{m}_x = f(x)$. Consequently $\hat{f}\phi$ is continuous for every $f \in \mathcal{M}(S)$, *i.e.*, ϕ is continuous.

Now all reduces to check that ϕ is a morphism of Riemann surfaces. Once this proved it is automatically open and surjective and it is enough to use 0.1.8. Let us take $x \in S$ and a generator f_x of $\mathfrak{m}_x$. We choose charts (U^x, ϕ_x), $x \in U^x$, (U^y, ψ_y), $y = \phi(x) \in U^y$ such that $\phi(U^x) \subseteq U^y$. Of course, $\psi_y = \hat{f}_x | U^y$ as constructed in A.15. Then the map $F : \phi_x(U^x) \longrightarrow \mathbb{C} : \phi_x(z) \longmapsto f_x(z)$ is analytic and the following diagram commutes:

$$\begin{array}{ccc} U^x & \xrightarrow{\ \phi\ } & U^y \\ \downarrow{\scriptstyle \phi_x} & & \downarrow{\scriptstyle \psi_y} \\ \phi_x(U^x) & \xrightarrow{\ F\ } & \mathbb{C} \end{array}$$

Case 2. Let us suppose now that $\sqrt{-1} \notin \mathcal{M}(S)$. Let us consider the double cover (S_c, π, τ) of S, *i.e.*, $\tau^2 = 1$, $\pi\tau = \pi$, $\tau \in \mathrm{Aut}(S_c)$, $\pi : S_c \longrightarrow S$.

We know by A.8 that $\mathcal{M}(S_c) = \pi^*(\mathcal{M}(S))(\sqrt{-1})$, and in particular $\sqrt{-1} \in \mathcal{M}(S_c)$. From the first case we can produce an isomorphism of Klein surfaces

$$\phi_c : S_c \longrightarrow X_c = S(\mathcal{M}(S_c)|\mathbb{R}).$$

We forget from now on the inclusion π^* and so we write $\mathcal{M}(S_c) = \mathcal{M}(S)(\sqrt{-1})$. Hence, with the notations in A.18 we have a morphism $p : X_c \longrightarrow X$. Now we define $\phi : S \longrightarrow X$ to make commutative the diagram

$$\text{(A.20.2)} \qquad \begin{array}{ccc} S_c & \xrightarrow{\ \phi_c\ } & X_c \\ \downarrow{\scriptstyle \pi} & & \downarrow{\scriptstyle p} \\ S & \xrightarrow{\ \phi\ } & X \end{array}$$

and we will show this is the searched isomorphism.

Claim 1. ϕ is a well defined map. Let us consider, with the notations in A.18, (1) and (2), the involution $\sigma \in \mathrm{Aut}(X_c)$ that verifies $p\sigma = p$. It is enough to see the equality $\sigma\phi_c = \phi_c\tau$. Once this is done, given $x, y \in S_c$, with $x \neq y$ but $\pi(x) = \pi(y)$ we know $y = \tau(x)$ and so $p\phi_c(y) = p\phi_c\tau(x) = p\sigma\phi_c(x) = p\phi_c(x)$, as desired. We prove the equality $\sigma\phi_c = \phi_c\tau$. Let us take $x \in S_c$. Then

$$\sigma\phi_c(x) = \{h = f + \sqrt{-1}\, g \in \mathcal{M}(S_c) | \ x \notin P(f - \sqrt{-1}\, g)\}.$$

On the other hand, since $\tau^*(-1) = -1$ but $\sqrt{-1} \notin \mathcal{M}(S)$ we get $\tau^*(\sqrt{-1}) = -\sqrt{-1}$.

Moreover, by A.8, $\tau^*(f)=f$ for $f\in\mathcal{M}(S)$. Hence $\tau^*(f+\sqrt{-1}\,g)=f-\sqrt{-1}\,g$ if $f,g\in\mathcal{M}(S)$. Thus, with the notations in A.5,

$$(f+\sqrt{-1}\,g)(\tau(x))=[\tau^*(f+\sqrt{-1}\,g)]^{\sim}(x)=(f-\sqrt{-1}\,g)(x),$$

and so

$$\sigma\phi_c(x)=\{h=f+\sqrt{-1}\,g\in M(S_c)|\tau(x)\notin P(f+\sqrt{-1}\,g)\}=\phi_c\tau(x).$$

Now by using 0.1.7 and A.20.2 it follows that ϕ is a surjective morphism. Also ϕ is a closed map, because S is compact and X is Hausdorff. Thus, applying 0.1.8 all reduces to see:

Claim 2. ϕ is injective. Let us take $u_1=\pi(x_1)$ and $u_2=\pi(x_2)$ in S with $\phi(u_1)=\phi(u_2)$. Then $p\phi_c(x_1)=p\phi_c(x_2)$ and by (3) in A.18, either $\phi_c(x_1)=\phi_c(x_2)$, or $\phi_c(x_1)=\sigma\phi_c(x_2)=\phi_c\tau(x_2)$. The injectivity of ϕ_c implies either $x_1=x_2$ or $x_1=\tau(x_2)$. In both cases $u_1=\pi(x_1)=\pi(x_2)=u_2$.

Comment. The last two propositions show that the functors

$$\begin{cases} \mathcal{F}_{\mathbb{R}} \longrightarrow \mathcal{K} : E|\mathbb{R} \longmapsto S(E|\mathbb{R}) \\ \mathcal{K} \longrightarrow \mathcal{F}_{\mathbb{R}} : S \longmapsto \mathcal{M}(S)|\mathbb{R} \end{cases}$$

are mutually inverse when acting on objects. Moreover, taking into account A.20.1 it is easy to check they are also inverse when acting on morphisms. This was our main purpose at the beginning. Now we shall obtain some information about the topology of $S\in\mathcal{K}$ by means of $\mathcal{M}(S)$. First of all:

Corollary A.21. *Let* S_1 *and* S_2 *be compact Klein surfaces. Then* S_1 *and* S_2 *are isomorphic if and only if* $\mathcal{M}(S_1)$ *and* $\mathcal{M}(S_2)$ *are* $\mathbb{R}$*-isomorphic.*

Proof. If $f:S_1\longrightarrow S_2$ is an isomorphism with inverse $g=f^{-1}$, the $\mathbb{R}$-homomorphism $f^*:\mathcal{M}(S_2)\longrightarrow\mathcal{M}(S_1)$ is an isomorphism with inverse g^*, by lemma A.5. Conversely, if $\mathcal{M}(S_1)|\mathbb{R} \cong \mathcal{M}(S_2)|\mathbb{R}$ we get

$$S_1\approx S(\mathcal{M}(S_1)|\mathbb{R})\approx S(\mathcal{M}(S_2)|\mathbb{R})\approx S_2.$$

At this point we are forced to introduce some elementary "real algebra".

Definition 6. Let E be a field. We say E is *real* if -1 is not a sum of squares in E.

Definition 7. An *ordering* in a field E is a subset $P\subseteq E$ such that:

(1) $P+P\subseteq P$; $P.P\subseteq P$.

(2) $P\cap(-P)=0$; $P\cup(-P)=E$, where $-P=\{-x|x\in P\}$.

Remarks A.22. (1) It is obvious that the relation $\leq_P$ on E defined by $x\leq_P y$ if and only if $y-x\in P$, is a total order relation on E compatible with field operations, *i.e.*,

$$\text{if } 0\le_P x,\ 0\le_P y, \text{ then } 0\le_P (x+y),\ 0\le_P xy.$$

(2) Since $x^2=(-x)^2$ and $P.P\subseteq P$, $P\cup(-P)=E$, it follows that $x^2\in P$ for any $x\in E$. In particular, $0\le_P 1$.

Definition 8. Let A be a subring of a field E. Let P be an ordering in E. We say that A is *convex with respect to* P if $0\le_P x\le_P y$, $y\in A$, $x\in E$, imply $x\in A$.

Proposition A.23. *Let* P *be an ordering in a field* E. *Let* V *be a valuation ring of* E *(i.e.,* $V\subseteq E$, $V\neq E$, *and* $f\in E\setminus V$ *implies* $f^{-1}\in V$*). The following statements are equivalent:*

(1) V *is convex with respect to* P.

(2) If $x\in \mathfrak{m}_V$, *then* $0<_P(1+x)$.

(3) For every $x,y\in V\cap P$ *such that* $x+y\in\mathfrak{m}_V$, *it is* $x,y\in\mathfrak{m}_V$.

Proof. Along the proof we write $<$ and $\le$ instead of $<_P$ and $\le_P$, respectively.

(1) ⇒ (2) Suppose $1+x\le 0$. Then $x\le -1<0$, and so $x^{-1}+1\ge 0$, *i.e.*, $0<-x^{-1}\le 1\in V$. Hence $-x^{-1}\in V$ and $x\notin\mathfrak{m}_V$.

(2) ⇒ (3) Let us suppose $x\notin\mathfrak{m}_V$. Thus $-(1+x^{-1}y)=-(x+y)x^{-1}\in\mathfrak{m}_V V\subseteq\mathfrak{m}_V$. By the hypothesis, $0<1-(1+x^{-1}y)=-x^{-1}y$. Since $x\neq 0$, by A.22(2), $xy=x^2(x^{-1}y)<0$. This is false, because $x,y\in P$.

(3) ⇒ (1) Assume there exist $x\in E\setminus V$, $y\in V$ such that $0\le x\le y$. Then $x^{-1}\in\mathfrak{m}_V$. Consequently $(x^{-1}y-1)+1=x^{-1}y\in\mathfrak{m}_V$ and $x^{-1}y-1\in V\cap P$, $1\in V\cap P$. This implies $1\in\mathfrak{m}_V$, absurd.

We are ready to state the key result in order to establish a relation between the topology of a Klein surface S and the algebraic properties of $\mathcal{M}(S)$. We adapt to our particular setting much more general arguments which appear, *e.g.*, in **[8]**.

Proposition A.24. *Let* $E|\mathbb{R}$ *be an algebraic function field in one variable. The following statements are equivalent:*

(1) E *is a real field.*

(2) E *admits an ordering.*

(3) There exist an ordering P *in* E *and* $V\in S(E|\mathbb{R})$ *such that* V *is convex with respect to* P.

(4) There exists $V\in S(E|\mathbb{R})$ *such that* $\kappa(V)=\mathbb{R}$ *(i.e.,* $V\in\partial S(E|\mathbb{R})$*).*

Proof. (1) ⇒ (2) This is a classical result of Artin-Schreier theory. For the sake of completeness we outline here the proof. Let S stand for the set of elements in E which can be written as sum of squares of elements in E. Let us consider the family

$$\mathscr{L}=\{M\subseteq E \mid S\subseteq M,\ -1\notin M,\ M+M\subseteq M,\ M.M\subseteq M\}.$$

Since E is real, $S\in\mathscr{L}$ and so, by Zorn's lemma, there exists a maximal element $P\in\mathscr{L}$. We claim it is an ordering in E. Since $-1\notin P$, it follows $P\cap(-P)=0$. All reduces to check $P\cup(-P)=E$. Let us take $a\in E\setminus(-P)$. Then it is easily seen that

$$P[a]=\{x+ay \mid x,y\in P\}\in\mathscr{L}.$$

Since $P\subseteq P[a]$ we get $P=P[a]$ by the maximality of P, and so $a\in P$.

(2) ⇒ (3) By assumption there exists some ordering P in E. Take

$$V=\{f\in E \mid \text{there exists } r\in\mathbb{R} \text{ with } 0\le|f|\le r\},$$

where $\le$ means $\le_P$ and $|f|=f$ or $-f$ according with $f\in P$ or $-f\in P$. Of course $\mathbb{R}\subseteq V$. Since $\mathbb{R}\subseteq E$, $\mathbb{R}\neq E$, the extension field $E|\mathbb{R}$ is not archimedean, and so $V\neq E$. Now given $f\in E\setminus V$, it must be $|f|>1$. Thus $0<|f^{-1}|<1$, and so $f^{-1}\in V$. Hence $V\in S(E|\mathbb{R})$. Finally, if $0\le f\le g\in V$ it is $0\le f\le g\le r$ for some $r\in\mathbb{R}$, *i.e.*, $f\in V$. Thus V is convex with respect to P.

(3) ⇒ (4) Let us take the valuation ring V given by the hypothesis. We claim that $\kappa(V)=\mathbb{R}$. Otherwise $\kappa(V)=\mathbb{C}$ and there exists $u\in\kappa(V)$ with $u^2=-1$. Let us write $u=(f+\mathfrak{m}_V)(g+\mathfrak{m}_V)^{-1}$ with $f,g\in V$, $g\notin\mathfrak{m}_V$. Then $f^2+g^2\in\mathfrak{m}_V$, $f^2,g^2\in P\cap V$. Since V is convex with respect to P, we deduce from (3) in A.23 that $g^2\in\mathfrak{m}_V$, a contradiction.

(4)⇒(1) Let us suppose $-1=f_1^2+\ldots+f_n^2$, $f_i\in E$. Then, for $f_0=1$ it is $f_0^2+\ldots+f_n^2=0$, and there exists $0\le i\le n$ such that each $g_j=f_jf_i^{-1}\in V$, $0\le j\le n$. Hence $-1=\sum_{j\neq i} g_j^2$ and so if $g_j+\mathfrak{m}_V=x_j\in\mathbb{R}$, $-1=\sum x_j^2$, absurd.

Corollary A.25. *Let S be a compact Klein surface.*

(1) S has nonempty boundary if and only if $\mathcal{M}(S)$ is a real field.

(2) Let us assume that ∂S is empty. Then S is orientable if and only if $\sqrt{-1}$ belongs to $\mathcal{M}(S)$.

Proof. By A.20, S is isomorphic as Klein surface to $X=S(\mathcal{M}(S)|\mathbb{R})$.

(1) If $\mathcal{M}(S)$ is real there exists, by (1)⇒(4) above, some $V\in X$ with $\kappa(V)=\mathbb{R}$, and by A.18, $V\in\partial X$. Hence ∂S is nonempty. The converse follows similarly, using now (4) ⇒ (1) in A.24.

(2) This was proved in A.1.(3).

Example A.24. We are going to construct a nonorientable compact Klein surface with empty boundary. Take $f=X^2+Y^2+1\in\mathbb{R}[X,Y]$.

(i) The ideal $\mathfrak{p}=f\mathbb{C}[X,Y]$ is prime, as a consequence of Eisenstein irreducibility criterion.

(ii) In particular, $\mathfrak{q}=f\mathbb{R}[X,Y]$ is a prime ideal and $E=q.f.(\mathbb{R}[X,Y]/\mathfrak{q})$ is an algebraic function field in one variable over $\mathbb{R}$. Thus $S=S(E|\mathbb{R})$ is a compact Klein surface. If $u=X+\mathfrak{q}$ and $v=Y+\mathfrak{q}$, it is $-1=u^2+v^2$, $u,v\in E$. Hence $\mathcal{M}(S)\cong E$ is not real, and so ∂S is empty. To prove that S is nonorientable we must check that $\sqrt{-1}\notin E$. Otherwise, we could find polynomials $g,h\in\mathbb{R}[X,Y]$, $h\notin\mathfrak{q}$, such that $(g+\mathfrak{q})^2(h+\mathfrak{q})^{-2}=-1$, *i.e.*, $g^2+h^2\in\mathfrak{q}$. Hence $(g+\sqrt{-1}\,h)(g-\sqrt{-1}\,h)\in\mathfrak{q}\subseteq\mathfrak{p}$ and $\mathfrak{p}$ being prime we can assume without loss of generality that $g+\sqrt{-1}\,h\in\mathfrak{p}$. That means

$$g+\sqrt{-1}\,h=f(u+\sqrt{-1}\,v) \text{ for certain } u,v\in\mathbb{R}[X,Y]$$

and, in particular, $h=fv\in\mathfrak{q}$, a contradiction. (Of course it is easy to check that $E|\mathbb{R}\cong\mathcal{M}(\mathbb{P}_2(\mathbb{R}))$ and so, by A.21, $S=\mathbb{P}_2(\mathbb{R})$).

In order to describe the boundary of a bordered compact Klein surface in a more geometric way, we stand first some standard terminology.

Definition 9. A *real* (irreducible) *algebraic curve* is a subset C of some $\mathbb{R}^n$ verifying:

(1) There exists $f\in A=\mathbb{R}[X_1,\dots,X_n]$ such that $C=\{x\in\mathbb{R}^n|f(x)=0\}$.

(2) If $\mathfrak{J}(C)=\{h\in A|h(x)=0$ for every $x\in C\}$, the quotient ring $P[C]=A/\mathfrak{J}(C)$ is a domain of dimension 1. This is equivalent to say that the quotient field $\mathbb{R}(C)$ of $P[C]$ is an algebraic function field in one variable over $\mathbb{R}$.

We say that C is *smooth* if $P[C]$ is integrally closed. The only topology we consider on C is the *euclidean* topology induced by the one of $\mathbb{R}^n$.

A well known result in real algebraic geometry is the following:

Proposition A.26. *Let* $E|\mathbb{R}$ *be an algebraic function field in one variable. Assume* E *is a real field. Then there exists a smooth compact model of* $E|\mathbb{R}$, *i.e., a compact and smooth real algebraic curve* $C\subseteq\mathbb{R}^n$ *such that* $E|\mathbb{R}\cong\mathbb{R}(C)|\mathbb{R}$.

Sketch of the proof. Let $\mathfrak{p}$ be a prime ideal of $\mathbb{R}[X,Y]$ such that E is the quotient field of $\mathbb{R}[X,Y]/\mathfrak{p}$. Let $M=\{(x,y)\in\mathbb{R}^2|f(x,y)=0$ for every $f\in\mathfrak{p}\}$. By the real Nullstellensatz, [8], Chapter IV, since E is real, we get $\mathfrak{p}=\mathfrak{J}(M)$. Hence $E|\mathbb{R}$ is isomorphic to $\mathbb{R}(M)|\mathbb{R}$. On the other hand, let $C\subseteq\mathbb{P}_n(\mathbb{R})$ be a projective normalization of M, [112], Chapter II. Hence C is compact and smooth. But the real projective space is affine, [8], Chapter III, and so $C\subseteq\mathbb{R}^m$ for some m. Since $\mathbb{R}(M)|\mathbb{R}\cong\mathbb{R}(C)|\mathbb{R}$ we are done.

We are now ready to state

Theorem A.27. *Let* S *be a compact Klein surface with nonempty boundary. Let* $C \subseteq \mathbb{R}^n$ *be a compact smooth model of* $\mathcal{M}(S)|\mathbb{R}$. *Then* C *and* ∂S *are homeomorphic.*

Proof. Let $A = P[C]$. For each maximal ideal $\mathfrak{m}$ in A, the localization $A_{\mathfrak{m}}$ is a noetherian and integrally closed local domain of dimension one, containing $\mathbb{R}$, whose quotient field is $\mathbb{R}(C) \cong \mathcal{M}(S)$. Thus $A_{\mathfrak{m}} \in S(\mathcal{M}(S)|\mathbb{R}) = X$. In particular, for each $a \in C$, the ideal $\mathfrak{m}_a = \{f \in A \mid f(a) = 0\}$ is the kernel of the epimorphism $f \longmapsto f(a)$ from A onto $\mathbb{R}$. So $\mathfrak{m}_a$ is a maximal ideal and $\kappa(A_{\mathfrak{m}_a}) = \mathbb{R}$, *i.e.*, $A_{\mathfrak{m}_a} \in \partial X$. Moreover, if $A_{\mathfrak{m}} \in \partial X$, there exists $a \in C$ such that $\mathfrak{m} = \mathfrak{m}_a$. In fact we get a canonical epimorphism $A \xrightarrow{\pi} A/\mathfrak{m} = \mathbb{R}$ and if we write $A = \mathbb{R}[X_1, \ldots, X_n]/\mathfrak{J}(C)$, $u_i = x_i + \mathfrak{J}(C) \in A$, $1 \le i \le n$, the point $a = (\pi(u_1), \ldots, \pi(u_n)) \in C$. To see that, take $F \in \mathbb{R}[X_1, \ldots, X_n]$ with $C = \{x \in \mathbb{R}^n \mid F(x) = 0\}$. Clearly $F \in \mathfrak{J}(C)$ and so $\pi(F + \mathfrak{J}(C)) = 0$. That means $F(a) = 0$, and so $a \in C$. Of course $\mathfrak{m} = \ker \pi \subseteq \mathfrak{m}_a$ and, $\mathfrak{m}$ being maximal, $\mathfrak{m} = \mathfrak{m}_a$.

On the other hand $\partial S \overset{top}{\approx} \partial X$ and all reduces now to check that the well defined map

$$\alpha : C \longrightarrow \partial X : a \longmapsto A_{\mathfrak{m}_a}$$

is a homeomorphism.

Claim 1. α is injective and continuous. Indeed, given two distinct points a,b in C with coordinates $a = (a_1, \ldots, a_n)$, $b = (b_1, \ldots, b_n)$, the function

$$f = \sum_{i=1}^{n} (u_i - a_i)^2 \in A$$

verifies $f(a) = 0$, $f(b) \neq 0$. Thus $f^{-1} \in A_{\mathfrak{m}_b} \setminus A_{\mathfrak{m}_a}$ which proves that α is injective.

Let us consider a subbasic open set M in ∂X, given as $M = (\hat{l})^{-1}(U) \cap \partial X$ for some open subset U of the Riemann sphere Σ and some $l \in \mathcal{M}(S) = \mathbb{R}(C)$. Then l can be viewed as a continuous function $l : C \longrightarrow \mathbb{R} \cup \{\infty\} \subseteq \Sigma$ and we claim that $\alpha^{-1}(M) = l^{-1}(U)$, which proves the continuity of α. In fact

$$a \in \alpha^{-1}(M) \Leftrightarrow A_{\mathfrak{m}_a} \in (\hat{l})^{-1}(U) \Leftrightarrow l + \mathfrak{m}_a \in U \Leftrightarrow l(a) \in U \Leftrightarrow a \in l^{-1}(U).$$

Now, since C is compact and ∂X is Hausdorff, it is enough to prove

Claim 2. α is surjective. Given $V \in \partial X$ there exists, by A.24, an ordering P in $\mathcal{M}(S)$ such that V is convex with respect to P. We must prove that $A \subseteq V$. Once this is done, take $\mathfrak{m} = \mathfrak{m}_V \cap A$, which is a prime ideal of A, and $A_{\mathfrak{m}} \subseteq V$. Since $V \neq \mathcal{M}(S)$, $\mathfrak{m}$ is not the zero ideal, and so, A being one dimensional, $\mathfrak{m}$ is maximal. As it was proved before, $A_{\mathfrak{m}} \in X$. Since $A_{\mathfrak{m}} \subseteq V$ with $\mathfrak{m}_V \cap A_{\mathfrak{m}} = \mathfrak{m}A_{\mathfrak{m}}$, then $A_{\mathfrak{m}} = V$. In this situation we already proved that $\mathfrak{m} = \mathfrak{m}_a$ for some $a \in C$, *i.e.*, $V = \alpha(a)$.

Since $A = \mathbb{R}[u_1, \ldots, u_n]$ and $\mathbb{R} \subseteq V$, all reduces to see that V contains each u_i and, V being integrally closed, it is enough to see $u_i^2 \in V$. Let $r \in \mathbb{R}$, $r > 0$ be such that C is contained in the open ball B of center the origin and radius r. We are going to see that $0 \leq_P u_i^2 <_P r^2 \in \mathbb{R} \subseteq V$. By convexity this implies $u_i^2 \in V$. Let $g = r^2 - u_i^2 \in A$, and let K be a real closure of $\mathcal{M}(S)$ with respect to P. If $g \leq_P 0$, there exists $h \in K$ such that $h^2 = -g$. Since $A[h]$ is an $\mathbb{R}$-algebra of finite type and the inclusion $A[h] \hookrightarrow K$ is a homomorphism of $\mathbb{R}$-algebras, there exists, by Artin - Lang theorem, [8], Chapter IV, a homomorphism of $\mathbb{R}$-algebras ψ : $A[h] \longrightarrow \mathbb{R}$. Let us consider the point $b = (\psi(u_1), \ldots, \psi(u_n)) \in \mathbb{R}^n$. In fact $b \in C$ because

$$F(b) = \psi(F(x_1, \ldots, x_n) + \mathfrak{I}(C)) = \psi(0) = 0.$$

However, $\psi(u_1)^2 + \ldots + \psi(u_n)^2 \geq \psi(u_i)^2 = r^2 - \psi(g) = r^2 + \psi(h)^2 \geq r^2$, *i.e.*, $b \notin B$ what is absurd.

We are now ready to get also a characterization of orientability of bordered compact Klein surfaces:

Proposition A.28. *Let* S *be a compact Klein surface with nonempty boundary. Let* (S_c, π, τ) *be its double cover and let* C *be a compact smooth model of* $\mathcal{M}(S)|\mathbb{R}$. *Let* $C_{\mathbb{C}}$ *be the complexification of* C. *The following statements are equivalent:*

(1) S *is orientable.*

(2) $S_c \backslash \pi^{-1}(\partial S)$ *is non connected.*

(3) $C_{\mathbb{C}} \backslash C$ *is non connected.*

Proof. Let $X = S \backslash \partial S$ which is a Klein surface without boundary. We can construct its double cover X_c as in 0.1.12. The question is that in case X is orientable, *i.e.*, X is a Riemann surface itself, then X_c consists of two disjoint copies of X, and in particular X_c is non connected. Of course, $X_c = S_c \backslash \pi^{-1}(\partial S)$. Hence

$$\text{S is orientable} \Leftrightarrow \text{X is orientable} \Leftrightarrow X_c = S_c \backslash \pi^{-1}(\partial S) \text{ is non connected.}$$

This proves (1) $\Leftrightarrow$ (2). On the other hand, it is well known that if A is the coordinate ring $P[C_{\mathbb{C}}]$ of $C_{\mathbb{C}}$, then the map

$$\beta : C_{\mathbb{C}} \longrightarrow S(\mathcal{M}(S_c)|\mathbb{C}) \overset{top}{\approx} S_c : a \longmapsto A_{\mathfrak{m}_a}$$

is a homeomorphism. The proof of the last theorem shows that $\beta(C) = \pi^{-1}(\partial S)$. Hence

$$S_c \backslash \pi^{-1}(\partial S) \overset{top}{\approx} C_{\mathbb{C}} \backslash C$$

and this proves (2) $\Leftrightarrow$ (3).

In order to apply the results in this book to study groups of birational isomorphisms of real algebraic curves we only need:

Proposition A.29. *Let* S *be a compact Klein surface, and* $E = \mathcal{M}(S)$. *Let* $\mathrm{Aut}(E|\mathbb{R})$

be the group of field $\mathbb{R}$*-automorphisms of* E. *Let* Aut(S) *be the group of automorphisms of* S. *Then:*

(1) The groups Aut(S) *and* Aut(E|$\mathbb{R}$) *are (anti-)isomorphic.*

(2) If ∂S *is nonempty and* $C\subseteq\mathbb{R}^n$ *is a model of* E|$\mathbb{R}$, *the group* B(C) *of birational isomorphisms* $C\longrightarrow C$ *is isomorphic to* Aut(S).

Proof. (1) By A.5 and A.6, the map ε : $\mathrm{Aut(S)}\longrightarrow\mathrm{Aut(E|\mathbb{R})}$: $f\longmapsto f^*$ is an (anti-)homomorphism of groups. Given $f\in\mathrm{Aut(S)}$, $f\neq 1_S$, there exists $x\in S$ with $y=f(x)\neq x$. Since $S=S(E|\mathbb{R})$ we apply A.14 to deduce that there exists $g\in E$ with $Z(g)=\{y\}$. Hence, with the notations in A.5,

$$[f^*(g)]^\sim(x)=g(f(x))=g(y)=0\neq g(x)=[1_E^*(g)]^\sim(x).$$

Thus $f^*\neq 1_E$, and so ε is injective. Finally we prove it is surjective. Given $F\in\mathrm{Aut(E|\mathbb{R})}$ and $V\in S(E|\mathbb{R})$, it is clear that $F^{-1}(V)\in S(E|\mathbb{R})$. Hence we define a map f : $S=S(E|\mathbb{R})\longrightarrow S:V\longmapsto F^{-1}(V)$. It is a morphism of Klein surfaces because given $V\in S$ and $h\in V$ with $\mathfrak{m}_V=hV$, then $\mathfrak{m}_{F^{-1}(V)}=F^{-1}(h)F^{-1}(V)$. Moreover, it is continuous since, given $f_1,\dots,f_l\in E$, open subsets $U_1,\dots,U_l$ in Σ and $A=(\hat{f}_1)^{-1}(U_1)\cap\dots\cap(\hat{f}_l)^{-1}(U_l)$, then $g_j=F(f_j)\in E=\mathcal{M}(S)$ and it is easily seen that $f^{-1}(A)=(\hat{g}_1)^{-1}(U_1)\cap\dots\cap(\hat{g}_l)^{-1}(U_l)$. Hence, by using 0.1.8, in order to prove that $f\in\mathrm{Aut(S)}$, only bijectivity is needed. But given $x\in V_1\backslash V_2$, then $F^{-1}(x)\in f(V_1)\backslash f(V_2)$, whilst $F^{-1}(F(V))=V$, $F(V)\in S$, for V, V_1, V_2 in S. Now all reduces to check that $\varepsilon(f)=F$. In fact, given $V\in S$, we have a field isomorphism (already used to compute $f^{-1}(A)$), between $\kappa(V)$ and $\kappa(F^{-1}(V))$ given by

$$\kappa(F^{-1}(V))\longrightarrow\kappa(V) : g+\mathfrak{m}_{F^{-1}(V)}\longmapsto F(g)+\mathfrak{m}_V.$$

Thus, given $g\in E$ and $V\in S$ with $g\in F^{-1}(V)$, we get

$$[\varepsilon(f)(g)]^\sim(V)=[f^*(g)]^\sim(V)=g(f(V))=g(F^{-1}(V))=g+\mathfrak{m}_{F^{-1}(V)}=F(g)+\mathfrak{m}_V=[F(g)]^\sim(V).$$

In case $g\notin F^{-1}(V)$ we deduce $[\varepsilon(f)(g^{-1})]^\sim(V)=[F(g^{-1})]^\sim(V)$. Since $\varepsilon(f(g))=(\varepsilon(f)(g^{-1}))^{-1}$ and $F(g)=(F(g^{-1}))^{-1}$, we conclude $[\varepsilon(f)(g)]^\sim=[F(g)]^\sim$ for every $g\in E$, and so by A.5, $\varepsilon(f)=F$.

(2) Since C is a model of E, then $B(C)\cong\mathrm{Aut(E|\mathbb{R})}\cong\mathrm{Aut(S)}$.

Final Remark A.30. We can understand in the language of real algebraic curves the meaning of those integers which appear as proper periods or periods in period-cycles in the signature of a given NEC group. Let S be a compact Klein surface of algebraic genus $p(S)\geq 2$. It can be written as a quotient $S=H/\Gamma$, Γ being a surface NEC group. Let us consider another NEC group Λ with signature

$$\tau=(0;+;[m_1,\dots,m_r];\{(n_1,\dots,n_s)\}),$$

containing Γ as a normal subgroup. Let Γ^+ and Λ^+ stand for the associated fuchsian groups and $S_c=H/\Gamma^+$ for the double cover of S. Then $G_1=\Lambda^+/\Gamma^+$ and $G_2=\Lambda/\Gamma$ are groups of automorphisms of S_c and S respectively. Moreover,

$S_c/G_1 = H/\Lambda^+$ has genus 0, *i.e.*, $S_c/G_1 = \Sigma$. Let us denote by R_π the set of ramification points of the projection $\pi: S_c \longrightarrow S_c/G_1 = \Sigma$ and $\{p_1,\dots,p_t\} = \pi(R_\pi)$. Let us also call $\upsilon_1,\dots,\upsilon_t$ the multiplicities of π at the branch points over $p_1,\dots,p_t$, respectively. Then, if

$$p : H \longrightarrow H/\Gamma^+ = S_c \text{ and } H^* = H \backslash (\pi p)^{-1}(\{p_1,\dots,p_t\})$$

the group Λ^+ is the covering transformation group of the covering $\pi p: H^* \longrightarrow \Sigma^*$, where $\Sigma^* = \Sigma \backslash \{p_1,\dots,p_t\}$, and so $\sigma(\Lambda^+) = (0;+;[\upsilon_1,\dots,\upsilon_t];\{-\})$. On the other hand, using that $\tau = \sigma(\Lambda)$ and example 2.2.5, we know

$$\sigma(\Lambda^+) = (0;+;[m_1,m_1,\dots,m_r,m_r,n_1,\dots,n_s];\{-\}).$$

Consequently π ramifies over $t=2r+s$ points with multiplicities given by $\{m_1,\dots,m_r,n_1,\dots,n_s\}$.

From the algebraic point of view, it follows at once from Example A.10.(1) that $\Sigma = S(\mathbb{C}(T)|\mathbb{R})$ and $\Delta = S(\mathbb{R}(T)|\mathbb{R})$, since irreducible polynomials in $\mathbb{C}[T]$ have the form T-a with $a \in \mathbb{C}$, whilst, in the real case they are either of type T-a, $a \in \mathbb{R}$, or $(T-a)^2+b^2$ with $a,b \in \mathbb{R}$, $b \neq 0$. But $\mathbb{C}(T) = \mathbb{R}(T)(\sqrt{-1})$ and so, by A.18, Σ is the double cover of Δ. Thus we get a diagram

$$\begin{array}{ccc} S_c & \xrightarrow{\quad\pi\quad} & \Sigma \\ \downarrow{\scriptstyle p_1} & & \downarrow{\scriptstyle p_2} \\ S & \xrightarrow{\quad\tilde{\pi}\quad} & \Delta \end{array}$$

where p_1 and p_2 are double covers. Hence $\tilde{\pi}$ ramifies over $r+s$ points with multiplicities $m_1,\dots,m_r,n_1,\dots,n_s$ and if $\{q_1,\dots,q_r,l_1,\dots,l_s\}$ is the projection under $\tilde{\pi}$ of its branch locus, $q_1,\dots,q_r \in \Delta \backslash \partial\Delta = \mathbb{C}\backslash\mathbb{R}$, and each $l_i \in \mathbb{R} \cup \{\infty\}$. In terms of real algebraic curves, $\mathbb{P}_1(\mathbb{C})$ and $\mathbb{P}_1(\mathbb{R})$ are, respectively, compact and smooth algebraic models of $\mathbb{C}(T)$ and $\mathbb{R}(T)$. Let C be a compact and smooth model of $\mathcal{M}(S)|\mathbb{R}$ and $C_{\mathbb{C}}$ its complexification. Then, from the above and A.26 ($\partial S \approx C$) we get a diagram

$$\begin{array}{ccc} C_{\mathbb{C}} & \xrightarrow{\quad\pi\quad} & \mathbb{P}_1(\mathbb{C}) \\ \uparrow{\scriptstyle j_1} & & \uparrow{\scriptstyle j_2} \\ C & \xrightarrow{\quad\tilde{\pi}|C\quad} & \mathbb{P}_1(\mathbb{R}) \end{array}$$

Of course $\tilde{\pi}|C$ ramifies over s points, $l_1,\dots,l_s \in \mathbb{P}_1(\mathbb{R})$, with multiplicities $n_1,\dots,n_s$ and π ramifies over $s+2r$ points with multiplicities $n_1,\dots,n_s,m_1,m_1,\dots,m_r,m_r$.

A. NOTES

The functorial equivalence between $\mathcal{K}$ and $\mathcal{F}_{\mathbb{R}}$ was stated by Alling-Greenleaf [5]. We follow here a more direct approach, inspired in the same line of ideas. The key point in order to simplify arguments in [5] is proposition A.18. The implication $\mathcal{M}(S)$ real $\Rightarrow$ ∂S is nonempty, and proposition A.27 were proved, respectively, in [5] and [4]. However we present here new proofs. In particular, for the first, we avoid the use of a deep result of Witt [129], employed in [5]. Another proof of A.28 can be found in [29], lemma 2.2.

The excellent books of Forster [49], Chevalley [36] and Bochnak-Coste-Roy [8] supply the necessary background on classical Riemann surfaces, valuation theory on algebraic function fields in one variable and topics in real algebra as real valuations or Artin-Lang theorem, respectively.

References

[1] Accola,R.D.M. On the number of automorphisms of a closed Riemann surface. Trans. Amer. Math. Soc. **131** (1968) 398-408.

[2] Accola,R.D.M. Riemann surfaces with automorphism groups admitting partitions. Proc. Amer. Math. Soc. **21** (1969) 477-482.

[3] Ahlfors,L.V., Sario,L. *Riemann surfaces.* Princeton University Press, Princeton (1960).

[4] Alling,N.L. *Real elliptic curves.* Math. Studies 54, Notas de Mat. 81. North-Holland, Amsterdam (1981).

[5] Alling,N.L., Greenleaf,N. *Foundations of the theory of Klein surfaces.* Lect. Notes in Math., 219, Springer-Verlag, Berlin, etc. (1971).

[6] Beardon,A.F. *A primer on Riemann surfaces.* London Math. Soc. Lect. Note Series 78 (1984).

[7] Behnke,H., Sommer,F. *Theorie der analytischen Funktionen einer komplexen Veränderlichen.* Springer-Verlag, Berlin, etc. (1955).

[8] Bochnak,J., Coste,M., Roy,M.F. *Géométrie algébrique réelle.* Ergeb. der Math., 12, Springer-Verlag, Berlin, etc. (1987).

[9] Brandt,R., Stichtenoth,H. Die automorphismengruppen hyperelliptischer Kurven. Manuscripta Math. **55** (1986) 83-92.

[10] Brumfiel,G.W. *Partially ordered rings and semialgebraic geometry.* Lect. Note Series of London Math. Soc.37 Cambridge Univ. Press (1979).

[11] Bujalance,E. Normal subgroups of NEC groups. Math. Z. **178** (1981) 331-341.

[12] Bujalance,E. Proper periods of normal NEC subgroups with even index. Rev. Mat. Hisp.-Amer. (4) **41** (1981) 121-127.

[13] Bujalance,E. Normal NEC signatures. Illinois J. Math. **26** (1982) 519-530.

[14] Bujalance,E. Cyclic groups of automorphisms of compact non-orientable surfaces without boundary. Pacific J. Math. **109** (1983) 279-289.

[15] Bujalance,E. Automorphism groups of compact Klein surfaces with one boundary component. Math. Scand. **59** (1986) 45-58.

[16] Bujalance,E. Automorphism groups of compact planar Klein surfaces. Manuscripta Math. **56** (1986) 105-124.

[17] Bujalance,E. Una nota sobre el grupo de automorfismos de una superficie de Klein compacta. Rev. R. Acad. Ci. Madrid **81** (1987) 565-569.

[18] Bujalance,E., Bujalance,J.A., Martínez,E. On the automorphism group of hyperelliptic Klein surfaces. Michigan Math. J. **35** (1988) 361-368.

[19] Bujalance,E., Etayo,J.J. Hyperelliptic Klein surfaces with maximal symmetry. London Math. Soc. Lect. Note Series **112** (1986) 289-296.

[20] Bujalance,E., Etayo,J.J. Large automorphism groups of hyperelliptic Klein surfaces. Proc. Amer. Math. Soc. **103** (1988) 679-686.

[21] Bujalance,E., Etayo,J.J., Gamboa,J.M. Hyperelliptic Klein surfaces. Quart. J. Math. Oxford. (2) **36** (1985) 141-157.

[22] Bujalance,E., Etayo,J.J., Gamboa,J.M. Groups of automorphisms of hyperelliptic Klein surfaces of genus 3. Michigan Math. J. **33** (1986) 55-74.

[23] Bujalance,E., Etayo,J.J., Gamboa,J.M. Algebraic function fields in one variable over real closed ground fields. Travaux en cours **24**. Hermann. (1987) 125-129.

[24] Bujalance,E., Etayo,J.J., Gamboa,J.M., Martens,G. Minimal genus of Klein surfaces admitting an automorphism of given order. Arch. Math. **52** (1989) 191-202.

[25] Bujalance,E., Etayo,J.J., Martínez,E. Automorphism groups of hyperelliptic Riemann surfaces. Kodai Math. J. **10** (1987) 174-181.

[26] Bujalance,E., Gamboa,J.M. Automorphism groups of algebraic curves of $\mathbb{R}^n$ of genus 2. Arch. Math. **42** (1984) 229-237.

[27] Bujalance,E., Gromadzki,G. On nilpotent groups of automorphisms of compact Klein surfaces. Proc. Amer. Math. Soc. (to appear).

[28] Bujalance,E., Martínez,E. A remark on NEC groups of surfaces with boundary. Bull. London Math. Soc. **21** (1989) 263-266.

[29] Bujalance,E., Singerman,D. The symmetry type of a Riemann surface. Proc. London Math. Soc. (3) **51** (1985) 501-519.

[30] Bujalance,J.A. Normal subgroups of even index in an NEC group. Arch. Math. **49** (1987) 470-478.

[31] Bujalance,J.A. Topological types of Klein surfaces with maximum order automorphism. Glasgow Math. J. **30** (1988) 87-96, 369.

[32] Cazacu,C.A. On the morphisms of Klein surfaces. Rev. Roum. Math. Pure et Appl. **31** (1986) 461-470.

[33] Chetiya,B.P. Groups of automorphisms of compact Riemann surfaces. Ph.D. thesis, Birmingham Univ. (1971).

[34] Chetiya,B.P. On genuses of compact Riemann surfaces admitting solvable automorphism groups. Indian J. pure appl. Math. **12** (1981) 1312-1318.

[35] Chetiya,B.P., Patra,K. On metabelian groups of automorphisms of compact Riemann surfaces. J. London Math. Soc. (2) **33** (1986) 467-472.

[36] Chevalley,C. *Introduction to the theory of algebraic functions in one variable.* Math. Surveys 6, Amer. Math. Soc., New York (1951).

[37] Cohen,J. On Hurwitz extensions by $PSL_2(7)$. Math. Proc. Cambridge Phil. Soc. **86** (1979) 395-400.

[38] Cohen,J. On non Hurwitz groups and non-congruence subgroups of the modular group. Glasgow Math. J. **22** (1981) 1-7.

[39] Conder,M.D.E. Generators for alternating and symmetric groups. J. London Math. Soc. (2) **22** (1980) 75-86.

[40] Conder,M.D.E. The genus of compact Riemann surfaces with maximal automorphism group. J. Algebra **108** (1987) 204-247.

[41] Delfs,H., Knebusch,M. *Locally semialgebraic spaces.* Lect. Notes in Math., 1173. Springer-Verlag, Berlin, etc. (1985).

[42] Duma,A. Holomorphe Differentiale höherer Ordnung auf kompakten Riemannschen Flächen. Schriftenreihe d. Univ. Münster, 2 series, **14** (1978).

[43] Etayo,J.J. Klein surfaces with maximal symmetry and their groups of automorphisms. Math. Ann. **268** (1984) 533-538.

[44] Etayo,J.J. NEC subgroups and Klein surfaces. Bol. Soc. Mat. Mex. **29** (1984) 35-41.

[45] Etayo,J.J. On the order of automorphisms groups of Klein surfaces. Glasgow Math. J. **26** (1985) 75-81.

[46] Etayo,J.J., Pérez-Chirinos,C. Bordered and unbordered Klein surfaces with maximal symmetry. J. Pure App. Algebra **42** (1986) 29-35.

[47] Farkas,H.M. Remarks on automorphisms of compact Riemann surfaces, in: Discontinuous groups and Riemann surfaces. Princeton Univ. Press (1974).

[48] Fennessey,E., Pride,S.J. Equivalences of two-complexes, with applications to NEC-groups. Math. Proc. Camb. Phil. Soc. **106** (1989) 215-228.

[49] Forster,O. *Lectures on Riemann surfaces.* Grad. Texts in Math. 81, Springer-Verlag, Berlin, etc. (1977).

[50] Fricke,R., Klein,F. *Vorlesungen über die Theorie der automorphen Funktionen* (2 vols.). B.G. Teubner, Leipzig (1897 and 1912).

[51] Gordan,P. Über endliche Gruppen linearer Transformationen einer Veränderlichen. Math. Ann. **12** (1877) 23-46.

[52] Greenberg,L. Maximal Fuchsian groups. Bull. Amer. Math. Soc. **69** (1963) 569-573.

[53] Greenberg,L. Maximal groups and signatures. Ann. of Math. Studies **79** (1974) 207-226.

[54] Greenberg,L. Finiteness theorems in Fuchsian and Kleinian groups, in: Discrete groups and automorphic functions, W.J. Harvey (ed.), Acad. Press, New York (1977) 218-221.

[55] Greenleaf,N., May,C.L. Bordered Klein surfaces with maximal symmetry. Trans. Amer. Math. Soc. **274** (1982) 265-283.

[56] Gromadzki,G. Maximal groups of automorphisms of Riemann surfaces in various classes of finite groups. Rev. R. Acad. Ci. Madrid **82** (1988) 267-276.

[57] Gromadzki,G. On soluble groups of automorphisms of Riemann surfaces. Can. Math. Bull. (to appear).

[58] Gromadzki,G. Abelian groups of automorphisms of compact non-orientable Klein surfaces without boundary. Commentationes Mathematicae **28** (1989) 197-217.

[59] Gromadzki,G., Maclachlan,C. Supersoluble groups of automorphisms of compact Riemann surfaces. Glasgow Math. J. **31** (1989) 321-327.

[60] Hall, W. Automorphisms and coverings of Klein surfaces. Ph.D. thesis. Southampton Univ. (1977).

[61] Harvey,W.J. Cyclic groups of automorphisms of a compact Riemann surface. Quart. J. Math. Oxford (2) **17** (1966) 86-97.

[62] Harvey,W.J. On branch loci in Teichmüller space. Trans. Amer. Math. Soc. **153** (1971) 387-399.

[63] Heins,M. On the number of 1-1 directly conformal maps which a multiply-connected plane region of finite connectivity p $(p>2)$ admits onto itself. Bull. Amer. Math. Soc. **52** (1946) 454-457.

[64] Hoare,A.H.M. Subgroups of NEC groups and finite permutation groups. Quart J. Math. Oxford (2) **41** (1990) 45-59.

[65] Hoare,A.H.M., Karras,A., Solitar,D. Subgroups of infinite index in Fuchsian groups. Math. Z. **125** (1972) 59-69.

[66] Hoare,A.H.M., Karras,A., Solitar,D. Subgroups of NEC groups. Comm. in Pure and Appl. Math. **26** (1973) 731-744.

[67] Hoare,A.H.M., Singerman,D. Subgroups of plane groups. London Math. Soc. Lect. Note Series **71** (1982) 221-227.

[68] Homma,M. Automorphisms of prime order of curves. Manuscripta Math. **33** (1980) 99-109.

[69] Hurwitz,A. Über algebraische Gebilde mit eindeutigen Transformationen in sich. Math. Ann. **41** (1893) 402-442

[70] Kato,T. On the number of automorphisms of a compact bordered Riemann surface. Kodai Math. Sem. Rep. **24** (1972) 224-233.

[71] Kato,T. On the order of automorphism group of a compact bordered Riemann surface of genus four. Kodai Math. J. **7** (1984) 120-132.

[72] Keen,L. Canonical polygons for finitely generated fuchsian groups. Acta Math. **115** (1965) 1-16.

[73] Keen,L. On Fricke moduli. Advances in the theory of Riemann surfaces. Ann. of Math. Studies **66** (1971) 205-224.

[74] Knebusch,M. *Weakly semialgebraic spaces*. Lect. Notes in Math. 1367 Springer Verlag. Berlin, etc. (1988).

[75] Kravetz,S. On the geometry of Teichmüller spaces and the structure of their modular groups. Ann. Acad. Sci. Fenn. A **278** (1959) 35pp.

[76] Kuribayashi,A. On analytic families of compact Riemann surfaces with non-trivial automorphisms. Nagoya Math. J. **28** (1966) 119-165.

[77] Kuribayashi,I. Hyperelliptic AM curves of genus three and associated representations. Preprint.

[78] Lehner,J. *A short course on automorphic functions.* Holt, Reinhart and Winston. New York (1966).

[79] Lehner,J., Newman,M. On Riemann surfaces with maximal automorphism groups. Glasgow Math. J. **8** (1967) 102-112.

[80] Lewittes,J. Automorphisms of compact Riemann surfaces. Amer. J. Math. **85** (1963) 734-752.

[81] Lønsted,K., Kleiman,S.L. Basics on families of hyperelliptic curves. Compos. Math. **38** (1979) 83-111.

[82] Macbeath,A.M. *Discontinuous groups and birational transformations.* Proc. of Dundee Summer School, Univ. of St. Andrews (1961).

[83] Macbeath,A.M. On a theorem of Hurwitz. Proc. Glasgow Math. Assoc. **5** (1961) 90-96.

[84] Macbeath,A.M. Groups of homeomorphisms of a simply connected space. Ann. of Math. (2) **79** (1964) 473-488.

[85] Macbeath,A.M. On a curve of genus 7. Proc. London Math. Soc. (3) **15** (1965) 527-542.

[86] Macbeath,A.M. The classification of non-euclidean crystallographic groups. Can. J. Math. **19** (1967) 1192-1205.

[87] Macbeath,A.M. Generators of the linear fractional groups. Number theory, Proc. Symp. Pure Math. **12** (1967) 14-32.

[88] Macbeath,A.M., Singerman,D. Spaces of subgroups and Teichmüller space. Proc. London Math. Soc. **31** (1975) 211-256.

[89] Maclachlan,C. A bound for the number of automorphisms of a compact Riemann surface. J. London Math. Soc. **44** (1969) 265-272.

[90] Maclachlan,C. Maximal normal Fuchsian groups. Illinois J. Math. **15** (1971) 104-113.

[91] Maclachlan,C. Smooth coverings of hyperelliptic surfaces. Quart. J. Math. Oxford (2) **22** (1971) 117-123.

[92] Maclachlan,C., Harvey,W.J. On mapping-class groups and Teichmüller spaces. Proc. London Math. Soc. (3) **30** (1975) 496-512.

[93] Magnus,W., Karras,A., Solitar,D. *Combinatorial group theory. Presentations of groups in terms of generators and relations.* Interscience Publishers. John Wiley & Sons, Inc. New York (1966).

[94] Martínez,E. Convex fundamental regions for NEC groups. Arch. Math. **47** (1986) 457-464.

[95] May,C.L. Automorphisms of compact Klein surfaces with boundary. Pacific J. Math. **59** (1975) 199-210.

[96] May,C.L. Cyclic automorphism groups of compact bordered Klein surfaces. Houston J. Math. **3** (1977) 395-405.

[97] May,C.L. A bound for the number of automorphisms of a compact Klein surface with boundary. Proc. Amer. Math. Soc. **63** (1977) 273-280.

[98] May,C.L. Large automorphisms groups of compact Klein surfaces with boundary I. Glasgow Math. J. **18** (1977) 1-10.

[99] May,C.L. Maximal symmetry and fully wound coverings. Proc. Amer. Math. Soc. **79** (1980) 23-31.

[100] May,C.L. A family of M^*-groups. Can. J. Math. **38** (1986) 1094-1109.

[101] May,C.L. Nilpotent automorphism groups of bordered Klein surfaces. Proc. Amer. Math. Soc. **101** (1987) 287-292.

[102] May,C.L. Supersolvable M^*-groups. Glasgow Math. J. **30** (1988) 31-40.

[103] Nakagawa,K. On the orders of automorphisms of a closed Riemann surface. Pacific J. Math. **115** (1984) 435-443.

[104] Nakagawa,K. On the automorphism groups of a compact bordered Riemann surface of genus 5. Kodai Math. J. **9** (1986) 206-214.

[105] Oikawa,K. Notes on conformal mappings of a Riemann surface onto itself. Kodai Math. Sem. Rep. **8** (1956) 23-30, 115-116.

[106] Poincaré,H. Sur l'uniformization des fonctions analytiques. Acta Math. **31** (1908) 1-63.

[107] Potyagaylo,L.D. Fundamental polygonal groups of non-Euclidean crystallography. Siberian Math. J. **26** (1986) 586-591.

[108] Preston,R. Projective structures and fundamental domains on compact Klein surfaces. Ph.D. thesis. Univ. of Texas (1975).

[109] Rotman,J.J. *An introduction to the theory of groups.* Allyn & Bacon. Boston (1973).

[110] Sah,C.H. Groups related to compact Riemann surfaces. Acta Math. **123** (1969) 13-42.

[111] Schwarz,H.A. Über diejenigen algebraischen Gleichungen zwischen zwei Veränderlichen Grossen, welche eine Schaar rationaler eindeutig umkehrbarer Transformationen in sich selbst zulassen. J. reine und angew. Math. **87** (1879) 139-145.

[112] Shafarevich,I.R. *Basic Algebraic Geometry.* Grundlehren der Math. 213. Springer-Verlag. Berlin, etc. (1977).

[113] Siegel,C.L. Some remarks on discontinuous groups. Ann. of Math. **46** (1945) 708-718.

[114] Singerman,D. Automorphisms of compact non-orientable Riemann surfaces. Glasgow Math. J. **12** (1971) 50-59.

[115] Singerman,D. Finitely generated maximal Fuchsian groups. J. London Math. Soc. (2) **6** (1972) 29-38.

[116] Singerman,D. Symmetries of Riemann surfaces with large automorphism group. Math. Ann. **210** (1974) 17-32.

[117] Singerman,D. On the structure of non-euclidean crystallographic groups. Proc. Camb. Phil. Soc. **76** (1974) 233-240.

[118] Singerman,D. Automorphisms of maps, permutation groups and Riemann surfaces. Bull. London Math. Soc. **8** (1976) 65-68.

[119] Singerman,D. PSL(2,q) as an image of the extended modular group with applications to group actions on surfaces. Proc. Edinburgh Math. Soc. **30** (1987) 143-151.

[120] Suzuki,M. *Group theory* (2 vols.) Springer-Verlag. Berlin, etc. (1978 and 1982).

[121] Timmann,S. A bound for the number of automorphisms of a finite Riemann surface. Kodai Math. Sem. Rep. **28** (1976) 104-109.

[122] Tsuji,R. On conformal mapping of a hyperelliptic Riemann surface onto itself. Kodai Math. Sem. Rep. **10** (1958) 127-136.

[123] Tsuji,R. Conformal automorphisms of a compact bordered Riemann surface of genus three. Kodai Math. Sem. Rep. **27** (1976) 271-290.

[124] Uzzell,B. Groups acting on hyperbolic 3-space. Ph.D. thesis. Univ. of Birmingham (1981).

[125] Weil,A. On discrete subgroups of Lie groups I. Ann. of Math. (2) **72** (1960) 369-384.

[126] Wilkie,H.C. On non-euclidean crystallographic groups. Math. Z. **91** (1966) 87-102.

[127] Wiman,A. Über die hyperelliptischen Kurven und diejenigen vom Geschlecht p=3, welche eindeutige Transformationen in sich zulassen. Bihang Till. Kongl. Svenska Vetenskaps-Akademiens Handlingar **21**, 1, no.1 (1895) 23pp.

[128] Wiman,A. Über die algebraischen Kurven von den Geschlechten p=4,5 und 6, welche eindeutige Transformationen in sich besitzen. Bihang Till. Kongl. Svenska Vetenskaps-Akademiens Handlingar **21**, 1, no.3 (1895) 41pp.

[129] Witt,E. Zerlegung reeller algebraischer Funktionen in Quadrate Schiefkörper über reellem Funktionenkörper. J. reine und angew. Math. **171** (1935) 4-11.

[130] Zieschang,H. *Surfaces and planar discontinuous groups.* Lect. Notes in Math. 835. Springer-Verlag. Berlin, etc. (1980).

[131] Zomorrodian,R. Nilpotent automorphism groups of Riemann surfaces. Trans. Amer. Math. Soc. **288** (1985) 241-255.

[132] Zomorrodian,R. Classification of p-groups of automorphisms of a Riemann surface and their lower series. Glasgow Math. J. **29** (1987) 237-244.

Subject index

Symbol index

Usual notations

The letters $\mathbb{N}$, $\mathbb{Z}$, $\mathbb{Q}$, $\mathbb{R}$, $\mathbb{C}$ denote the sets of natural, integer, rational, real and complex numbers, respectively. The real and complex projective spaces of dimension n are denoted by $\mathbb{P}_n(\mathbb{R})$ and $\mathbb{P}_n(\mathbb{C})$. The cyclic group of order N, written additively, is $\mathbb{Z}_N$ and the dihedral group of order 2N is D_N. The localization of a ring A at a prime ideal $\mathfrak{p}$ is $A_{\mathfrak{p}}$ and tr.deg E|k is the transcendence degree of the field extension E|k. The semidirect product of two groups is denoted by G:H; if H is a subgroup of G the index of H in G is [G:H], while $H \triangleleft G$ means that H is a normal subgroup of G. The commutator subgroup of G is $G'=[G,G]$; $G^{(n)}=[G^{(n-1)}]'$ and G_{ab} is the quotient G/G'. The order of G is $|G|$, and the order of an element x in G is #(x). The commutator $aba^{-1}b^{-1}$ of two elements a and b of G is [a,b] and $a^b=bab^{-1}$. If G acts on a surface S, then $St_G(p)$ is the set of $f \in G$ such that f(p)=p. Finally, 1_X is the identity map on a set X, detA and trA are the determinant and trace of a square matrix A, $\{m_1,\ldots,\hat{m}_i,\ldots,m_n\}$ is the result of the deletion of m_i in the set $\{m_1,\ldots,m_n\}$, and the abbreviations $\cong$, $\approx$, *g.c.d.* and *l.c.m.* mean, respectively, algebraic isomorphism, topological or dianalytic isomorphism, greatest common divisor and lowest common multiple.

Other notations

LECTURE NOTES IN MATHEMATICS

Edited by A. Dold, B. Eckmann and F. Takens

Some general remarks on the publication of monographs and seminars

In what follows all references to monographs, are applicable also to multiauthorship volumes such as seminar notes.

§1. Lecture Notes aim to report new developments - quickly, informally, and at a high level. Monograph manuscripts should be reasonably self-contained and rounded off. Thus they may, and often will, present not only results of the author but also related work by other people. Furthermore, the manuscripts should provide sufficient motivation, examples and applications. This clearly distinguishes Lecture Notes manuscripts from journal articles which normally are very concise. Articles intended for a journal but too long to be accepted by most journals, usually do not have this "lecture notes" character. For similar reasons it is unusual for Ph.D. theses to be accepted for the Lecture Notes series.

Experience has shown that English language manuscripts achieve a much wider distribution.

§2. Manuscripts or plans for Lecture Notes volumes should be submitted (preferably in duplicate) either to one of the series editors or to Springer- Verlag, Heidelberg. These proposals are then refereed. A final decision concerning publication can only be made on the basis of the complete manuscripts, but a preliminary decision can usually be based on partial information: a fairly detailed outline describing the planned contents of each chapter, and an indication of the estimated length, a bibliography, and one or two sample chapters - or a first draft of the manuscript. The editors will try to make the preliminary decision as definite as they can on the basis of the available information. We generally advise authors not to prepare the final master copy of their manuscript (cf. §4) beforehand.

§3. Final manuscripts should contain at least 100 pages of mathematical text and should include

- a table of contents;
- an informative introduction, perhaps with some historical remarks: it should be accessible to a reader not particularly familiar with the topic treated;
- a subject index: this is almost always genuinely helpful for the reader.

§4. Lecture Notes are printed by photo-offset from the master-copy delivered in camera-ready form by the authors. Springer-Verlag provides technical instructions for the preparation of manuscripts, for typewritten manuscripts special stationery, with the prescribed typing area outlined, is available on request. Careful preparation of the manuscripts will help keep production time short and ensure satisfactory appearance of the finished book. For manuscripts typed or typeset according to our instructions, Springer-Verlag will, if necessary, contribute towards the preparation costs at a fixed rate.

The actual production of a Lecture Notes volume takes 6-8 weeks.

§5. Authors receive a total of 50 free copies of their volume, but no royalties. They are entitled to purchase further copies of their book for their personal use at a discount of 33.3 %, other Springer mathematics books at a discount of 20 % directly from Springer-Verlag.

Commitment to publish is made by letter of intent rather than by signing a formal contract. Springer-Verlag secures the copyright for each volume.

Addresses:

Professor A. Dold, Mathematisches Institut, Universität Heidelberg, Im Neuenheimer Feld 288, 6900 Heidelberg, Federal Republic of Germany

Professor B. Eckmann, Mathematik, ETH-Zentrum
8092 Zürich, Switzerland

Prof. F. Takens, Mathematisch Instituut, Rijksuniversiteit Groningen, Postbus 800, 9700 AV Groningen, The Netherlands

Springer-Verlag, Mathematics Editorial, Tiergartenstr. 17,
6900 Heidelberg, Federal Republic of Germany, Tel.: (06221) 487-410

Springer-Verlag, Mathematics Editorial, 175 Fifth Avenue,
New York, New York 10010, USA, Tel.: (212) 460-1596

LECTURE NOTES

ESSENTIALS FOR THE PREPARATION OF CAMERA-READY MANUSCRIPTS

Springer
Springer-Verlag
Berlin Heidelberg New York
London Paris Tokyo Hong Kong

The preparation of manuscripts which are to be reproduced by photo-offset require special care. Manuscripts which are submitted in technically unsuitable form will be returned to the author for retyping. There is normally no possibility of carrying out further corrections after a manuscript is given to production. Hence it is crucial that the following instructions be adhered to closely. If in doubt, please send us 1 - 2 sample pages for examination.

General. The characters must be uniformly black both within a single character and down the page. Original manuscripts are required: photocopies are acceptable only if they are sharp and without smudges.

On request, Springer-Verlag will supply special paper with the text area outlined. The standard TEXT AREA (OUTPUT SIZE if you are using a 14 point font) is 18 x 26.5 cm (7.5 x 11 inches). This will be scale-reduced to 75% in the printing process. If you are using computer typesetting, please see also the following page.

Make sure the TEXT AREA IS COMPLETELY FILLED. Set the margins so that they precisely match the outline and type right from the top to the bottom line. (Note that the page number will lie outside this area). Lines of text should not end more than three spaces inside or outside the right margin (see example on page 4).

Type on one side of the paper only.

Spacing and Headings (Monographs). Use ONE-AND-A-HALF line spacing in the text. Please leave sufficient space for the title to stand out clearly and do NOT use a new page for the beginning of subdivisons of chapters. Leave THREE LINES blank above and TWO below headings of such subdivisions.

Spacing and Headings (Proceedings). Use ONE-AND-A-HALF line spacing in the text. Do not use a new page for the beginning of subdivisons of a single paper. Leave THREE LINES blank above and TWO below headings of such subdivisions. Make sure headings of equal importance are in the same form.

The first page of each contribution should be prepared in the same way. The title should stand out clearly. We therefore recommend that the editor prepare a sample page and pass it on to the authors together with these instructions. Please take the following as an example. Begin heading 2 cm below upper edge of text area.

MATHEMATICAL STRUCTURE IN QUANTUM FIELD THEORY

John E. Robert
Mathematisches Institut, Universität Heidelberg
Im Neuenheimer Feld 288, D-6900 Heidelberg

Please leave THREE LINES blank below heading and address of the author, then continue with the actual text on the same page.

Footnotes. These should preferable be avoided. If necessary, type them in SINGLE LINE SPACING to finish exactly on the outline, and separate them from the preceding main text by a line.

Symbols. Anything which cannot be typed may be entered by hand in BLACK AND ONLY BLACK ink. (A fine-tipped rapidograph is suitable for this purpose; a good black ball-point will do, but a pencil will not). Do not draw straight lines by hand without a ruler (not even in fractions).

Literature References. These should be placed at the end of each paper or chapter, or at the end of the work, as desired. Type them with single line spacing and start each reference on a new line. Follow "Zentralblatt für Mathematik"/"Mathematical Reviews" for abbreviated titles of mathematical journals and "Bibliographic Guide for Editors and Authors (BGEA)" for chemical, biological, and physics journals. Please ensure that all references are COMPLETE and ACCURATE.

IMPORTANT

Pagination. For typescript, number pages in the upper right-hand corner in LIGHT BLUE OR GREEN PENCIL ONLY. The printers will insert the final page numbers. For computer type, you may insert page numbers (1 cm above outer edge of text area).

It is safer to number pages AFTER the text has been typed and corrected. Page 1 (Arabic) should be THE FIRST PAGE OF THE ACTUAL TEXT. The Roman pagination (table of contents, preface, abstract, acknowledgements, brief introductions, etc.) will be done by Springer-Verlag.

If including running heads, these should be aligned with the inside edge of the text area while the page number is aligned with the outside edge noting that right-hand pages are odd-numbered. Running heads and page numbers appear on the same line. Normally, the running head on the left-hand page is the chapter heading and that on the right-hand page is the section heading. Running heads should not be included in proceedings contributions unless this is being done consistently by all authors.

Corrections. When corrections have to be made, cut the new text to fit and paste it over the old. White correction fluid may also be used.

Never make corrections or insertions in the text by hand.

If the typescript has to be marked for any reason, e.g. for provisional page numbers or to mark corrections for the typist, this can be done VERY FAINTLY with BLUE or GREEN PENCIL but NO OTHER COLOR: these colors do not appear after reproduction.

COMPUTER-TYPESETTING. Further, to the above instructions, please note with respect to your printout that
- the characters should be sharp and sufficiently black;
- it is not strictly necessary to use Springer's special typing paper. Any white paper of reasonable quality is acceptable.

If you are using a significantly different font size, you should modify the output size correspondingly, keeping length to breadth ratio 1 : 0.68, so that scaling down to 10 point font size, yields a text area of 13.5 x 20 cm (5 3/8 x 8 in), e.g.

Differential equations.: use output size 13.5 x 20 cm.

Differential equations.: use output size 16 x 23.5 cm.

Differential equations.: use output size 18 x 26.5 cm.

Interline spacing: 5.5 mm base-to-base for 14 point characters (standard format of 18 x 26.5 cm).
If in any doubt, please send us 1 - 2 sample pages for examination. We will be glad to give advice.

Vol. 1259: F. Cano Torres, Desingularization Strategies for Three-Dimensional Vector Fields. IX, 189 pages. 1987.

Vol. 1260: N.H. Pavel, Nonlinear Evolution Operators and Semigroups. VI, 285 pages. 1987.

Vol. 1261: H. Abels, Finite Presentability of S-Arithmetic Groups. Compact Presentability of Solvable Groups. VI, 178 pages. 1987.

Vol. 1262: E. Hlawka (Hrsg.), Zahlentheoretische Analysis II. Seminar, 1984–86. V, 158 Seiten. 1987.

Vol. 1263: V.L. Hansen (Ed.), Differential Geometry. Proceedings, 1985. XI, 288 pages. 1987.

Vol. 1264: Wu Wen-tsün, Rational Homotopy Type. VIII, 219 pages. 1987.

Vol. 1265: W. Van Assche, Asymptotics for Orthogonal Polynomials. VI, 201 pages. 1987.

Vol. 1266: F. Ghione, C. Peskine, E. Sernesi (Eds.), Space Curves. Proceedings, 1985. VI, 272 pages. 1987.

Vol. 1267: J. Lindenstrauss, V.D. Milman (Eds.), Geometrical Aspects of Functional Analysis. Seminar. VII, 212 pages. 1987.

Vol. 1268: S.G. Krantz (Ed.), Complex Analysis. Seminar, 1986. VII, 195 pages. 1987.

Vol. 1269: M. Shiota, Nash Manifolds. VI, 223 pages. 1987.

Vol. 1270: C. Carasso, P.-A. Raviart, D. Serre (Eds.), Nonlinear Hyperbolic Problems. Proceedings, 1986. XV, 341 pages. 1987.

Vol. 1271: A.M. Cohen, W.H. Hesselink, W.L.J. van der Kallen, J.R. Strooker (Eds.), Algebraic Groups Utrecht 1986. Proceedings. XII, 284 pages. 1987.

Vol. 1272: M.S. Livšic, L.L. Waksman, Commuting Nonselfadjoint Operators in Hilbert Space. III, 115 pages. 1987.

Vol. 1273: G.-M. Greuel, G. Trautmann (Eds.), Singularities, Representation of Algebras, and Vector Bundles. Proceedings, 1985. XIV, 383 pages. 1987.

Vol. 1274: N. C. Phillips, Equivariant K-Theory and Freeness of Group Actions on C*-Algebras. VIII, 371 pages. 1987.

Vol. 1275: C.A. Berenstein (Ed.), Complex Analysis I. Proceedings, 1985–86. XV, 331 pages. 1987.

Vol. 1276: C.A. Berenstein (Ed.), Complex Analysis II. Proceedings, 1985–86. IX, 320 pages. 1987.

Vol. 1277: C.A. Berenstein (Ed.), Complex Analysis III. Proceedings, 1985–86. X, 350 pages. 1987.

Vol. 1278: S.S. Koh (Ed.), Invariant Theory. Proceedings, 1985. V, 102 pages. 1987.

Vol. 1279: D. Ieşan, Saint-Venant's Problem. VIII, 162 Seiten. 1987.

Vol. 1280: E. Neher, Jordan Triple Systems by the Grid Approach. XII, 193 pages. 1987.

Vol. 1281: O.H. Kegel, F. Menegazzo, G. Zacher (Eds.), Group Theory. Proceedings, 1986. VII, 179 pages. 1987.

Vol. 1282: D.E. Handelman, Positive Polynomials, Convex Integral Polytopes, and a Random Walk Problem. XI, 136 pages. 1987.

Vol. 1283: S. Mardešić, J. Segal (Eds.), Geometric Topology and Shape Theory. Proceedings, 1986. V, 261 pages. 1987.

Vol. 1284: B.H. Matzat, Konstruktive Galoistheorie. X, 286 pages. 1987.

Vol. 1285: I.W. Knowles, Y. Saitō (Eds.), Differential Equations and Mathematical Physics. Proceedings, 1986. XVI, 499 pages. 1987.

Vol. 1286: H.R. Miller, D.C. Ravenel (Eds.), Algebraic Topology. Proceedings, 1986. VII, 341 pages. 1987.

Vol. 1287: E.B. Saff (Ed.), Approximation Theory, Tampa. Proceedings, 1985–1986. V, 228 pages. 1987.

Vol. 1288: Yu. L. Rodin, Generalized Analytic Functions on Riemann Surfaces. V, 128 pages, 1987.

Vol. 1289: Yu. I. Manin (Ed.), K-Theory, Arithmetic and Geometry. Seminar, 1984–1986. V, 399 pages. 1987.

Vol. 1290: G. Wüstholz (Ed.), Diophantine Approximation and Transcendence Theory. Seminar, 1985. V, 243 pages. 1987.

Vol. 1291: C. Mœglin, M.-F. Vignéras, J.-L. Waldspurger, Correspondances de Howe sur un Corps p-adique. VII, 163 pages. 1987

Vol. 1292: J.T. Baldwin (Ed.), Classification Theory. Proceedings, 1985. VI, 500 pages. 1987.

Vol. 1293: W. Ebeling, The Monodromy Groups of Isolated Singularities of Complete Intersections. XIV, 153 pages. 1987.

Vol. 1294: M. Queffélec, Substitution Dynamical Systems – Spectral Analysis. XIII, 240 pages. 1987.

Vol. 1295: P. Lelong, P. Dolbeault, H. Skoda (Réd.), Séminaire d'Analyse P. Lelong – P. Dolbeault – H. Skoda. Seminar, 1985/1986. VII, 283 pages. 1987.

Vol. 1296: M.-P. Malliavin (Ed.), Séminaire d'Algèbre Paul Dubreil et Marie-Paule Malliavin. Proceedings, 1986. IV, 324 pages. 1987.

Vol. 1297: Zhu Y.-l., Guo B.-y. (Eds.), Numerical Methods for Partial Differential Equations. Proceedings. XI, 244 pages. 1987.

Vol. 1298: J. Aguadé, R. Kane (Eds.), Algebraic Topology, Barcelona 1986. Proceedings. X, 255 pages. 1987.

Vol. 1299: S. Watanabe, Yu.V. Prokhorov (Eds.), Probability Theory and Mathematical Statistics. Proceedings, 1986. VIII, 589 pages. 1988.

Vol. 1300: G.B. Seligman, Constructions of Lie Algebras and their Modules. VI, 190 pages. 1988.

Vol. 1301: N. Schappacher, Periods of Hecke Characters. XV, 160 pages. 1988.

Vol. 1302: M. Cwikel, J. Peetre, Y. Sagher, H. Wallin (Eds.), Function Spaces and Applications. Proceedings, 1986. VI, 445 pages. 1988.

Vol. 1303: L. Accardi, W. von Waldenfels (Eds.), Quantum Probability and Applications III. Proceedings, 1987. VI, 373 pages. 1988.

Vol. 1304: F.Q. Gouvêa, Arithmetic of *p*-adic Modular Forms. VIII, 121 pages. 1988.

Vol. 1305: D.S. Lubinsky, E.B. Saff, Strong Asymptotics for Extremal Polynomials Associated with Weights on ℝ. VII, 153 pages. 1988.

Vol. 1306: S.S. Chern (Ed.), Partial Differential Equations. Proceedings, 1986. VI, 294 pages. 1988.

Vol. 1307: T. Murai, A Real Variable Method for the Cauchy Transform, and Analytic Capacity. VIII, 133 pages. 1988.

Vol. 1308: P. Imkeller, Two-Parameter Martingales and Their Quadratic Variation. IV, 177 pages. 1988.

Vol. 1309: B. Fiedler, Global Bifurcation of Periodic Solutions with Symmetry. VIII, 144 pages. 1988.

Vol. 1310: O.A. Laudal, G. Pfister, Local Moduli and Singularities. V, 117 pages. 1988.

Vol. 1311: A. Holme, R. Speiser (Eds.), Algebraic Geometry, Sundance 1986. Proceedings. VI, 320 pages. 1988.

Vol. 1312: N.A. Shirokov, Analytic Functions Smooth up to the Boundary. III, 213 pages. 1988.

Vol. 1313: F. Colonius, Optimal Periodic Control. VI, 177 pages. 1988.

Vol. 1314: A. Futaki, Kähler-Einstein Metrics and Integral Invariants. IV, 140 pages. 1988.

Vol. 1315: R.A. McCoy, I. Ntantu, Topological Properties of Spaces of Continuous Functions. IV, 124 pages. 1988.

Vol. 1316: H. Korezlioglu, A.S. Ustunel (Eds.), Stochastic Analysis and Related Topics. Proceedings, 1986. V, 371 pages. 1988.

Vol. 1317: J. Lindenstrauss, V.D. Milman (Eds.), Geometric Aspects of Functional Analysis. Seminar, 1986–87. VII, 289 pages. 1988.

Vol. 1318: Y. Felix (Ed.), Algebraic Topology – Rational Homotopy. Proceedings, 1986. VIII, 245 pages. 1988

Vol. 1319: M. Vuorinen, Conformal Geometry and Quasiregular Mappings. XIX, 209 pages. 1988.

Vol. 1320: H. Jürgensen, G. Lallement, H.J. Weinert (Eds.), Semigroups, Theory and Applications. Proceedings, 1986. X, 416 pages. 1988.

Vol. 1321: J. Azéma, P.A. Meyer, M. Yor (Eds.), Séminaire de Probabilités XXII. Proceedings. IV, 600 pages. 1988.

Vol. 1322: M. Métivier, S. Watanabe (Eds.), Stochastic Analysis. Proceedings, 1987. VII, 197 pages. 1988.

Vol. 1323: D.R. Anderson, H.J. Munkholm, Boundedly Controlled Topology. XII, 309 pages. 1988.

Vol. 1324: F. Cardoso, D.G. de Figueiredo, R. Iório, O. Lopes (Eds.), Partial Differential Equations. Proceedings, 1986. VIII, 433 pages. 1988.

Vol. 1325: A. Truman, I.M. Davies (Eds.), Stochastic Mechanics and Stochastic Processes. Proceedings, 1986. V, 220 pages. 1988.

Vol. 1326: P.S. Landweber (Ed.), Elliptic Curves and Modular Forms in Algebraic Topology. Proceedings, 1986. V, 224 pages. 1988.

Vol. 1327: W. Bruns, U. Vetter, Determinantal Rings. VII,236 pages. 1988.

Vol. 1328: J.L. Bueso, P. Jara, B. Torrecillas (Eds.), Ring Theory. Proceedings, 1986. IX, 331 pages. 1988.

Vol. 1329: M. Alfaro, J.S. Dehesa, F.J. Marcellan, J.L. Rubio de Francia, J. Vinuesa (Eds.): Orthogonal Polynomials and their Applications. Proceedings, 1986. XV, 334 pages. 1988.

Vol. 1330: A. Ambrosetti, F. Gori, R. Lucchetti (Eds.), Mathematical Economics. Montecatini Terme 1986. Seminar. VII, 137 pages. 1988.

Vol. 1331: R. Bamón, R. Labarca, J. Palis Jr. (Eds.), Dynamical Systems, Valparaiso 1986. Proceedings. VI, 250 pages. 1988.

Vol. 1332: E. Odell, H. Rosenthal (Eds.), Functional Analysis. Proceedings, 1986–87. V, 202 pages. 1988.

Vol. 1333: A.S. Kechris, D.A. Martin, J.R. Steel (Eds.), Cabal Seminar 81–85. Proceedings, 1981–85. V, 224 pages. 1988.

Vol. 1334: Yu.G. Borisovich, Yu. E. Gliklikh (Eds.), Global Analysis – Studies and Applications III. V, 331 pages. 1988.

Vol. 1335: F. Guillén, V. Navarro Aznar, P. Pascual-Gainza, F. Puerta, Hyperrésolutions cubiques et descente cohomologique. XII, 192 pages. 1988.

Vol. 1336: B. Helffer, Semi-Classical Analysis for the Schrödinger Operator and Applications. V, 107 pages. 1988.

Vol. 1337: E. Sernesi (Ed.), Theory of Moduli. Seminar, 1985. VIII, 232 pages. 1988.

Vol. 1338: A.B. Mingarelli, S.G. Halvorsen, Non-Oscillation Domains of Differential Equations with Two Parameters. XI, 109 pages. 1988.

Vol. 1339: T. Sunada (Ed.), Geometry and Analysis of Manifolds. Procedings, 1987. IX, 277 pages. 1988.

Vol. 1340: S. Hildebrandt, D.S. Kinderlehrer, M. Miranda (Eds.), Calculus of Variations and Partial Differential Equations. Proceedings, 1986. IX, 301 pages. 1988.

Vol. 1341: M. Dauge, Elliptic Boundary Value Problems on Corner Domains. VIII, 259 pages. 1988.

Vol. 1342: J.C. Alexander (Ed.), Dynamical Systems. Proceedings, 1986–87. VIII, 726 pages. 1988.

Vol. 1343: H. Ulrich, Fixed Point Theory of Parametrized Equivariant Maps. VII, 147 pages. 1988.

Vol. 1344: J. Král, J. Lukeš, J. Netuka, J. Veselý (Eds.), Potential Theory – Surveys and Problems. Proceedings, 1987. VIII, 271 pages. 1988.

Vol. 1345: X. Gomez-Mont, J. Seade, A. Verjovski (Eds.), Holomorphic Dynamics. Proceedings, 1986. VII, 321 pages. 1988.

Vol. 1346: O. Ya. Viro (Ed.), Topology and Geometry – Rohlin Seminar. XI, 581 pages. 1988.

Vol. 1347: C. Preston, Iterates of Piecewise Monotone Mappings on an Interval. V, 166 pages. 1988.

Vol. 1348: F. Borceux (Ed.), Categorical Algebra and its Applications. Proceedings, 1987. VIII, 375 pages. 1988.

Vol. 1349: E. Novak, Deterministic and Stochastic Error Bounds in Numerical Analysis. V, 113 pages. 1988.

Vol. 1350: U. Koschorke (Ed.), Differential Topology. Proceedings, 1987. VI, 269 pages. 1988.

Vol. 1351: I. Laine, S. Rickman, T. Sorvali, (Eds.), Complex Analysis, Joensuu 1987. Proceedings. XV, 378 pages. 1988.

Vol. 1352: L.L. Avramov, K.B. Tchakerian (Eds.), Algebra – Some Current Trends. Proceedings, 1986. IX, 240 Seiten. 1988.

Vol. 1353: R.S. Palais, Ch.-l. Terng, Critical Point Theory and Submanifold Geometry. X, 272 pages. 1988.

Vol. 1354: A. Gómez, F. Guerra, M.A. Jiménez, G. López (Eds.), Approximation and Optimization. Proceedings, 1987. VI, 280 pages. 1988.

Vol. 1355: J. Bokowski, B. Sturmfels, Computational Synthetic Geometry. V, 168 pages. 1989.

Vol. 1356: H. Volkmer, Multiparameter Eigenvalue Problems and Expansion Theorems. VI, 157 pages. 1988.

Vol. 1357: S. Hildebrandt, R. Leis (Eds.), Partial Differential Equations and Calculus of Variations. VI, 423 pages. 1988.

Vol. 1358: D. Mumford, The Red Book of Varieties and Schemes. V, 309 pages. 1988.

Vol. 1359: P. Eymard, J.-P. Pier (Eds.), Harmonic Analysis. Proceedings, 1987. VIII, 287 pages. 1988.

Vol. 1360: G. Anderson, C. Greengard (Eds.), Vortex Methods. Proceedings, 1987. V, 141 pages. 1988.

Vol. 1361: T. tom Dieck (Ed.), Algebraic Topology and Transformation Groups. Proceedings, 1987. VI, 298 pages. 1988.

Vol. 1362: P. Diaconis, D. Elworthy, H. Föllmer, E. Nelson, G.C. Papanicolaou, S.R.S. Varadhan. École d'Été de Probabilités de Saint-Flour XV–XVII, 1985–87. Editor: P.L. Hennequin. V, 459 pages. 1988.

Vol. 1363: P.G. Casazza, T.J. Shura. Tsirelson's Space. VIII, 204 pages. 1988.

Vol. 1364: R.R. Phelps, Convex Functions, Monotone Operators and Differentiability. IX, 115 pages. 1989.

Vol. 1365: M. Giaquinta (Ed.), Topics in Calculus of Variations. Seminar, 1987. X, 196 pages. 1989.

Vol. 1366: N. Levitt, Grassmannians and Gauss Maps in PL-Topology. V, 203 pages. 1989.

Vol. 1367: M. Knebusch, Weakly Semialgebraic Spaces. XX, 376 pages. 1989.

Vol. 1368: R. Hübl, Traces of Differential Forms and Hochschild Homology. III, 111 pages. 1989.

Vol. 1369: B. Jiang, Ch.-K. Peng, Z. Hou (Eds.), Differential Geometry and Topology. Proceedings, 1986–87. VI, 366 pages. 1989.

Vol. 1370: G. Carlsson, R.L. Cohen, H.R. Miller, D.C. Ravenel (Eds.), Algebraic Topology. Proceedings, 1986. IX, 456 pages. 1989.

Vol. 1371: S. Glaz, Commutative Coherent Rings. XI, 347 pages. 1989.

Vol. 1372: J. Azéma, P.A. Meyer, M. Yor (Eds.), Séminaire de Probabilités XXIII. Proceedings. IV, 583 pages. 1989.

Vol. 1373: G. Benkart, J.M. Osborn (Eds.), Lie Algebras, Madison 1987. Proceedings. V, 145 pages. 1989.

Vol. 1374: R.C. Kirby, The Topology of 4-Manifolds. VI, 108 pages. 1989.

Vol. 1375: K. Kawakubo (Ed.), Transformation Groups. Proceedings, 1987. VIII, 394 pages, 1989.

Vol. 1376: J. Lindenstrauss, V.D. Milman (Eds.), Geometric Aspects of Functional Analysis. Seminar (GAFA) 1987–88. VII, 288 pages. 1989.

Vol. 1377: J.F. Pierce, Singularity Theory, Rod Theory, and Symmetry-Breaking Loads. IV, 177 pages. 1989.

Vol. 1378: R.S. Rumely, Capacity Theory on Algebraic Curves. III, 437 pages. 1989.

Vol. 1379: H. Heyer (Ed.), Probability Measures on Groups IX. Proceedings, 1988. VIII, 437 pages. 1989